INSTRUMENTATION AND PROCESS CONTROL

Prof. D.C. Sikdar
Deptt. of Chemical Engineering,
Dayananda Sagar College of Engineering,
Shavige Malleswara Hills,
Kumaraswamy Layout
BANGALORE

KHANNA BOOK PUBLISHING CO. (P) LTD.
Publisher of Science, Technology and Engineering Books
4C/4344, Ansari Road, Darya Ganj, New Delhi-110002
Phone: 011-23244447-48 **Mobile:** +91-99109 09320
E-mail: contact@khannabooks.com
Website: www.khannabooks.com

Dear Readers,

To prevent the piracy, this book is secured with HIGH SECURITY HOLOGRAM on the front title cover. In case you don't find the hologram on the front cover title, please write us to at contact@khannabooks.com or whatsapp us at +91-99109 09320 and avail special gift voucher for yourself.

Specimen of Hologram on front Cover title:

Moreover, there is a SPECIAL DISCOUNT COUPON for you with EVERY HOLOGRAM.

How to avail this SPECIAL DISCOUNT:

Step 1: Scratch the hologram
Step 2: Under the scratch area, your "coupon code" is available
Step 3: Logon to www.khannabooks.com
Step 4: Use your "coupon code" in the shopping cart and get your copy at a special discount
Step 5: Enjoy your reading!

Price: ₹ 399.00

ISBN: 978-93-82609-04-9

Instrumentation and Process Control
by D.C. Sikdar

First Edition: 2016
Revised Edition: 2018
Reprint: 2022

To view complete list of books, Please scan the QR Code: KPH

Published by:
Khanna Book Publishing Co. (P) Ltd.
CIN: U22110DL1998PTC095547

Visit us at: **www.khannabooks.com**
Write us at: **contact@khannabooks.com**

Printed in India by: Vaibhav Products, Delhi.

Dedicated to My Parents

Wife BULU, Son ATANU

&

Members of the Family

TOMY and HONEY

For their Love, Understanding and Support

Dedicated to My Parents

Wife [illegible]

&

Members of the Family

[illegible] and [illegible]

For their Love, Understanding and Support

PREFACE

The Instrumentation and Process Control is an Engineering subject that deals with art of measurements, architect plan, mechanism and algorithm for maintaining the output of a specific process within a desired range. For example, consider the maintenance of temperature of a chemical reactor. It may be controlled to obtain a constant product output. Process control is extensively used in industry to enable mass production of consistent products from continuously operated processes such as oil refinery, cement manufacturing unit and many others. Process control enables us to operate a complex process from a central control room.

Instrumentation and Process Control is one of the core courses at the undergraduate level of Chemical Engineering curriculum. In this course, more emphasis is given to the fundamental of measuring instrument, Laplace Transform, Basic concept of Process Control, First order and Second order system, Control of Industrial Bio-processes, Controller and Final control elements, Block diagram reduction techniques, Determination of Stability of a process, Advanced control techniques and Control Structure of unit operations, all coming under the realm of Instrumentation and Process Control. The conventional textbook on Instrumentation and Process Control are inadequate to meet these demand. Hence, there is a need for student friendly textbook on Instrumentation and Process Control that can bridge the gap between the theoretical learning and practical implementation.

During this two and half decades of my experience in teaching to the undergraduate and post graduate students, I have come across the students who have strong feeling of liking and disliking towards the Instrumentation and Process Control subject. Unfortunately, disliking frequently outnumbered the liking. The nature of the subject and determination of stability is partly to be blamed for this scenario. The main reason is the scarcity of the student friendly textbook. These points have been kept in mind while preparing the context of the book.

This book is intends to serve as a simple textbook for the undergraduate and post graduate students of Chemical Engineering and other related branches such as Medical Electronics, Aeronautical Engineering, Polymer Science and Engineering, Bio-technology as well as Diploma in Chemical Engineering. In this book the chapters and topics are organized in a systematic way in order that students get a thorough knowledge and complete understanding of the subject. The usefulness of the book has been further enhanced by the inclusion of a large number of illustrative figures, solved examples, review questions and exercises at the end of each chapter. In this book only SI unit is used. An attempt has been made to avoid all intricacies that usually make the subject dry.

This book is the outgrowth of the several lectures, I have delivered to the Chemical Engineering students of undergraduate and post graduate at Dayananda Sagar College of Engineering, Bangalore. The response from the students who benefitted from these lectures is the main source of inspiration for writing this textbook. The comments and suggestions for improving the quality of the book would be gratefully accepted.

D.C. Sikdar

ACKNOWLEDGEMENT

I recall with gratitude the stimulating experiences I spent at NITK Surathkal during my undergraduate study under the supervision of Dr. P. N. Singh as a Professor & Head of the Department, which contributed much in sustaining my interest in the subject of Instrumentation and Process Control.

I express my deep sense of gratitude to Dr. Hemachandra Sagar, Chairman and Dr. Premachandra Sagar, Vice-Chairman and CEO, Dayananda Sagar Institutions, who has enthused me with the spirit of writing the book that would be useful for the student community.

I am thankful to Sri Galiswamy, Secretary, Dayananda Sagar Institutions, Dr. C. P. S. Prakash, Principal, Dayananda Sagar College of Engineering and Dr. R. Ravishankar, Head of the Department of Chemical Engineering, Dayananda Sagar College of Engineering, Bangalore for their sustained support rendered during the writing of this book.

I am also thankful to Dr. S. Murthyshekhar, Prof. G. K. Mahadevaraju and Prof. B. S. Thirumalesh, Department of Chemical Engineering, Dayananda Sagar College of Engineering, Bangalore, who has been of great assistance in shaping this book.

Thanks are due to Dr. G. Mugeraya, Department of Chemical Engineering, National Institute of Technology, Surathkal, Dr. P. Ravichandran, Principal, St. Joseph's College of Engineering, Chennai, Dr. G. P. Desai, Department of Chemical Engineering, Bapuji Institute of Engineering and Technology, Davangere, Dr. Harish Kumar, Department of Chemical Engineering, Manipal Institute of Technology, Manipal, Prof. S. Murthy, Department of Chemical Engineering, Siddaganga Institute of Technology, Tumkur, Dr. Chetan Nayak, Department of Chemical Engineering, B. M. S. College of Engineering, Bangalore, Dr. Jagadish Patil, Department of Chemical Engineering, R. V. College of Engineering, Bangalore, for their support and encouragement during the writing of this book.

I thank my family and relatives for their direct and indirect support. My wife Bulu, son Atanu, family members Tomy and Honey, needs a special mention for their patience and graceful support during the writing of this book.

Finally, Last but not least, I express my thanks to management and editorial board and production department of Khanna publishers, New Delhi for their careful evaluation and constructive suggestions during the writing this book. It is my duty to thank Mr. Mukul Seth managing director of khanna publisher for publishing my book.

I hope that this book will be very useful to the readers in all aspects. I welcome helpful comments and suggestions from the readers for its further improvements.

D.C. Sikdar

CONTENTS

LIST OF SYMBOLS

a	=	constant
a	=	Cross sectional area
a_1	=	Cross sectional area
a_2	=	Cross sectional area
a	=	Element of Routh table
A	=	Constant
A	=	Heat transfer surface area
A	=	Magnitude of step change
A	=	Position of float
A	=	Reactant
A_1	=	Cross sectional area of tank1
A_2	=	Cross sectional area
b	=	Element of Routh table
b	=	proportional band
B	=	Bulb
B	=	Constant
B	=	Length of breadth
B	=	Magnetic Flux
B	=	Position of float
B	=	Product
B_1	=	Constant
c	=	Constant
c	=	Element of Routh table
c_o	=	Exit concentration
c_i	=	Inlet concentration
C	=	Controlled variable
C	=	Damper constant
C	=	Velocity coefficient
$C_i(t)$	=	Deviation Variable
C_1	=	Constant
C_2	=	Constant
C_3	=	Constant
C_D	=	Coefficient of discharge
Cp	=	Specific heat at constant pressure
Cp_{kf}	=	Specific heat of kettle fluid
Cp_m	=	Specific heat of mercury
d	=	Element of Routh table
d	=	Inner diameter
D	=	Denominator
e	=	Capacitance
e	=	Element of Routh table
e	=	Inverse of logarithm
f	=	Element of Routh table
f	=	Sustained frequency
f_n	=	Natural frequency
F	=	Volumetric flow ate of inlet
F(t)	=	Applied force
f(t)	=	Function
g	=	Acceleration due to gravity
g	=	Element of Routh table
G_c	=	Transfer function of the process
G_p	=	Transfer function of the controller
G_v	=	Transfer function of the valve
h	=	Difference in manometric fluid
h	=	Empty height of capillary
h	=	Heat transfer coefficient
h	=	Level of liquid in tank
h	=	Liquid column height
h_c	=	Height of the capillary column
h_1	=	Height of the capillary filled with air
h_{jf}	=	Heat transfer coefficient of jacketed fluid
h_s	=	Steady level of water in tank
h_{1s}	=	Steady state level of water in tank1
h_{2s}	=	Steady state level of water in tank2
h_m	=	Heat transfer coefficient of mercury
H	=	Transfer Function of measuring element

$H(t)$ = Deviation variable
$H_i(t)$ = Deviation variable
$H_2(t)$ = Deviation variable

i = i^{th} term of Numerical number
I_m = Imaginary axis
j = j^{th} term of Numerical number

k = Constant
k = Reaction rate constant
k = Spring constant
k = Steady state gain
k^l = Constant
k_c = Controller gain
k_c = Steady state gain
k_p = Steady state gain of process
k_v = Steady state gain of process

l = Length of the conductor
L = Laplace Transform
L = Length of the manometer
L = Length of the thermometer bulb
L_w = Breadth of the weirs

m = Mass of mercury
m_{kf} = Mass of kettle fluid
M = Mass of a body

n = Number of rows
n = Valve flow coefficient
N = Net encirclement
N = Numerator

o = Zero

P = Deviation variable
p = Output pressure signal
P = Pressure
P_a = Absolute pressure
P_{gtm} = Atmospheric pressure

p_1 = Applied pressure at section 1
p_2 = Applied pressure at section 2
P_g = Gauge pressure
p_1 = Pole1
p_2 = Pole2
p_3 = Pole3
p_4 = Pole4
P_1 = Pressure at section 1
P_2 = Pressure at section 2
p_s = Steady state pressure
P_v = Vacuum pressure

q = Inlet flow rate to tank1
q_1 = Outlet flow rate from tank1
q = Volumetric flow rate
q_1 = Inlet flow rate to tank2
q_2 = Outlet flow rate from tank2
q_s = Steady state flow rate
q_{1s} = Steady state inlet flow rate to tank1
q_{2s} = Steady state inlet flow rate to tank2
q_{os} = Steady state outlet flow rate
q_{1os} = Steady state outlet flow rate from tank1
q_{2os} = Steady state outlet flow rate from tank1
$Q(s)$ = Deviation variable
$Q_1(S)$ = Deviation variable
$Q_2(S)$ = Deviation variable

r = Centre of gravity
r = Polar coordinate
r = radial distance
r = Rate of reaction
R = resistance
R = Rangeability
R = Set point
R_1 = Resistance of tank1
R_2 = Resistance of tank2
Re = Real value

S = Break away point
S = function variable

t = Time
T = Period of Oscillation
T = Turn down ratio
T_1 = Initial temperature
T_2 = Final temperature
T_1 = Temperature of the source
T_2 = Temperature of the sink

u = Velocity of the conductive material

v = Velocity of the flowing fluid
v = Voltage
v = Volume of the bulb
V = Volume
V = load variable
V_1 = Initial volume
V_2 = Final volume
V_s = Steady state voltage

w = Watts

x = Concentration at any time
x = real value
$X(t)$ = Deviation variable
$x(t)$ = Inlet concentration
x_s = Steady state concentration

y = Concentration
y = Real value
y = Concentration at the outlet at any time
$y(t)$ = Exit concentration
y_s = Steady state concentration at the outlet
$Y(t)$ = Deviation variable

Z = Complex variable
Z = Characteristic equation
Z = number of zero
Z_1 = Zero1
Z_2 = Zero2
$Z(t)$ = Deviation variable

α = Angle of pole
α_1 = Angle of pole1
α_2 = Angle of pole2
α_3 = Angle of pole3
β = Angle of zero
β_1 = Angle of zero1
β_2 = Angle of zero2
δ = Change
ρ = Density
μ = Viscosity
σ = Boltzmann constant
ω = Radian frequency
ω_c = Corner frequency
ω_{co} = Cross over frequency
π = Pai
τ = Time constant
τ_I = Integral time constant
τ_D = Derivative time constant
τ_1 = Time constant of tank1
τ_2 = Time constant of tank2
τ_P = Time constant of the process
τ_V = Time constant of the valve
τ_{oil} = Time constant of oil
τ_{air} = Time constant of air
θ = Angle
θ_1 = Angle of pole1
θ_2 = Angle of pole2
θ_3 = Angle of pole3
θ_d = Angle of departure (Negative)
θ_d^* = Angle of departure (Positive)
Φ = Phase angle
$\in$ = Damping coefficient
AR = Amplitude Ratio
GM = gain Margin
DR = Decay ratio
OS = Over shoot
PM = Phase Margin

Chapter 1

Instrumentation Fundamental

1.1 INTRODUCTION

Instrumentation is the art of measuring the value of some plant parameters such as temperature, pressure, level, flow and supplying a signal that is proportional to the measured parameter to provide the indication or automatic control. The perfection of the instrument depends upon the measurement. There are two basic method of measurement.

(i) **Direct Method :** Length measurement, *e.g.*, all physical dimensions.

(ii) **Indirect Method :** To determine the quality of product, *e.g.*, milk, bacteria dimination in milk.

There are three or four phases in most of the measuring system, each phases being made up with a distinct component or group of components which perform required and definite steps in measurement.

1.2 CRITERIA OF SELECTION OF MEASURING INSTRUMENTS

The following factors to be considered for selection of a measuring instrument :

(i) How accurate is the measurement to be made, *i.e.*, the accuracy from the measuring instrument.

(ii) When the final data reqnired, *i.e.*, at the time of taking the measurements or latter on.

(iii) The cost criterion, *i.e.*, how expensive can the measuring process be?

(iv) In what form the data should be displayed, *i.e.*, indicating, recording, integrating or photographic.

(v) Whether quantity to be measured has constant value or is it a time variant? If time variant then whether variation is linear, parabolic or some other type?

1.3 STATIC CHARACTERISTICS OF AN INSTRUMENTS

The static characteristics of an instrument are in general those that must be considered when the instrument is used to measure a condition not varying with time. The static characteristics of an instruments are :

(i) Accuracy

(ii) Reproducibility
(iii) Sensitivity
(iv) Static error
(v) Drift
(vi) Dead zone

1.3.1 Accuracy

It is the degree of correctness with which a measuring means yields the true value referred to accepted engineering standards, such as standard meter. Accuracy is determined as the maximum amount by which the result differs from the true values. It is influenced by the factors like static error, dynamic error, reproducibility and dead zone.

1.3.2 Reproducibility

It is the degree of closeness with which a given value may be repeatedly measure. It may be specified in terms of units for a given period of time. Perfect reproducibility means that the instrument has no drift.

1.3.3 Sensitivity

It is defined as the ratio of output response to a specified change in input.

1.3.4 Static Error

It is the difference between the true value of a quantity not changing with time and the value indicated by the instrument.

$$\text{Instrument reading} = \text{True value} + \text{Scale error}$$

For plus static error the instrument read high and for minus static error the instruments read low.

1.3.5 Drift

It is an insidious quality in industrial instrument because it is rarely apparent and must be carefully guarded against by continuous inspection and maintenance.

1.3.6 Dead Zone

It is the largest range of values of a measured variable to which the instruments does not responds. This is some times called dead spot and hysteresis. It is the range within which variable can vary without being detected. Dead zone usually occurs with friction in an indicating or recording instrument, particularly the latter. It may also be found in certain kinds of mechanisms that can only indicate small and discrete changes in value of a measured variable.

1.4 DYNAMIC CHARACTERISTICS OF AN INSTRUMENT

Instrument rarely responds instantaneously to change in the measured variable. Instead, they exhibit a characteristic slowness,or sluggishness due to such things as mass, thermal capacitance, fluid capacitance etc. Furthermore, pure delay in time is often encountered where the instrument wait for some reactions to take place.

The dynamic behaviour of an instrument is determined by subjecting its primary element to some known and predetermined variation in measured quantity. The most common variations are :

(i) Step change
(ii) Linear change
(iii) Sinusoidal change
(iv) Speed of response
(v) Fidelity
(vi) Measuring lag

1.4.1 Step Change

It is a change in which the primary element is subjected to an instantaneous and finite change in measured variable.

1.4.2 Linear Change

It is a change in which the primary element is following a measured variable, changing linearly with time.

1.4.3 Sinusoidal Change

It is a change in which the primary element is following measured variable, the magnitude of which changes in accordance with a sinusoidal function of constant amplitude. The physical conditions of imposing these changes on an instrument may cause great difficulties.

1.4.4 Speed of Response

It is the quickness of an instrument to read the measured variable.

1.4.5 Fidelity

It is the degree to which an instrument indicates the changes in measured variable without dynamic error.

1.4.6 Measuring Lag

It is a retardation or delay in the response of an instrument to changes in the measured quality.

1.5 TRANSDUCER

Modern technology requires electrical instrumentation, not only in electrical works but also practically in all other kinds of works.

Most of the non-electrical applications, such as mechanical or thermal problems requires a device to convert the quantity or phenomenon to be observed into a voltage or current, which can then be observed on an oscilloscope, read on a meter or recorded on a chart. Such a device is called transducer. It is defined as a device actuated by power from one system and supplying power to another system, *e.g.*, a thermocouple is actuated by power in the form of heat and it supplies electrical power to a control or measuring circuit. Various type of transducers are available for use. They can be classified according to their function. Most of them are commonly used with electronics devices such as amplifiers. Some common transducers are :

(i) Pressure transducer
(ii) Level transducer
(iii) Velocity transducer

1.5.1 Pressure Transducer

A transducer is a device for changing the variation of a quantity to those of another. For example, pressure to voltage. A pressure sensor usually acts as a transducer and generate a electrical signal as a function of the pressure imposed. It is used for control and monitoring the thousands of everyday applications.

For example, it can be used indirectly to measure the variables such as fluid flow, water level and altitude. Hence, pressure sensor alternatively called pressure transducer or pressure transmitter. The cost of the pressure sensor depends upon the technology in design and its performances. Some pressure sensors are designed to measure in a dynamic mode for capturing very high speed change in pressure, *e.g.*, combustion pressure in gas turbine.

1.5.2 Level Transducer

A level sensor usually acts as a transducer and generate a electrical signal as a function of the level imposed. The level sensors can be continuous or point values. The substance to be measured either can be inside a container or it can be in natural form. The continuous level sensors measure level within a specified range and determine the exact amount of substances in a particular place, while the point value sensor indicate whether the substance is above or below the sensing point.

1.5.3 Velocity Transducer

Velocity sensor usually acts as a transducer and generate the electrical signal as a function of the velocity imposed. This information passed on to the VIC (Velocity Indicating Controller) and convert the electrical signal to pressure. Depending on the pressure pneumatic control valve work and allow the fluid to pass through with a required velocity. Velocity transducer generally used for flow control in industry.

1.6 INDICATOR

An indicating instrument is used when only the present value of the variable has meaning and the past record is of no consequence.

Indicators are of three types :
(i) Eccentric scale indicator
(ii) Concentric scale indicator
(iii) Linear scale indicator

1.6.1 Eccentric Scale Indicator

The eccentric scale indicator is commonly used on mechanical instruments such as pressure thermometers, flow meters, and pressure gauges. The scale length is usually about 6-8 inch, but may be much smaller. A simple eccentric scale indicator is shown in Fig. 1.1.

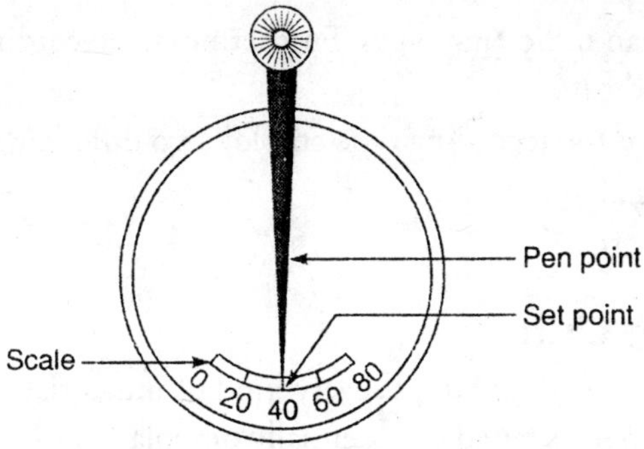

Fig. 1.1 Eccentric scale indicator.

1.6.2 Concentric Scale Indicator

The concentric scale indicator is used for nearly all kinds of instruments. The scale length is usually about 20 inch, in the precision type instruments. In simple dial type pressure gauge thermometers the scale length may be as small as 2 inch. A simple concentric scale indicator is shown in Fig. 1.2.

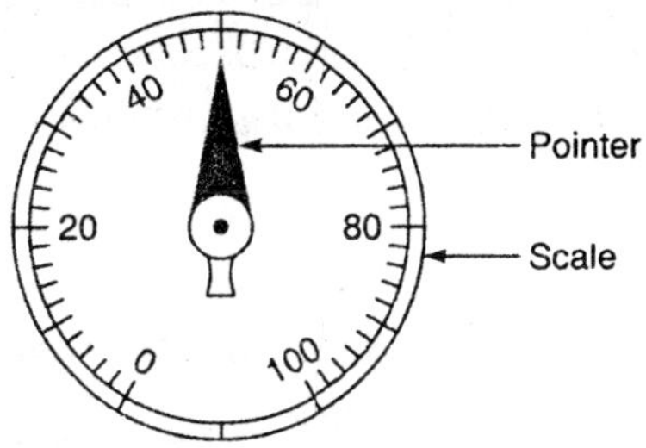

Fig. 1.2 Concentric scale indicator.

1.6.3 Linear Scale

The linear scale is commonly used only on the potentiometer resistance thermometer type instrument. For such instruments an indicator is made simply by omitting the strip chart, chart drive an pen. Galvanometer type millivolt meters employ a linear scale in a curved from.

The use of the various indicating instrument is generally restricted, because a free choice of scale type is not possible. For readability there can be no doubt that the concentric scale generally offers the greatest scale length in the smallest space.

1.7 RECORDER

Modern industrial plant relies on instrumentation as a regular part of its production equipment. In the vast majority, instrumentation is more elaborate, and in an increasing number of processes, complete instrumentation is a necessity without that plant can not operate efficiently.

A whole new field is gradually taking shape, and the instrumentation engineer is becoming an important factor in the engineering of newly designed manufacturing process.

Process instrumentation, whether it is the temperature control of milk, the operation control means for a distillery or the control means for a large steel plant is accomplished largely by experience.

A recording instrument should be employed when a permanent record of the variable is desirable. A record is useful for three reasons :

(i) The process operator may refer to the instrument to observe the trend of variable as a guide to process operation.

(ii) The record may be useful in locating trouble on the job.

(iii) The record may be required for reference to past performances.

A recording instrument should not be used when any of the foregoing functions are not required and when a simple indicator will suffice. Obviously recording instruments are more expensive and require more maintenance than indicating types. Furthermore, a recording instrument without a chart in place is not useful as an indicator.

Industrial instruments of the recording type employ two different styles of recording charts :

(i) Circular recording chart
(ii) Strip recording chart

1.7.1 Circular Recording Chart

The circular recording chart is used on almost every kind of industrial instrument like flow, pressure, temperature and level. This chart is based on essentially on polar co-ordinates with the exceptions that the zero is moved away from the geometric centre and the time lines are segments of an arc instead of straight lines. The chart is clamped at the geometric centre and rotates about the centre. The period of rotation of the circular chart is usually 8 hr, 24 hr or 7 days, although nearly any speed of rotation can be obtained by choosing a properly geared synchronous motor. A simple circular recording chart is shown in Fig. 1.3.

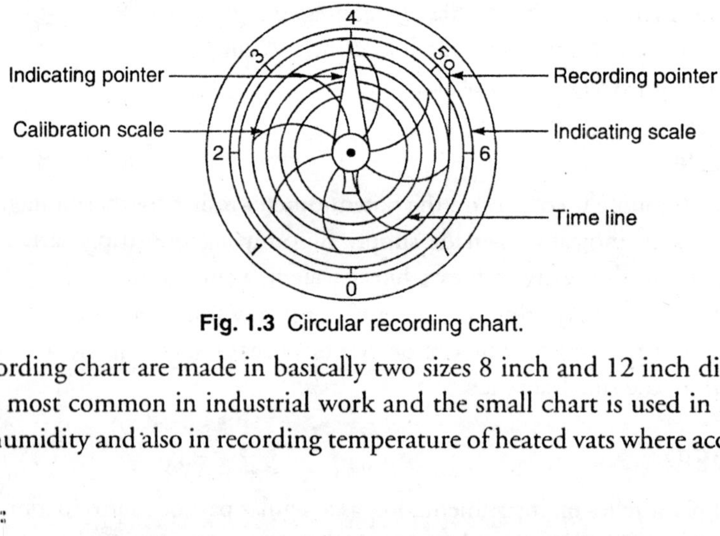

Fig. 1.3 Circular recording chart.

Circular recording chart are made in basically two sizes 8 inch and 12 inch diameter. The larger chart is by far the most common in industrial work and the small chart is used in recording ambient temperature and humidity and also in recording temperature of heated vats where accuracy of reading is not essential.

Advantages :

(i) The entire record of one process period is available at a glance.
(ii) Long chart length is available for one process period.
(iii) It is easily filed for future reference.

Disadvantages :

(i) Time values towards the centre are difficult to read.
(ii) The chart must be changed once every process period.

Circular chart is most advantageously used to record a mildly fluctuating variable at about 50% of scale span where the process period is not greater than 24 hrs.

The circular charts are printed under carefully controlled temperature and humidity conditions on high grade paper. Humidity variation under actual use causes the paper to expand and contract. Since the fibres in the chart to expand or contact to elliptical shape, the error in calibration is very small and is not corrected or compensated.

1.7.2 Strip Recording Chart

The strip recording chart is generally used on the potentiometer type instrument or on the resistance thermometer. The chart is driven from the supply roll to the reroll by a synchronous motor. The speed of the paper is usually given in inches per hr, 1, 2 and 4 inch per hr is very common. Speed upto 20 inch per hr are available. A simple strip chart recorder is shown in Fig. 1.4.

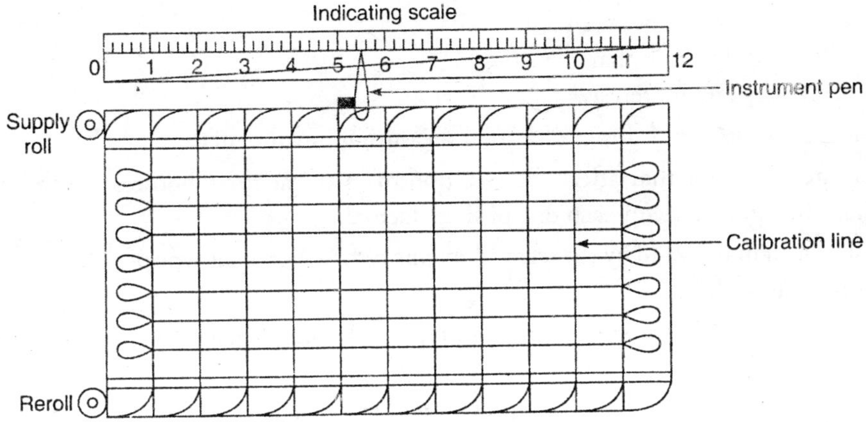

Fig. 1.4 Strip chart recorder.

The strip recording chart used with the recording pyrometer and resistance thermometer type instrument has rectangular coordinates and a width of approximately 10 inch. Strip recording charts for galvanometer type instruments are not so common in recordings. These are usually about 6 inch in width and the time lines consists of arc segments instead of straight lines because the pen on such instrument is arranged to move in arc.

Advantages :

(i) It does not require frequent changing.

(ii) Long process period are easily recorded.

(iii) Readability is same all values of scale span.

Disadvantages :

(i) A limited length of the chart is visible.

(ii) It is difficult to file and not easily used for reference.

The strip chart is most advantageously used to record a variable changing between wide limits and to records a variable over long process period.

REVIEW QUESTIONS

1.1 Explain the fundamental properties of a good measuring instrument.

1.2 Explain the static characteristics of an instrument.

1.3 Explain the dynamic characteristics of an instrument.

1.4 What are the basic requirements of recorders?

1.5 Write the working principle of strip chart recorder.

1.6 Write the working principle of circular chart recorder.

1.7 Explain dynamic accuracy.

1.8 Write notes on transducers.

1.9 Explain the following terms :
(i) Range and span
(ii) Accuracy and precision
(iii) Static error and dynamic error

1.10 Explain the difference between :
(i) Static and dynamic characteristics.
(ii) Sensitivity and hysteresis.

1.11 What are the essential features of a good measuring instruments

1.12 What are the factor that affect the selection of instruments? Choose a suitable example and explain the significance of each one of these factors.

1.13 What do you understand by the term "dynamic error"? For what type of change dynamic error is important and why?

1.14 What is indicator? What are the different types of indicator generally used in industry? Explain them in detail.

1.15 Explain the working principle of a level transducer in detail.

□□□

Chapter 2

Pressure Measurements

2.1 INTRODUCTION

Pressure is the force acting on unit area and is one of the important process variable. Usually pressure is expressed under the following names :

(i) Absolute pressure
(ii) Differential pressure
(iii) Gauge pressure
(iv) Vacuum pressure

2.1.1 Absolute Pressure

It is the total pressure exerted by the fluid. It is designated by P_a.

2.1.2 Differential Pressure

It is the algebraic difference between the two pressures.

2.1.3 Gauge Pressure

It is the special case of differential pressure where one pressure is atmospheric. It is the algebraic difference between the total pressure exerted by the fluid and the pressure exerted by the atmosphere. It is represented by the equation

$$P_g = P_a - P_{atm} \qquad \ldots(2.1)$$

where, P_g = Gauge pressure
P_a = Absolute pressure
P_{atm} = Atmospheric pressure

2.1.4 Vacuum Pressure

It is the another case of differential pressure and is defined as

$$P_v = P_{atm} - P_a \qquad \ldots(2.2)$$

where, P_v = Vacuum pressure
P_{atm} = Atmospheric pressure
P_a = Absolute pressure

Negative vacuum and positive gauge pressure are identical. Pressure is normally expressed as kg/cm^2 or pound per square inch (Psi). It can also be expressed as multiplication of liquid column height and its density. In SI unit it is expressed as N/m^2.

$$P = \rho g h \qquad ...(2.3)$$

where, P = Pressure, N/m^2

ρ = Density of liquid, kg/m^3

g = Acceleration due to gravity, m/s^2

h = Liquid column height, m

One standard atmosphere may be expressed as :

1 standard atm = 760 mmHg at 25°C

= 29.92 mmHg at 32°F

= 10.33 kg/cm^2

= 14.223 Psi

= 1.013×10^5 N/m^2

2.2 CLASSIFICATION OF PRESSURE MEASURING INSTRUMENT

The pressure measuring instrument may be classified as :

(i) Instrument which uses liquid column

(ii) Instrument which uses elastic pressure element

(iii) Instrument which uses electrical devices

(iv) Instrument which uses electronic devices.

2.2.1 Instrument which Uses Liquid Column

These are generally called as manometers. These are the simplest form of pressure measuring devices. Here pressure is measured in terms of head of liquid column (h) and is converted to pressure by multiplying density (ρ). It is represented by the equation

$$P = \rho\ h,\ kg/cm^2 \qquad ...(2.4)$$

Manometers are normally limited to measurement of differential pressure upto 0.2 kg/cm^2. The instrument which uses liquid column are as follows :

(i) U-tube manometer

(ii) Inclined tube manometer

(iii) Well type manometer

(iv) Cistern barometer

(v) Fixed cistern barometer

(vi) Mcleod gauge.

(i) U-Tube Manometer

It is the simplest of all the manometer. It consists of a U shaped glass tube and seale attached to measure the difference of liquid in the two limbs. A simple U-tube manometer is shown in Fig. 2.1.

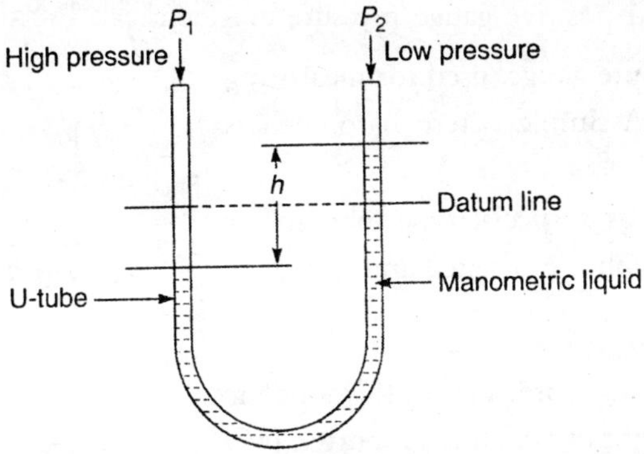

Fig. 2.1 U-tube manometer.

When the two different pressure sources are applied to two limbs, the differential pressure is given in terms of difference in liquid column height '*h*' which can be converted to pressure units in kg/cm^2 by multiplying density in corresponding units.

If one limb is open to atmosphere it gives gauge pressure. If one limb is evacuated and sealed it gives absolute pressure.

(ii) Inclined Tube Manometer

In this case one limb of the manometer is inclined to the vertical one. A simple inclined tube manometer is shown in Fig. 2.2.

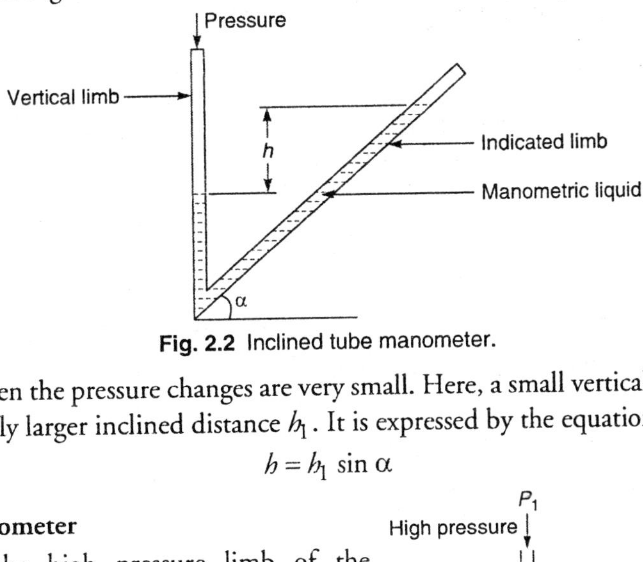

Fig. 2.2 Inclined tube manometer.

This is used when the pressure changes are very small. Here, a small vertical height is measured in terms of comparatively larger inclined distance h_1. It is expressed by the equation

$$h = h_1 \sin \alpha \quad \text{...(2.5)}$$

(iii) Well Type Manometer

In this case the high pressure limb of the manometer is connected to large reservoir. This design permits the pressure to be measure by measuring the rise in the level of narrow tube. A simple well type manometer is shown in Fig. 2.3.

The cross-sectional area of the narrow tube is very small compared to the reservoir. Hence, the level change in the reservoir is very less compared to the narrow tube.

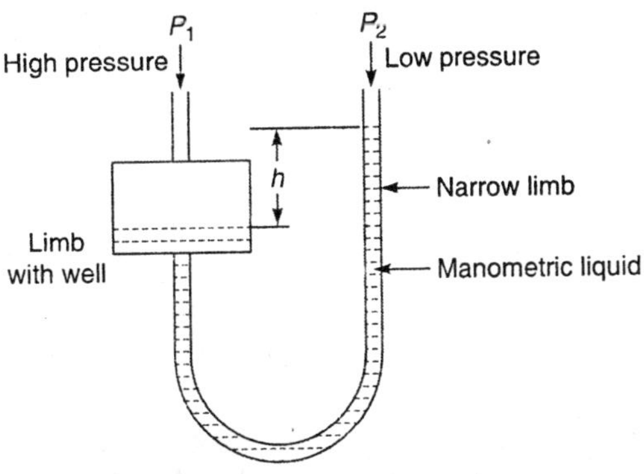

Fig. 2.3 Well type manometer.

(iv) Cistern Barometer

It is an absolute pressure gauge, used for measuring the atmospheric pressure. A simple cistern barometer is shown in Fig. 2.4.

It can be considered as a special type manometer where the narrow vertical limb is evacuated and sealed.

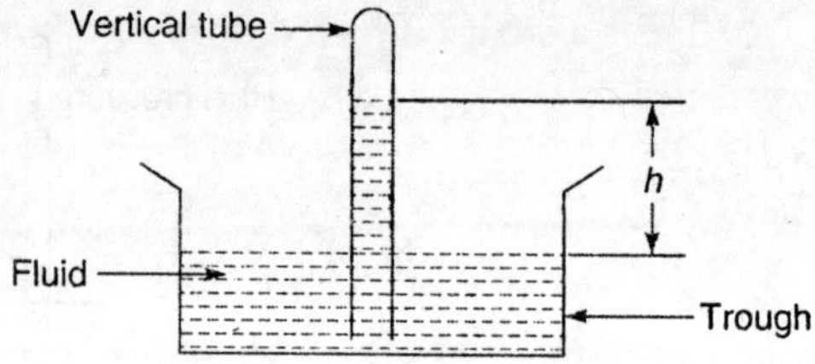

Fig. 2.4 Cistern barometer.

(v) Fixed Cistern Barometer

It is used to measure gauge pressure. In this type the liquid sealed bell is the weight of which is supported by a spring. A typical fixed cistern barometer is shown in Fig. 2.5.

The applied pressure rises the bell and is counteracted by spring tension. The motion of the bell is converted into circular motion of the pointer which is moving on a circular scale, is calibrated in terms of pressure.

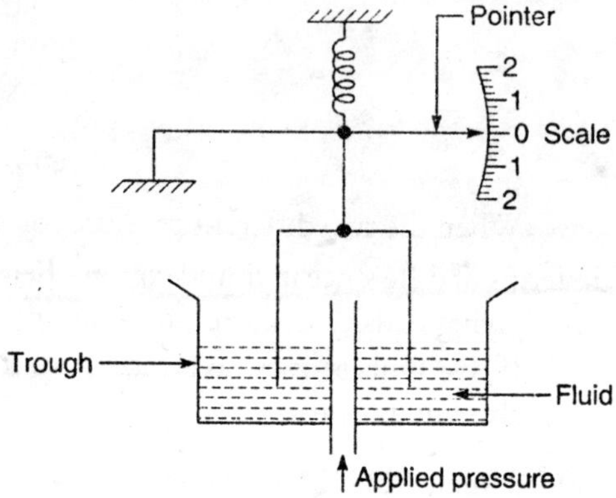

Fig. 2.5 Fixed cistern barometer.

(vi) Mcleod Gauge

This is a device used to measure very low pressure, *i.e.*, 0.1 micron to 50 mmHg. The principle used here is Boyle's law

$$P_1V_1 = P_2V_2 \qquad \ldots(2.6)$$

A known volume of gas V_1 whose unknown pressure P_1 is compressed to a known pressure P_2 and volume V_2. Then unknown pressure P_1 may be calculated using the equation

$$P_1 = \frac{P_2V_2}{V_1} \qquad \ldots(2.7)$$

At first mercury is filled upto the cut off point and the unknown pressure is connected at the reference point which is to be determined. The capillary above the cutoff point and bulb *B* is filled up with gas. A simple Mcleod gauge is shown in Fig. 2.6.

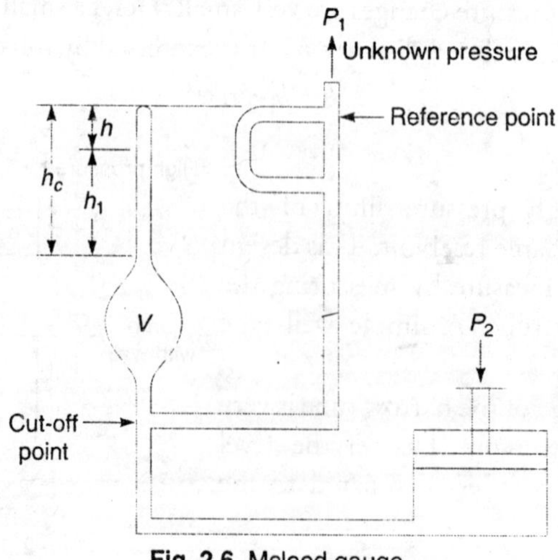

Fig. 2.6 Mcleod gauge.

Then mercury is forced upto zero level in the reference capillary and the unknown pressure P_1 is calculated using Boyle's law.

$$P_1 V_1 = P_2 V_2 \qquad \ldots(2.6)$$

where, P_1 = Unknown pressure
$V_1 = V + ah_c$, Initial volume of the gas intrapped
V = Volume of the bulb + Capillary tube upto the bulb from cut-off point
a = Area of cross-section of capillary
$P_2 = P_1 + h$, Final amplified pressure
$V_2 = ah$, Volume of the gas in the capillary

$$P_1 (V + ah_c) = (P_1 + h)\, ah \qquad \ldots(2.8)$$

$$P_1 V + P_1 ah_c = P_1 ah + ah^2$$

$$P_1 V + P_1 ah_c - P_1 ah = ah^2$$

$$P_1 (V + a(h_c - h)) = ah^2$$

But,

$$(h_c - h) = h_1$$

$\therefore$

$$P_1 (V + ah_1) = ah^2 \qquad \ldots(2.9)$$

$$V >> \ldots ah_1$$

Therefore, ah_1 can be neglected.

$$P_1 = \frac{ah^2}{V} \qquad \ldots(2.10)$$

2.2.2 Instrument which Uses Elastic Pressure Element

These pressure gauge have elastic element that convert pressure signal into proportional mechanical displacement. Hence, it is also called transducer.

The instrument which uses elastic pressure elements are as follows :

(i) Bourdon pressure gauge
(ii) Diaphragm pressure gauge
(iii) Bellows pressure gauge

(i) Bourdon Pressure Gauge

Bourdon pressure gauge consists of a thin tube of elliptical cross-section sent into an arc of a circle lengthwise fixed at one end of source which accept the applied pressure. Other end is free and closed. A typical Bourdon pressure gauge is shown in Fig. 2.7.

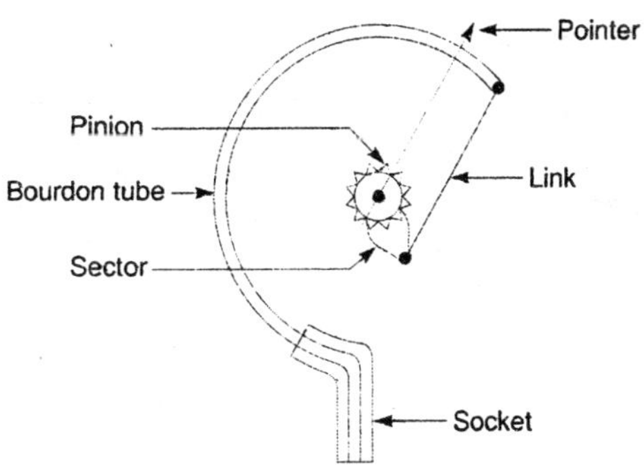

Fig. 2.7 Bourdon pressure gauge.

When the pressure is applied the oval shaped tube tend to become circular with subsequent increase in the radius of the circular arc. The movement of the free end of the tube is transmitted to a section through an adjustable connecting link. The sector is in a mesh with a pinion and a fixed pointer which rotate over a calibrated scale to give an indication of pressure. As the reference pressure in this case surrounding the Bourdon tube is usually atmospheric, the pressure measured is usually gauge pressure.

Requirements of the tube material :

(i) It must be hard enough to withstand the applied pressure without any part of it being stressed above the limit of proportionality.

(ii) It must be sufficiently stable to maintain its calibration indefinitely.

(iii) It must be immune to corrosion from the fluid inside the tube and from atmosphere outside it.

(iv) It must be easy to fabricate.

Gauge should be selected with a full scale pressure range of approximately twice the normal operating pressure and should not exceed 75% of the full scale. Some typical tube material and its pressure range is given in Table 2.1.

Table 2.1 : Material construction of bourdon tube.

S. No.	Material	Range (kg/cm^2)
1.	Phosphor bronze	1 – 70
2.	Berryllium copper	30 – 350
3.	Alloy steel	650 – 5500
4.	Stainless steel	2 – 70
5.	Monel	70 – 1350

There are some special type of Bourdon tube where Bourdon element is wound as helix or spiral.

Helix : It is a type of Bourdon element where the Bourdon tube is wound in the form of a helix to increase the tip movement. The design transmits only the circular component of the tip movement to open arm which is directly proportional to the pressure change.

Spiral : It is also a modification of Bourdon element used widely in industry. It is a thin walled tube flattened at both ends and is formed to spiral shape. The spiral arrangement provides larger movement to the free end which used for recording with suitable material. This type can be used to measure pressure in the range 0.7 to 260 kg/cm^2. As no sector and pinion are required, hence friction losses are reduced.

(ii) Diaphragm Pressure Gauge

Diaphragm is a circular element made of thin ductile metal which can be supported and the surface is free to move on applying force. The face of the diaphragm can be dished, flat or corrugated depending on strength and deflection desired. Some different faces of diaphragm are shown in Fig. 2.8.

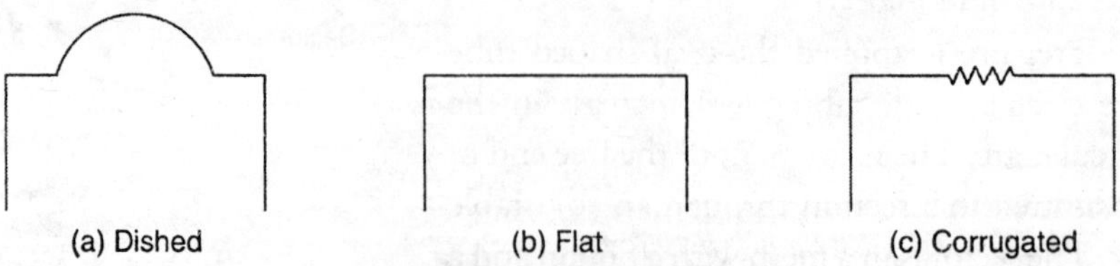

Fig. 2.8 Diaphragm faces.

In high precision instruments the diaphragm are used in pair back to back in the form of capsules and forming a battery of several capsules. A simple diaphragm gauge is shown in Fig. 2.9.

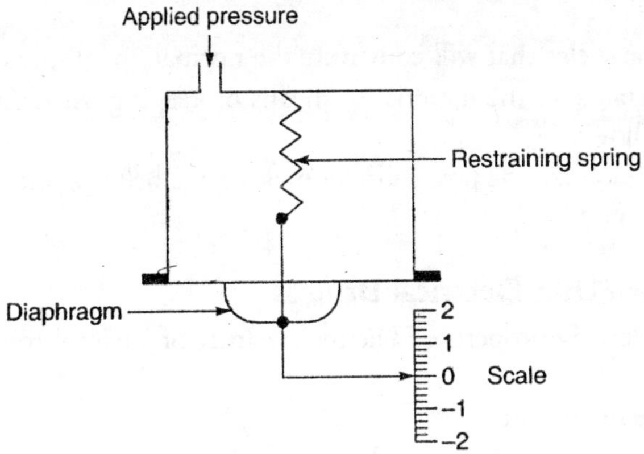

Fig. 2.9 Diaphragm pressure gauge.

The corrugated diaphragm will have 4 times the deflections of flat diaphragm subjected to the same pressure and its deflection pressure curve is linear in character. A typical material and its pressure range is given in Table 2.2.

Table 2.2 : Material construction of diaphragm.

S. No.	Material	Range (kg/cm^2)
1.	Brass	0.35 – 15

Requirements of diaphragm gauge :

(i) Dimension of the diaphragm and the total load applied on it must be compatible with physical properties of the materials used.

(ii) The flexibility must be large to provide the required sensitivity.

(iii) The dial scale should be calibrated to account for the accuracy.

(iv) The elastic modulus should not change within the working temperature.

(v) Output should be linear.

(iii) Bellows Pressure Gauge

The metallic bellows is a series of circular part so formed or joined that they can be expanded or contracted by the application of pressure or vacuum but not appreciable in other direction. A simple bellows pressure gauge is shown in Fig. 2.10.

By increasing the number of convolution in the bellows the axial movement can be increased. Increasing the diameter of the bellows will increase the force for a given pressure and makes the instrument more sensitive pressure gauge.

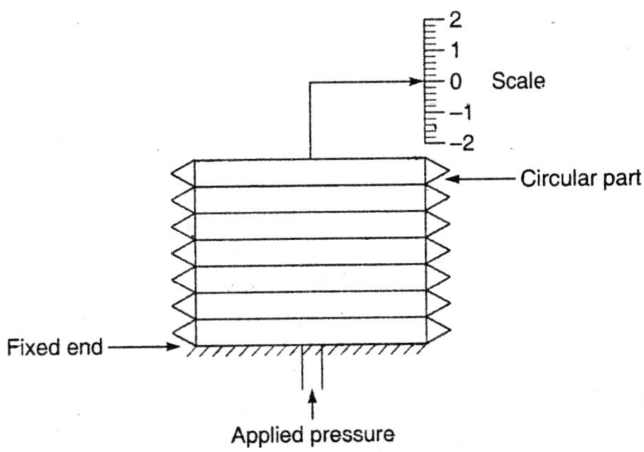

Fig. 2.10 Bellows pressure gauge.

Requirement of bellows material :

(i) Strength of pressure to be measured.

(ii) Hysteresis.

(iii) Fatigue, *i.e.*, the cycles that will constitute the normal life of units.

(iv) Corrosion resistance of the material or means of keeping the corrosive contaminants from reacting the bellows.

(v) Ease of fabrication, *i.e.*, the possibility of making the bellows out of the material that satisfy the condition 1 and 2.

2.2.3 Instrument which Uses Electrical Devices

It makes the use of the following properties. The temperature of a wire through which electric current flows depend on :

(i) The magnitude of current.

(ii) The rate at which heat is dissipated by conduction, convection and radiation.

If the current is kept constant then the temperature will depend on the rate at which heat is dissipated, which in term depend on the conductivity of media. Conductivity is again depend on the density of the media. If the pressure is reduced then conductivity is reduced and wire become hotter for the same current.

Thus we can conclude that for a given current the temperature of the wire directly dependent on the pressure of the surrounding medium. By measuring temperature of the wire or some measurable property dependent on temperature of the wire, with suitable calibration we can know the pressure of the medium. The instrument which uses electrical devices for the measurement of pressure are as follows :

(i) Pirani gauge

(ii) Thermocouple gauge

(i) Pirani Gauge

Principle used here is that the pressure is dependent on the temperature of the wire if the current is constant. But change in temperature changes the resistance of the wire. A simple pirani gauge is shown in Fig. 2.11.

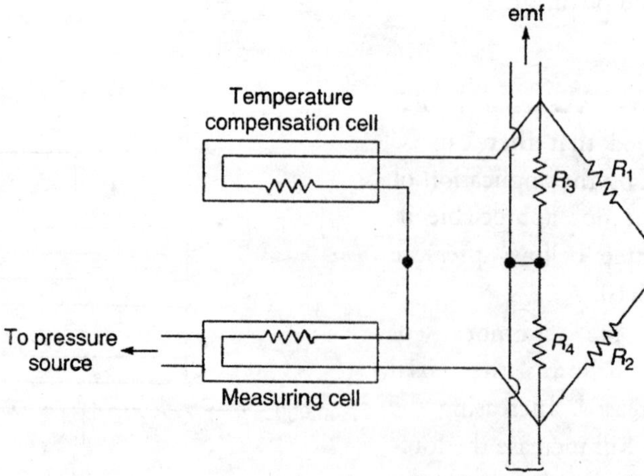

Fig. 2.11 Pirani gauge.

The open ended tube is exposed to the vacuum to be measured. The closed off tube is sealed at a pressure lower than 1 micron and acts to compensate for temperature variations.

The pirani gauge filament should not be used above 330°C. If organic vapour from grease, oil gasket materials are present in the system, at higher temperature decomposes organic vapours leaving a deposit on the filament causing a zero shift of calibration. The range is 20–1000 microns of Hg.

(ii) Thermocouple Gauge

It consists of two platinum resistance coil kept in two identical cells, one is evacuated and sealed and other is connected to the pressure source. These coils are connected to a regulated power supply to heat it. When the coils are heated the presence of any gas in the cell cools the resistor below the corresponding temperature in the evacuated cell. A simple thermocouple gauge is shown in Fig. 2.12.

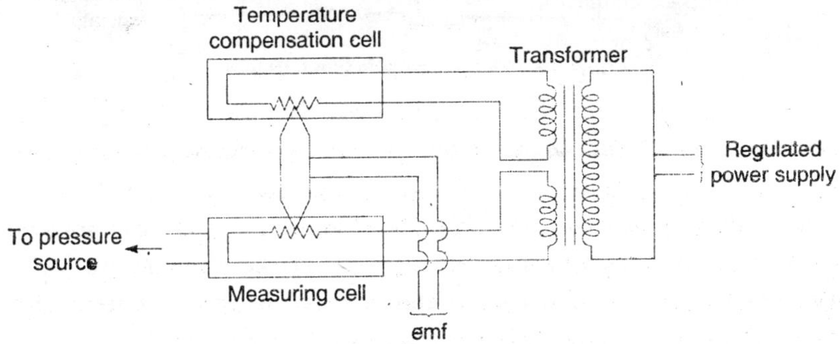

Fig. 2.12 Thermocouple gauge.

The thermocouple attached at the middle of each resistor will develop an emf which is proportional to the temperature difference of these junction.

This emf in terms is proportional to the pressure. The potentiometer which is used to measure the emf is calibrated in terms of pressure. The range, merit and demerits of this gauge is same as that of the pirani gauge.

2.2.4 Instrument which Uses Electronic Devices

It makes the use of following properties. Electron emitted at the cathode by the thermionic emission are attracted towards the grid, pass through the grid and start toward the plate. Ions are formed by collision of the electrons with molecules of the gas present in the tube. The instrument which uses electronics devices for the measurement of pressure are as follows :

(i) Hot filament ionization gauge
(ii) Cold cathode ionization gauge
(iii) Radio active ionization gauge

(i) Hot Filament Ionization Gauge

The hot filament ionization gauge is thermionic type. The tube is the form of triode, with cathode, grid and plate.

The electrons emitted at the hot cathode by thermionic emission are attracted towards the grid. Because of the velocity of electron it start towards the plate. Ions are formed by the collision of the electrons with the molecules of the gas present in the tube. A typical hot filament ionization gauge is shown in Fig. 2.13.

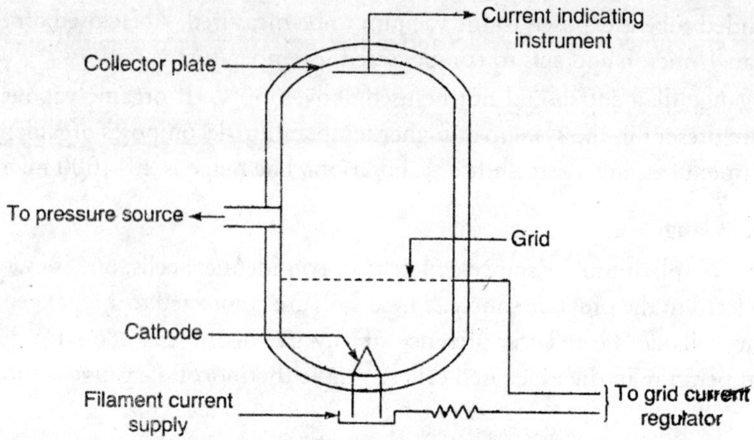

Fig. 2.13 Hot filament ionization gauge.

The +ve ions are collected on the plate and as a current +ve ion current exists. The amount of ion current is proportional to the amount of gas present provided the electron current is constant.

The electron to grid is maintained by a grid current. The ion current is maintained by instrument similar to the atomic balance potentiometer. This type of ionization gauge can not be used at pressure above 2 microns because of the probability of burning out the element. An ionization gauge in conjunction with a pirani gauge is used to measure absolute pressure above 2 microns. It can also be used for the measurement of absolute pressure of inert gases like O_2, CO_2 and H_2.

(ii) Cold Cathode Ionization Gauge

It consists of two parallel flat plates as cathode and a cylindrical anode placed in between them. When a very high voltage is maintained between anode or cathode, the electrodes will discharge ion. At low pressure the probability of these electrode colliding with the molecules and ionizing them are very less. And it is quite possible that birth rates of new electrons of ionization is smaller than death rates of electron. To over come this difficulties the gauge is placed in between the poles of an electromagnet so that magnetic field is at right angles to the plates. A typical cold cathode ionization gauge is shown in Fig. 2.14.

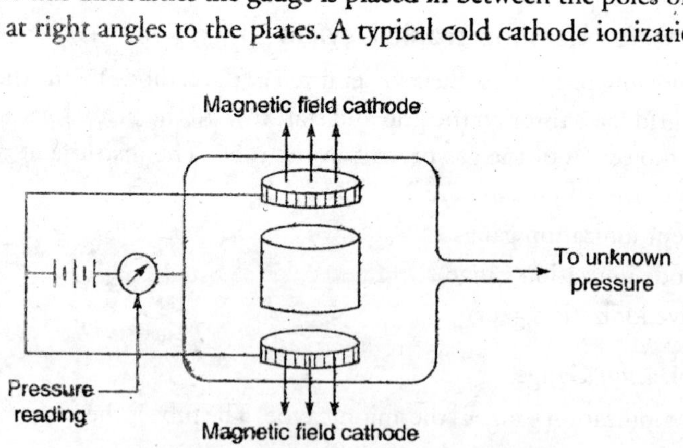

Fig. 2.14 Cold cathode ionization gauge.

The magnetic field generated by electromagnetic force the electron to move towards the anode cylinder in a helical path. This increased path length of electrons correspondingly increases the opportunity of its encounter with molecules and ionize molecules of gas even though they are very few in number.

This instrument may be used as leak detector. It has range ability 10^{-3} to 10^{3} mmHg.

(iii) Radio Active Ionization Gauge

If a source of radio active material is installed in the chamber (source), the alpha particles given off by the radio isotopes can ionize the gas molecules in the same way as the fast moving electrons could do in the thermionic gauge.

Radio active source is placed in the air tight chamber. The emitted alpha particle will collide with the gas molecules to produce ions.

The chamber is fed with DC source through partition insulation. The collector has –ve charge compared to the wall of the chamber. The +ve charged ions are attracted towards negatively charged ion collector, causing a current in the external circuit which can be amplified and measured in terms of pressure of the medium. The ionic current is of the order of 2×10^{-10} amp/mmHg.

It is very similar to the thermionic type with the following problem of thermionic gauge eliminated :

(i) The cathode can oxidize if appreciable amount of oxygen are present.

(ii) The gases being analyzed may be dissociated because of the high temperature.

(iii) The cathode may be attached by the gases or it may undergo undesirable chemical combination with them.

The gauge of operation of radio active ionization gauge is 10^{-3} to 10 microns of Hg.

REVIEW QUESTIONS

2.1 Explain the working principle of U-tube and inclined tube manometer.

2.2 Classify the pressure measuring instruments.

2.3 How do you calculate the pressure drop when water is flowing in tubes by using manometer?

2.4 Describe the principle of operation of Mcleod gauge with neat sketch.

2.5 What is the relationship between pressure and liquid level measurement? What pressure will be created by a column of liquid 5 m high, if the density of the liquid is 5000 kg/m^{3}?

2.6 Why are the elastic element pressure gauge preferable to liquid column manometer in industry? What materials are generally used for the elastic elements in pressure gauge? What characteristics should they posses? What is the range of operation of such pressure gauge?

2.7 Narrate the working of a pressure spring or bourdon tube pressure gauge.

2.8 Describe the working of a diaphragm pressure gauge with a neat sketch.

2.9 With a neat sketch explain the working of a bellows pressure gauge.

2.10 With the help of neat sketches explain the functioning and principle of operations of the following pressure measuring instruments :

(i) Pirani gauge

(ii) Ionization gauge

2.11 What are the various temperature measuring instruments used in industries to measure high temperatures? Explain the principle of operation and functioning of any two of them. Mention the range of operations.

2.12 What is the principle involved in pirani gauge. With a neat circuit diagram explain the working of pressure measurement using pirani gauge.

2.13 Draw a neat circuit diagram of thermocouple gauge and explain the working principle with range of operation.

2.14 What is the principle involved in ionization gauge? Draw a neat diagram of hot filament ionization gauge and explain its working principle.

2.15 With a neat diagram of cold cathode ionization gauge explain its working principle.

2.16 Write a notes on radio active ionization gauge.

2.17 With a neat diagram of thermocouple gauge explain its working principle.

□□□

Chapter 3

Temperature Measurements

3.1 INTRODUCTION

Temperature is the measurement of degree of hotness or coolness of a body. It is one of the most important variable in industrial processing. Primarily, because it is a fundamental condition characteristic of the thermal state of a body. In the case of industrial processes, it is this thermal state that determines whether desired condition or not. As in certain chemical processes, the physical conditions of a manufactured product. Temperature and heat flow are related quantitatively by the law of thermodynamics which express the fact that heat flows of it own accord from a body at higher temperature to a body at a lower temperature. It is therefore important in temperature measurement to remember that two bodies in intimate contact are at the same temperature only if there is no heat flow between them. Temperature measurement is based on the so called zeroth law of thermodynamics, which states that two bodies in thermal equilibrium with a third body is in thermal equilibrium with each other.

The SI unit for temperature is the base unit of thermodynamics temperature, namely the Kelvin, which is generally used in all calculation and specially for physico chemical purposes.

As a customary or practical unit, the degree celcius (°C) is retained. This is identical to the degree centigrade, the change in name being due to the fact that the grade and hence the centigrade has another meaning in some countries.

The degree celcius and degree Kelvin are identical units of temperature interval, since 0°C = 273 K, then T°C = T + 273 K. If follows, therefore that temperature intervals or temperature differences are the same in both cases.

3.2 TEMPERATURE MEASURING INSTRUMENTS

The various temperature measuring instruments are as follows :

(i) Thermometer
(ii) Bi-metallic thermometer
(iii) Bourdon tube thermometer
(iv) Thermocouple
(v) Thermopiles
(vi) Resistance thermometer
(vii) Pyrometer

3.2.1 Thermometer

It consists of a glass envelope, a responsive liquid normally mercury and an indicating scale. In the envelope there is a bulb and a capillary tube fused together. The liquid is filled in the bulb and a little portion of the capillary. As the temperature of the thermometric liquid increases, it expands and rises in the capillary. The level of liquid in the column rises to indicate the temperature of the medium surrounding the bulb. The scale may be marked directly on the capillary tube as in most of the laboratory or clinical type of thermometer. A simple mercury in glass thermometer is shown in Fig. 3.1.

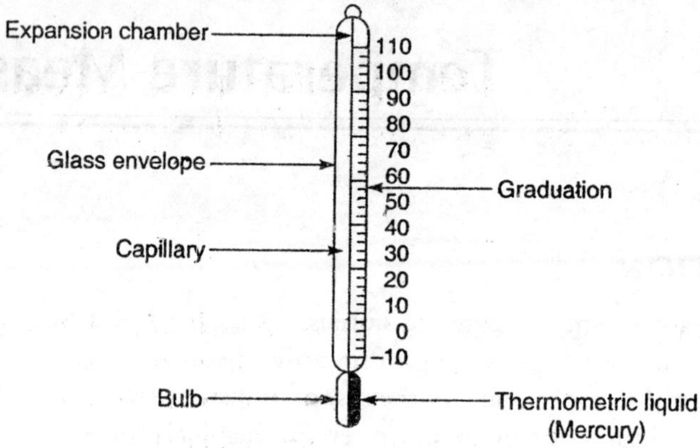

Fig. 3.1 Mercury in glass thermometer.

Advantages : Following are the advantages of mercury in glass thermometer :

(i) It is relatively low cost.
(ii) Simple in use.
(iii) Easy of checking for physical damage.
(iv) Absence of need for auxiliary power.

Disadvantages : Following are the disadvantages of mercury in glass thermometer :

(i) It is easily breakable.
(ii) Lack of adaptability to remove reading.
(iii) Surface temperature can not be measured.
(iv) Range of measurement is limited to 360°C.

3.2.2 Bi-metallic Thermometer

The construction of a bimetallic strip element provides a rough uneven surface yet accurate and simple device for the indication of temperature. A typical bimetallic thermometer is shown in Fig. 3.2.

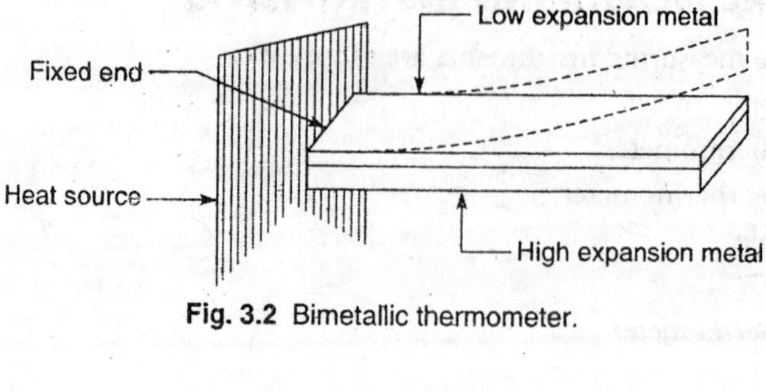

Fig. 3.2 Bimetallic thermometer.

A bimetallic thermometer is composed of two strips of metal welded together, each strip made from a metal having a different co-efficient of thermal expansion. When the strip is heated or cooled different expansion or contraction occurs causing the strip to bend. With one end of the strip is fixed the other end deflects in the direction proportional to temperature, square of the length and thickness. If the strip of bimetal is wound into helix or spiral and one end fixed, the other end will rotate when heat is supplied. Then the angular velocity is proportional to the temperature change, length of the strip and thickness of the material.

Invar (Iron 64% and Nickel 36%) is used as low expansion metal alloy and brass is used as high expansion metal alloy.

Range : –30 to 550°C. Generally thermometer of this type is not recommended for continuous use above 450°C.

Accuracy : ± 1% for low temperature and ± 2% for higher temperature. Changes in ambient temperature at the end do not seriously affect the accuracy.

Uses : Bimetallic thermometer is commonly used whenever the industrial mercury in glass thermometer is employed.

Advantages :

(i) Thermal expansion is linear.

(ii) Percentage of error is very less.

Disadvantages :

(i) Spot reading.

(ii) Stem of the thermometer must deep in fluid.

3.2.3 Bourdon Tube Thermometer

The bourdon tube thermometer may be broadly classified as :

(i) Liquid filled thermometer

(ii) Gas filled thermometer

(iii) Vapor pressure thermometer

(i) Liquid Filled Thermometer

It works on the principle of thermal expansion of liquid. When the bulb is immersed in a heated substance, the liquid expands and causes the pressure spring to unwind. The movement is converted into circular movement of lever which moves on a scale. This scale is calibrated in terms of temperature.

(ii) Gas Filled Thermometer

It works on the principle that if the volume is kept constant, pressure of a gas increases as the temperature is raised. From Charles law at constant volume, we have

$$\frac{T_1}{T_2} = \frac{P_1}{P_2} \qquad \ldots(3.1)$$

where, T_1 = Initial temperature

T_2 = Final temperature

P_1 = Initial pressure

P_2 = Final pressure

The system is filled with gas at high pressure so that the increase in the pressure for each degree temperature will be more. The deflection of bourdon tube is converted to circular movement of a lever on a scale which is calibrated in terms of temperature. Nitrogen is the gas often used because of its inertness.

(iii) Vapor Pressure Thermometer

Unlike liquid filled and gas filled thermometer which depends on their volumetric expansion for their operation, the vapour pressure thermometer depends on the partial pressure of the liquid which only partially filled the system. A typical vapour pressure bourdon tube thermometer is shown in Fig. 3.3.

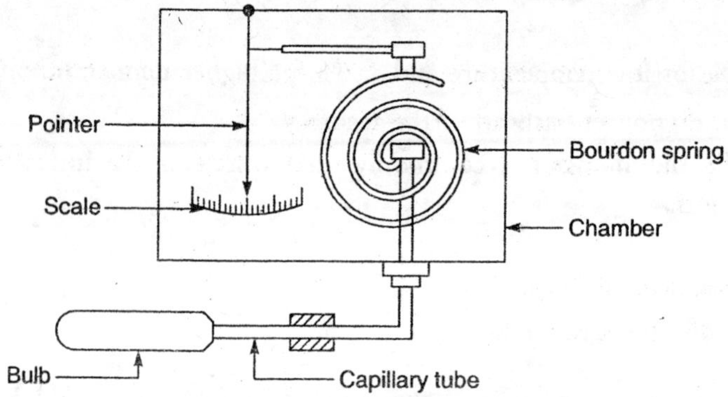

Fig. 3.3 Vapour pressure thermometer.

The property used here is the change in vapour pressure of the liquid with its temperature. The following points are to be noted in vapour pressure thermometer :

(i) Vapour pressure does not change linearly with temperature. At lower temperature change in vapour pressure is small and at higher temperature it is large for unit change in temperature.

(ii) Interchange of liquid and vapour in the system. When the temperature of the bulb is lower the liquid remains in the bulb and vapour occupies the tube and this is reversed when the bulb temperature is higher.

To overcome this limitation a modification is done and dual filled system in which two different liquids one volatile and other non-volatile are used.

The liquid used for vapour pressure thermometer are mercury, xylene, alcohol, ether etc. The liquid used for vapour pressure thermometer are methyl chloride, diethyl ether, ethyl alcohol, water, toluene, argon etc.

Maximum tube length for mercury filled is 150–200 ft and range of temperature for which it is used as – 40 to 540°C.

3.2.4 Thermocouple

It is a simple device made of wires of two dissimilar metals joined at their ends. It converts thermal energy into electrical energy.

Principle : If two different metals are joined at one end and a junction is made. If this junction is heated to a higher temperature with respect to free ends and if these two conductors of metals at free ends are connected by an instrument, an emf is setup and will establish a flow of current.

The flow of current from hot junction to free end of the wire at high temperature is called as see beck effect. The magnitude of the emf will depends on :

(i) The magnitude of difference between the temperature of these junctions.

(ii) The materials used as conductors.

Construction : It consists of two wires of suitable materials which are joined together at one end is called measuring junction made up by twisting together and joining the tips, and the other end is called reference junction or cold junction is connected to measuring unit through extension wire to complete the circuit.

For measuring the emf a millivoltmeter or for more accurate measurement a potentiometer system is used which is calibrated in terms of temperature.

To give protection to the junction from injurious gases a refractory sheath is used. This is also used to prevent mechanical damage. A simple thermocouple is shown in Fig. 3.4.

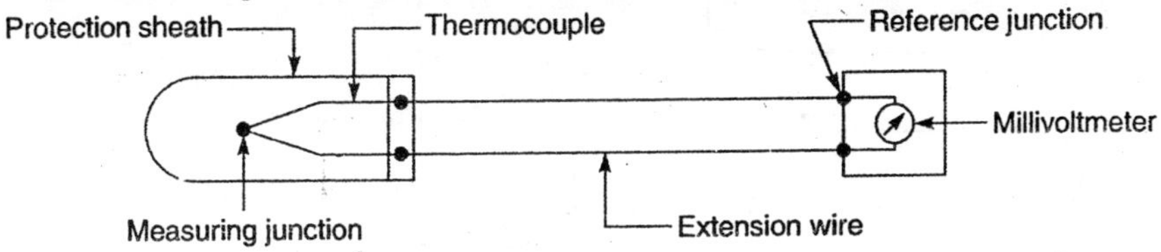

Fig. 3.4 Thermocouple.

Characteristics of thermocouple material : The characteristics of thermocouple materials are as follows :

(i) It should have linear temperature emf relationship.

(ii) Sufficient thermo emf per degree of temperature change to facilitate detection and measurement.

(iii) It should be physically able to withstand, sustained high temperature, rapid temperature changes and effect of corrosive atmosphere.

(iv) Thermocouple should be such that to follow only one characteristics whether the temperature rises or falls.

(v) Cost should be reasonable.

Some of the thermocouple and its characteristics are given in Table 3.1.

Table 3.1 : Thermocouple characteristics.

S. No.	Thermocouple	Sensitivity/°C	Normal Temperature Range (°C)	Maximum Temperature (°C)	Service Characterization
1.	Copper/constantan	0.03	–200 to 400	600	High resistance
2.	Iron/constantan	0.05	–200 to 800	1000	A low cost couple withstanding oxidation and reducing atmosphere
3.	Chromel/alumel	0.04	–200 to 1200	1400	Withstand oxidising atmosphere
4.	Platinum (50%) + Rhodium (10%)/ Platinum (100%)	0.006	–300 to 1400	1700	Withstand oxidation by emf effected in the presence of metal carbon and silicates

Also, there are tungsten–rhenium thermocouples in common use for measuring temperature upto 2100°C. These alloys have poor oxidation resistance and should be used in vacuum, hydrogen and inert atmosphere.

Peltier Effect : The absorption and evolution of heat at the junction of a thermocouple to the current flow in the circuit is called Peltier effect.

3.2.5 Thermopiles

The thermopiles are series of thermocouples connected side by side or in star formation. The output of a thermopile is equal to the sum of the emf and is much more sensitive than the thermocouple. A simple thermopile is shown in Fig. 3.5.

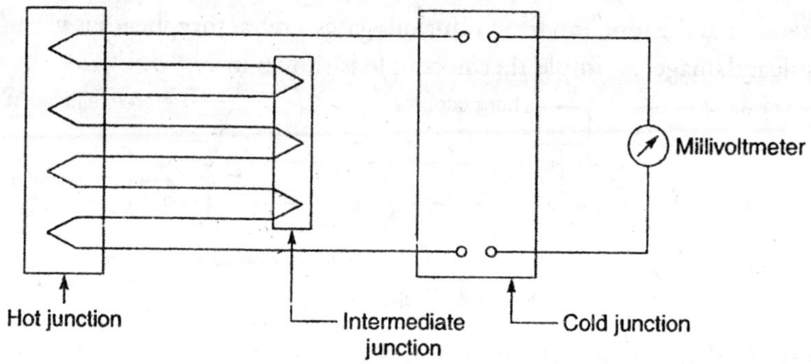

Fig. 3.5 Thermopile.

A chromel/constantan thermocouple having 14 couples gives about 1 mV/°C and achieve a sensitivity of 0.002°C. The hot junctions of individual couples must be carefully insulated from each other. Thermopiles arranged in parallel can be used to determine the average temperature of the system.

3.2.6 Resistance Thermometer

With increase in temperature the electrical resistance of most of the metals increases is almost directly proportionals to the rise in temperature. Therefore, if the electrical resistance of a wire of known calibrated material is measured, the temperature of the wire can be determined.

The sensing element which is used to measure the temperature is made by winding of resistance wire on a insulated material and may be placed in a stainless steel bulb to protect it from corrosive fluids and also damage to the wire. The temperature can be measured in terms of resistance of wire by using resistance bridge using the sensing element as one arm of the bridge. A simple resistance thermometer is shown in Fig. 3.6.

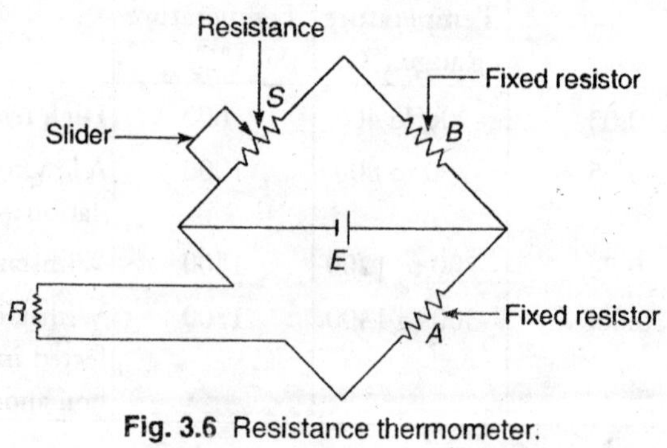

Fig. 3.6 Resistance thermometer.

The characteristic of resistance elements used for measuring temperature is given in the Table 3.2.

Table 3.2 : Characteristics of resistance element.

S. No.	Resistance wire	Rated resistance at 0°C (ohms)	Range of temperature (°C)
1.	Platinum	10	0 to 650
2.	Platinum	46	–200 to 500
3.	Platinum	100	–200 to 600
4.	Copper	50	–50 to 180
5.	Copper	100	–50 to 250

The resistance thermometer has an accuracy of ± 0.75% of scale range at temperature upto 600°C.

Advantages :

(i) This can be employed where the capillary length of liquid expansion thermometer would be excessive.

(ii) By suitable modification of the system, resistance thermometer can be used to determine temperature difference, average temperature etc.

3.2.7 Pyrometer

Temperature of an object can be measured by means of the quantity and characteristics of the energy which it radiate. It is generally called pyrometer. The broad classification of pyrometer is as follows :

(i) Radiation pyrometer

(ii) Optical pyrometer

(i) Radiation Pyrometer

It measures the rate of energy emission per unit area over a relatively broad range of wave length. The temperature is measured by measuring the intensity of radiation emitted by a hot body over a range of wavelength.

The Stefan Boltzmann equation states that the rate at which energy emitted from a unit area by a black body is a function of the difference of the fourth power of absolute temperature of a hot sources and a receiver. It may be represented by the equation

$$\frac{q}{A} = \sigma (T_1^4 - T_2^4) \qquad ...(3.2)$$

where, q = Amount of heat transfer

A = Heat transfer surface area

σ = Boltzmann constant

T_1 = Temperature of the source

T_2 = Temperature of the sink

Therefore, the temperature of a suitably arranged receiver may be theoretically related to the temperature of a hot target. A simple form of radiation pyrometer is shown in Fig. 3.7.

It consists of a lens or mirror optical system for focussing radiating energy on a detecting element and a detector which may be photocell on a thermopile or other element which varies as electrical quantity as a function of temperature. The optical system and detector must be protected with a housing and an auxiliary instrument provided for measuring the pyrometer signal, *i.e.*, millivoltmeter.

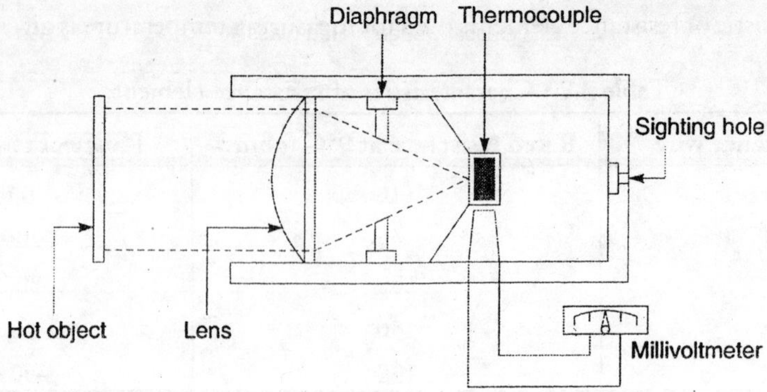

Fig. 3.7 Radiation pyrometer.

Range : It has range from 70 to 5500°C. An arbitrary classification for the range of operation over which a single type of instrument is recommended for use are :

(i) High range
(ii) Intermediate range
(iii) Low-intermediate range
(iv) Low range

Applications : These are particularly recommended for industrial temperature measurements under conditions which make difficult or impossible to use the other primary elements. These include :

(i) Temperature above the practical operating range of thermocouple.
(ii) Environments which contaminate or limit the life of thermocouple.
(iii) Moving targets.
(iv) Targets which would be accessible such as furnace interiors.
(v) Targets which would be damaged by contact with primary elements like thermocouple and resistance thermometer.
(vi) Average temperature of large surface area.

(ii) Optical Pyrometer

It is a type of pyrometer in which brightness (for a narrow wavelength interval) of a hot object is manually compared by the operators eye with that of a source of standard brightness.

The optical pyrometer is widely employed for accurate measurement of the temperature of furnaces, molten metals and other heated materials. It is primarily used in the range of 550 to 3000°C. Most of the optical pyrometers are manually operated and thereby, somewhat limited in their application. A typical optical pyrometer is shown in Fig. 3.8.

The optical pyrometers are manually operates with nearly monochromatic light. The wavelength is usually a band of about 0.01 micron wide at 0.65 micron in the red portion of the variable spectrum.

The pyrometer acts as a telescope. The lens focusses radiation from the source at the plane of the filament. The eye piece should also be focussed at the same plane, the red filter prevents all radiation from the hot sources except the red from reaching the eye. The intensity of the bulb is varied by the rheostat and the current flowing is indicated by millivoltmeter which is calibrated in terms of temperature.

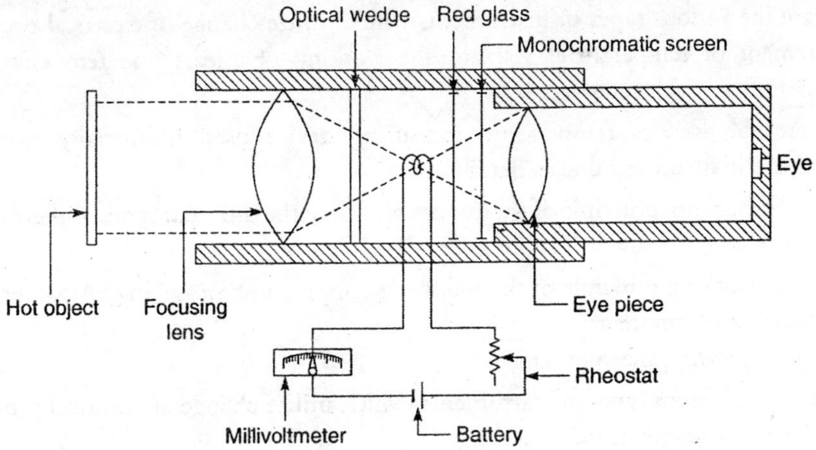

Fig. 3.8 Optical pyrometer.

By adjusting the variable resistance the brilliance of the lamp is matched with that of the object where upon the filament becomes invisible. Then the reading of the millivoltmeter is taken as the indication of the object temperature.

Applications : This instrument is extremely used by metallurgist both in industry and laboratory.

Range : It has range from 600 to 3000°C.

Characteristics : The characteristics of the optical pyrometers are as follows :

(i) No direct contact with the object whose temperature is to be measured.

(ii) Instrument can be used to measure-high temperature.

(iii) Measurement are practically independent of the distance of the operator from the heated body.

(iv) The temperature can be measured with great rapidity.

REVIEW QUESTIONS

3.1 What is the significance of temperature measurement?

3.2 Compute the temperature at which Fahrenheit and Centigrade scales coincide.

3.3 Develop an algebraic equation relating the degree Centigrade and Fahrenheit scale.

3.4 For an ordinary mercury in glass thermometer name the parts that constitute the primary element and the indicating elements.

3.5 With the help of a neat diagram explain the working of a radiation pyrometer. What are the possible errors in this type of instrument? What are the precautions to be taken?

3.6 Explain in detail the following temperature measuring instruments with neat sketch, working principle, range of operation and applications :

(i) Thermocouples

(ii) Thermopiles

3.7 Discuss the principles on which the bimetallic temperature measuring instrument and the mercury in glass thermometer work. Mention the range of temperature measured by them. Give their merits and demerits.

3.8 With a neat diagram explain the working of an optical pyrometer.

3.9 What are the various types of instruments which utilize change of electrical properties for the measurement of temperature? Explain the working of atleast one temperature measuring instrument from each category.

3.10 What are the various temperature measuring device used industrially to measure high temperature? Explain any one in detail.

3.11 What is the working principle of thermocouple? Give the different types of thermocouples with temperature ranges. How it is calibrated?

3.12 Explain the working principle of the following temperature measuring instruments :
(i) Resistance thermometer
(ii) Vapour pressure thermometer

3.13 Describe the different types of instruments, which utilize change of electrical properties for the measurement of temperature.

3.14 What is thermocouple? A large number of thermocouples are connected in parallel, then the generated emf will correspond to the mean of the temperature of individual junctions. Is this statement true? If yes give reasons, if no give reasons.

3.15 What is Peltier effect? Explain.

3.16 What are the various thermocouple materials? List them and mention temperature range for their applications.

3.17 What is thermowell? How will you select the materials of construction of thermowell?

□□□

Chapter 4

Flow Measurements

4.1 INTRODUCTION

The measurement of flow rate and flow quantity of material is made primarily for the purpose of mixing different fluids at specific proportion and to get maximum efficiency and quality control of materials involve in the process. Secondarily, flow measurement are made for the purpose of cost accounting and minimize the wastage of usually air, steam and water services.

4.2 CLASSIFICATION OF FLOW MEASURING INSTRUMENT

The flow measuring instruments may be broadly classified as :

(i) Rate meter
(ii) Quantity meter

4.2.1 Rate Meter

Rate meter is essential for getting information about the amount of material flowing in or out of process. For example domestic water meter which indicates the amount of water used at any particular time interval. The various rate meter are as follows :

(i) Head meter
(ii) Area meter
(iii) Head area meter
(iv) Velocity meter
(v) Thermal flow meter
(vi) Radio isotope tracer flow meter
(vii) Electro magnetic meter
(viii) Alum salt velocity meter

4.2.1.1 Head Meter

Head meter operates on the principle of converting kinetic energy of flowing fluid and energy due to static pressure. A head meter normally consists of a primary element which causes the restriction in the flow line to induce the head and a secondary element to measure the differential head as a means of determining flow rate. The various head meters are as follows :

(i) Orifice meter
(ii) Venturi meter
(iii) Flow nozzle
(iv) Pitot tube

(i) Orifice Meter

The orifice meter works on the principle of conversion of energy from static pressure to kinetic energy.

An orifice meter is a flat disc normally of 1.5 to 3 mm thick material with a circular hole and inserted between flanges in the pipe line. Usually the orifices are concentric but there are also eccentric or segmental type. A common type of orifice plates are shown in Fig. 4.1.

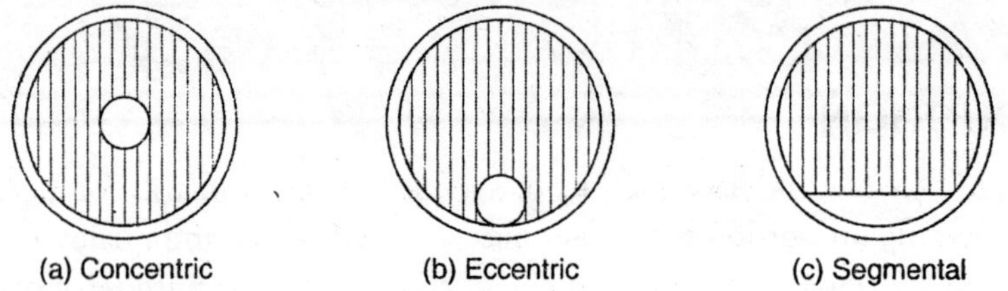

Fig. 4.1 Orifice plate.

The eccentric and segmental types are used for fluids consisting of suspended solids containing mixture of liquid and gas. Passage of solid material is not interfered with and there is no build up of solid matter against the upstream face of the orifice plate as would occur with normal concentric orifice. A simple orifice meter is shown in Fig. 4.2.

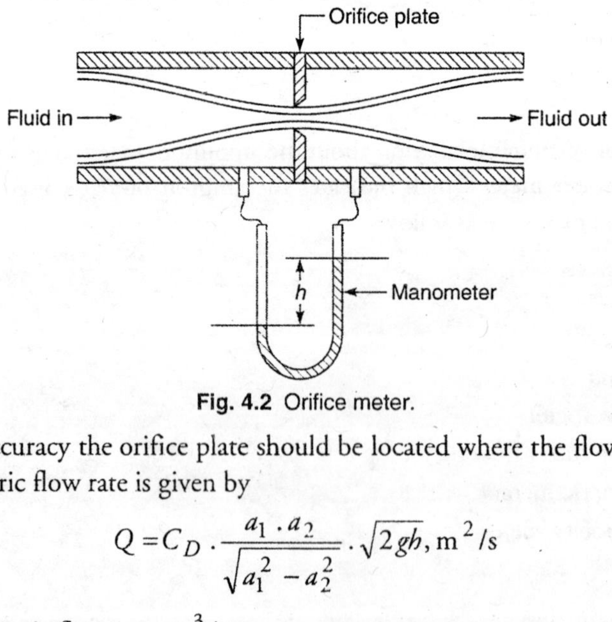

Fig. 4.2 Orifice meter.

For maximum accuracy the orifice plate should be located where the flow is most uniform. The expression for volumetric flow rate is given by

$$Q = C_D \cdot \frac{a_1 \cdot a_2}{\sqrt{a_1^2 - a_2^2}} \cdot \sqrt{2gh}, \text{ m}^2/\text{s} \qquad \ldots(4.1)$$

where, Q = Volumetric flow rate, m^3/s

a_1 = Cross-sectional area of pipe, m^2

a_2 = Cross-sectional area of orifice plate, m^2

C_D = Coefficient of discharge

g = Acceleration due to gravity, m/s^2

h = Difference in manometric fluid level, m.

(ii) Venturi Meter

The venturi meter works on the principle of conversion of energy from static pressure to kinetic energy.

It is mainly used for measuring high flow rates and the flow rates of liquids having suspended particles. Compare to orifice meter here pressure loss is minimum. The pressure recovery is attained by long conical exit section. A simple venturi meter is shown in Fig. 4.3.

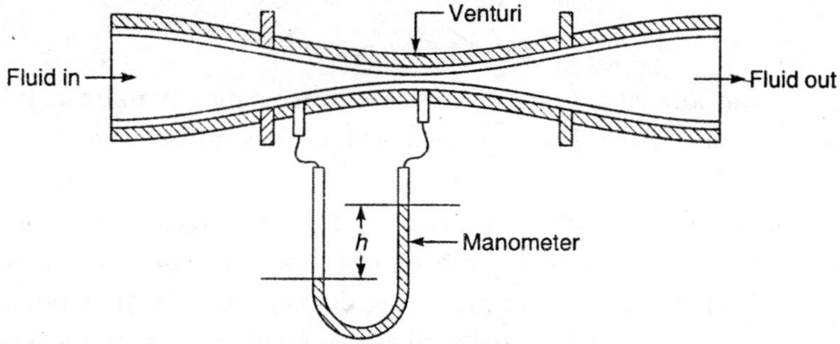

Fig. 4.3 Venturi meter.

In some cases to make possible large part of pressure a truncated recovery cone having an outlet diameter somewhat smaller than the inlet to the venturi is used. The shape and length of the outlet section has no effect on the coefficient of discharge of the tube.

When cost factor is not involved the most common reason for choice of venturi meter in preference to orifice meter is the better performance in the presence of solid in the flowing fluids suspension very quickly deposit on the upstream side of orifice plate and change of its coefficient of discharge. The venturi meter because of stream lined entrance and exit does not provide any dead pocket or impingements which permits these solids to settle out.

For Reynolds number above 10,000 based on upstream flow the discharge coefficient $C_D = 0.98$. The expression for volumetric flow rate is given by

$$Q = C_D \cdot \frac{a_1 \cdot a_2}{\sqrt{a_1^2 - a_2^2}} \cdot \sqrt{2gh}, \text{m}^3/\text{s} \qquad \ldots(4.2)$$

where, a_1 = Cross-sectional area of pipe, m^2

a_2 = Cross-sectional area of throat, m^2

(iii) Flow Nozzle

The flow nozzle works on the principle of conversion of energy from static pressure to kinetic energy.

It consists of a belt shaped approach section of elliptical profile followed by a cylindrical throat tangent to the ellipse. A simple flow nozzle is shown in Fig. 4.4.

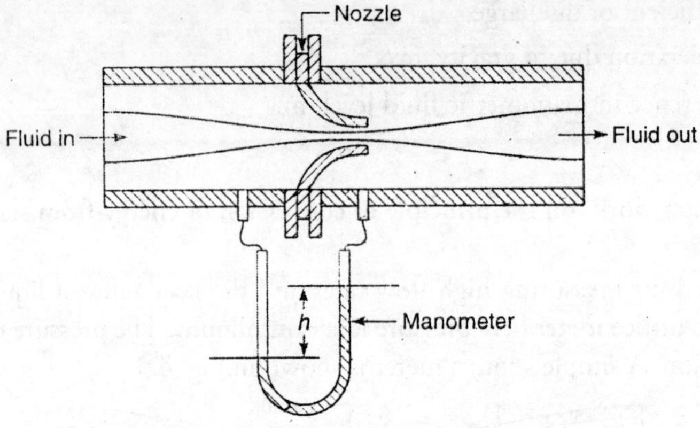

Fig. 4.4 Flow nozzle.

The pipe wall taps are located at 1 pipe dia upstream and 0.5 dia down stream from the inlet face of the nozzle to achieve the best result. The flow coefficient of flow nozzle is approximately 60% higher than that of the orifice plate.

Though the flow nozzle would appear to give a higher pressure recovery than an orifice, because of the flow coefficient higher than that of the orifice plate, lower diameter ratio is required to get the desired reading on the differential pressure measuring device. Since the flow nozzle requires smaller diameter ratio for getting the same differential head as that of an orifice plate for a given flow rate, the permanent pressure loss of a flow nozzle is of the same magnitude as that of an orifice.

The nozzle should be so installed that the flow is vertically up when there are entrained vapour or gases, because of the possibility of gas accumulation. For liquid suspended particles it should be installed vertically downwards so that the materials which dropout of suspension will be unable to alter approach conditions because there will be no good place for them to remain. The volumetric flow rate may be calculate as in the case of orifice meter.

(iv) Pitot Tube

The pitot tube works on the principle of conversion of kinetic energy to pressure at the impact tube.

It is a device to measure point velocity. The principle behind this instrument is that fluid flowing with certain velocity when strike the pitot tube, its forward motion is arrested and the impact results in the conversion of kinetic energy into potential energy. As a result of this an excess head is generated inside the tube. A simple pitot tube is shown in Fig. 4.5.

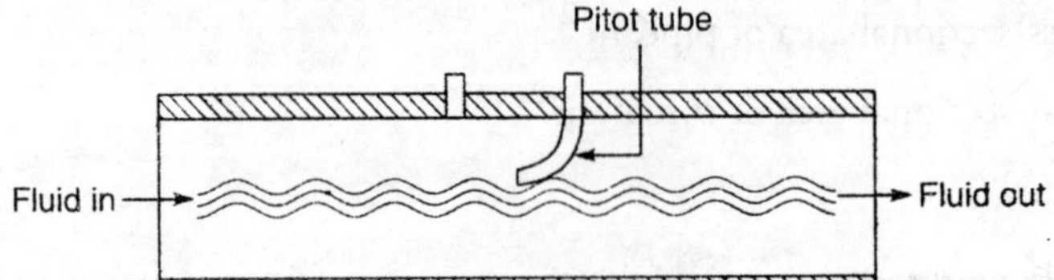

Fig. 4.5 Pitot tube.

A static pressure tapping is taken from the pipe and connected to the low pressure side of measuring instrument and high pressure side which given impact pressure to the pitot tube. The relation between the differential pressure and velocity at the tip of pitot tube is given by the relation

$$V = C \cdot \frac{\sqrt{\rho_m - \rho}}{\rho} \cdot \sqrt{2gh}, \text{m/s} \qquad \ldots(4.3)$$

where, V = Velocity of flowing fluid, m/s
C = Velocity coefficient
ρ_m = Density of manometric fluid, m^3/kg
ρ = Density of flowing fluid, m^3/kg
g = Acceleration due to gravity, m/s^2
h = Differential head of manometer, m

Then the velocity is converted into volumetric flow rate by multiplying cross-sectional area of the flow system and expressed as

$$Q = V \cdot A, \text{m}^3/\text{s} \qquad \ldots(4.4)$$

It has got a very limited application commercially as a primary element because of the following reasons :

(i) It has tendency to play when the flowing fluid contains small amounts of solids.
(ii) It has velocity range limitation when used with standard commercial instruments.
(iii) It gives only the point velocity, *i.e.*, it gives different velocities at different spot across the diameter.

4.2.1.2 Area Meter

In this meter the cross-sectional area for flow is varied to keep the constant pressure differential at the restriction. The flow rate is a function of cross-sectional area available for flow, *i.e.*, the position of the restriction. The general type of variable area meters are :

(i) Rota meter
(ii) Piston meter
(iii) Orifice and plug area meter

(i) Rota Meter

It is an indicating type of instrument. It consists of a long vertical tube having a uniform tapper, arranged with smaller section at the bottom and largest at the top. There is a free flowing float inside the tube, the position of which indicate the flow rate. In some design there are series of angled slots cut into the float so that a fluid flows past the float will be caused to spin to prevent the float from moving to the sides of the rota meter tube. A simple rota meter is shown in Fig. 4.6.

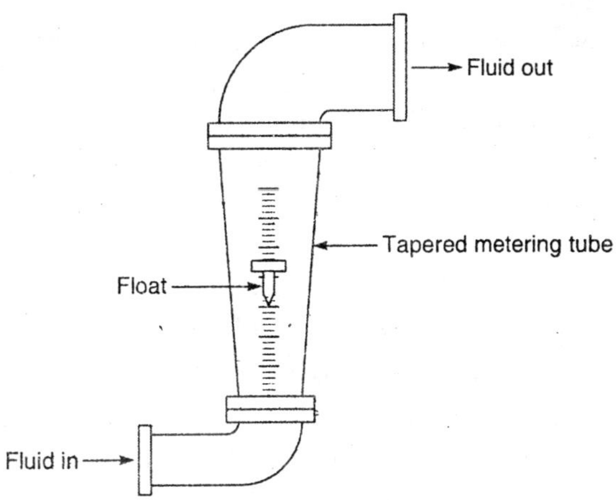

Fig. 4.6 Rotameter.

As the rate of flow through the rota meter increases, the float rises in the tube increasing the annulus between the float and the tube wall and thus maintain a constant pressure differential across the float. The position of float in the tube is a measure of volumetric flow rate and the tube may be calibrated in terms of volumetric flow rates. Rota meter may be used to measure the flow rates of both liquids and gases.

Rota meter usually made up of transparent material like glass and Acrolite. The choice of the material construction of rota meter depends upon the type of fluid to be used and its flow rates. Some examples are given below :

(i) **Glass rota meter :**

Liquid :	Low flow rate	30 – 300 ml/s
	High flow rate	0.5 – 5 lit/s
Gases :	Low flow rate	0.2 – 2 lit/s
	High flow rate	4 – 40 lit/s

(ii) **Metal rota meter :** Up to 120 lit/s

Precautions : It is important that the temperature of the fluid should remain constant, otherwise the viscosity and density of the fluid being metered may change and the temperature function may cause considerable errors particularly when gases are being measured. A further source of error is the tendency of the float to oscillate which make accurate scale reading difficult.

It should be noted that a rota meter can be installed only in a vertical position.

(ii) Piston Meter

In this type, a piston is accurately fitted inside a sleeve and is lifted by fluid pressure until sufficient port area in the sleeve is uncovered to permit passage of the flow. The flow is indicated by position of the piston. A simple piston meter is shown in Fig. 4.7.

This meter is quite similar to the rota meter. In this meter a constant pressure force must be exerted on the piston. If the pressure difference across the piston is more, it will push the piston up which decreases the differential pressure. In this manner piston is positioned such that it maintain a constant differential pressure.

The magnet attached at the top of the cylinder shows the position of piston. This in term gives the port area available for flow and the flow rate. The body material is cast iron or steel or other material depending on :

(i) The chemical activity of the liquid.

(ii) The strength of the material required.

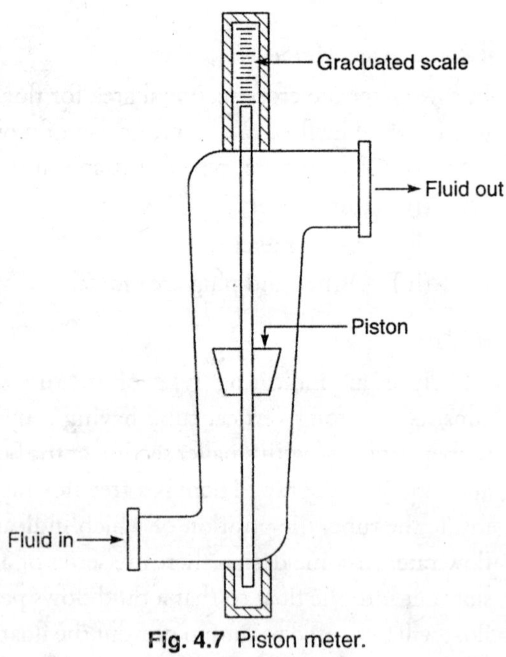

Fig. 4.7 Piston meter.

Characteristics of Piston Meter :

(i) These are well adopted to the measurement of highly variable flow rate.

(ii) Accuracy is generally 98%.

(iii) Corrosive liquids can be handled by suitably choosing the materials of construction.

(iii) Orifice and Plug Area Meter

It consists of a tappered plug and is of such a form that the area of the annular space between the orifice and the plug is proportional to the lift of the plug. A simple orifice and plug area meter is shown in Fig. 4.8.

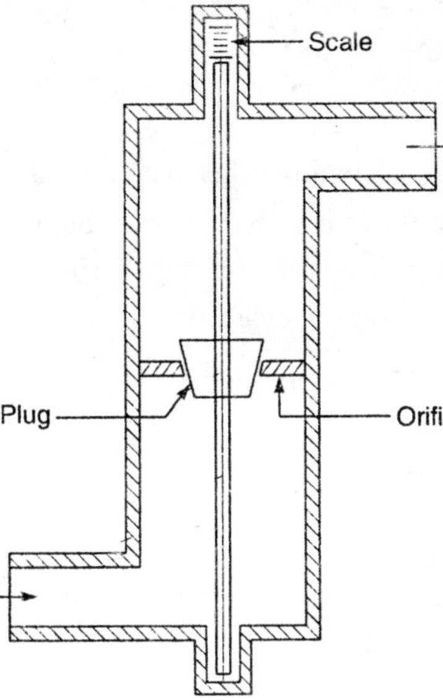

Fig. 4.8 Orifice and plug area meter.

The height by which the plug rises when a fluid flows past is a measure of the rate of flow.

4.2.1.3 Head Area Meter

Head area meters are often called open channel meter. It is also known as weirs. Here, both head and area of the stream vary with the flow and area of the stream. The flow depends on the head, *i.e.*, the level of liquid in the weir.

These are used for metering liquid, usually water in open channel or in conduit which do not flow full and do not flow under pressure. The general type of head area meters are as follows :

(i) Rectangular weir
(ii) Triangular weir
(iii) Trapezoidal weir
(iv) Flume

(i) Rectangular Weir

It provide restriction in the flow path by means of weirs opening in the dam through which it flows. A simple rectangular weir is shown in Fig. 4.9.

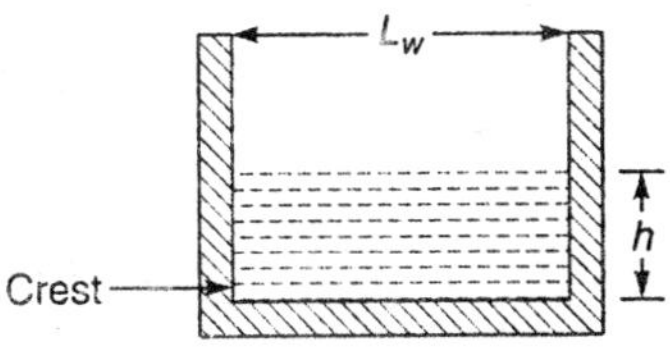

Fig. 4.9 Rectangular weirs.

This is most popular because of its easy construction, accuracy and simplicity. The expression for volumetric flow rate is given by

$$Q = \frac{2}{3} . C_D . (L_w - 0.2h) . \sqrt{2gh^3}, \text{m}^3/\text{s} \qquad ...(4.5)$$

where, Q = Volumetric flow rate, m^3/s
C_D = Coefficient of discharge
L_w = Breadth of weir, m
h = Height of the surface of liquid above the crest, m
g = Acceleration due to gravity, m/s^2

(ii) Triangular Weir

This has a greater practical range than any other type for a given size. This is more suitable for measuring flows less than 4 ft^3/s because of its greater head loss. A simple triangular weir is shown in Fig. 4.10.

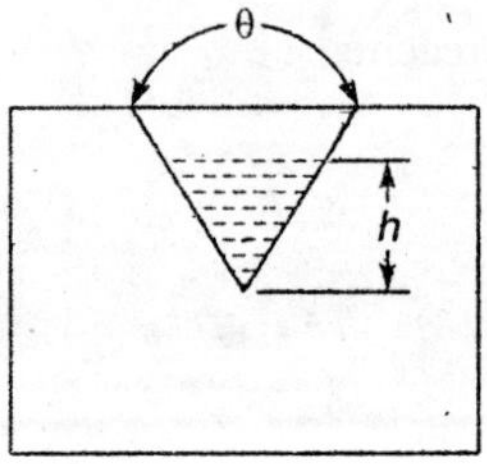

Fig. 4.10 Triangular weirs.

The flow rate can be calculated using the expression

$$Q = \frac{8}{15} . C_D . h . \tan \theta/2 . \sqrt{2gh^3}, m^3/s \quad ...(4.6)$$

where, θ = Angle between the sides.

The value of flow coefficient C_D varies with :

(i) Weirs angle
(ii) Head
(iii) Kinematic viscosity
(iv) Surface tension

C_D values varies from 0.58 to 0.7.

(iii) Trapezoidal Weir

This is not much in practical use because of its complexity. A simple trapezoidal weir is shown in Fig. 4.11.

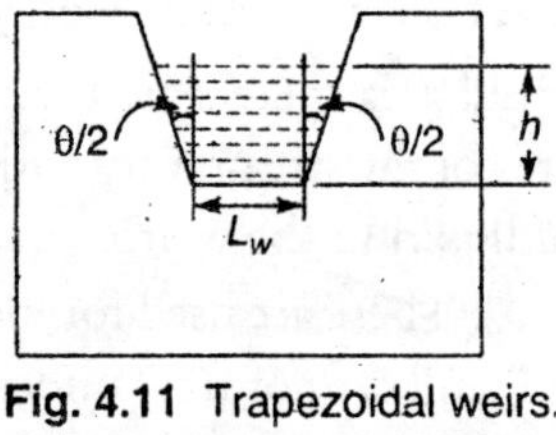

Fig. 4.11 Trapezoidal weirs.

The flow rate can be calculated using the expression

$$Q = \frac{2}{3} . C_D . L_w . \sqrt{2gh^3} \quad ...(4.7)$$

where, L_w = Length of the crest, m.

(iv) Flume

It is particularly used when large amount of liquid to be measured and also when large amount of materials are held in suspension as in irrigation water. A simple flume is shown in Fig. 4.12.

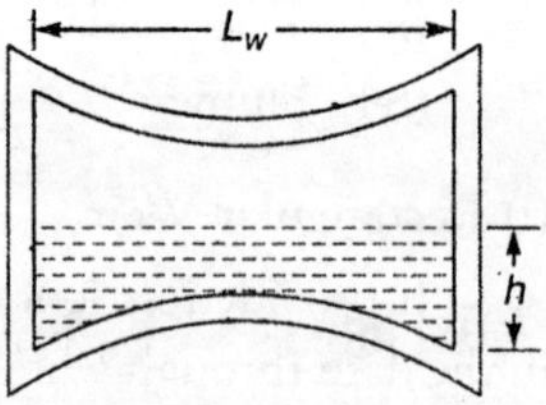

Fig. 4.12 Flume.

The expression for volumetric flow rate is given by

$$Q = \frac{2}{3} . C_D . L_w . \sqrt{2gh^3}, m^3/s$$

where, L_w = Length of the crest, m.

4.2.1.4 Velocity Meter

It is also called current meter. This is primarily an instrument for measuring the velocity of the flowing fluid. Flow rate can be measured from the velocity using the cross-sectional area of flow. The general type of velocity meter are as follows :

(i) Turbine meter
(ii) Cap anemometer
(iii) Vane anemometer
(iv) Speed anemometer

(i) Turbine Meter

A turbine wheel is set in the path of a fluid stream. The flowing fluid impinges on turbine blades imparting a force to blade surface and setting the rotor in motion. When a steady rotational speed is reached, it is proportional to the fluid velocity. A simple turbine meter is shown in Fig. 4.13.

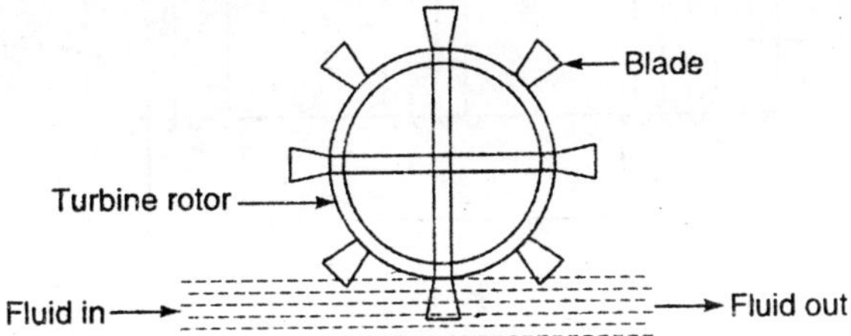

Fig. 4.13 Turbine meter.

(ii) Cap Anemometer

It consists of 3 ot 4 hemispherical hollow caps mounted on a horizontal plane with their tips vertical. As the flow passes over the caps more drag force is experienced by the cap with its open end facing into the stream than the caps with their rounded face towards the stream. The cap assembly is caused to spin around its spindle which drives a geared dial and register a reading approximately to flow velocity.

(iii) Vane Anemometer

It consists of a small wind mill which spin at different rate when subjected to varying velocities of flow. It should be correctly aligned to the flow direction if accurate reading is to be obtained. A simple vane anemometer is shown in Fig. 4.14.

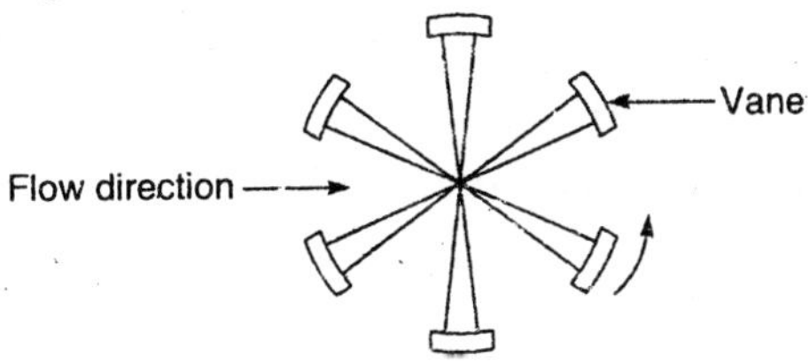

Fig. 4.14 Vane anemometer.

(iv) Speed Anemometer

It consists of a thin metal sheet. The deflection of which is measured when a fluid stream passes over it. The deflection of the reed tip can be read directly on a scale or the change in resistance of a strain gauge fixed to a thin cantilever beam at the base of the reed which is calibrated against in flow velocity. A simple speed anemometer is shown in Fig. 4.15.

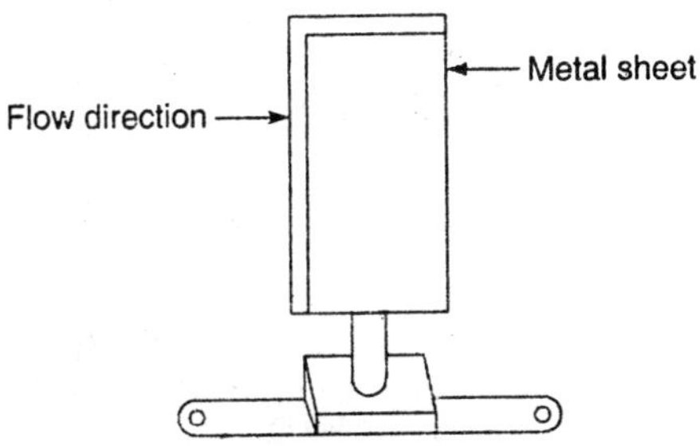

Fig. 4.15 Speed anemometer.

4.2.1.5 Thermal Flow Meter

It works on the principle that for any given fluid a definite amount of heat is required to increase the temperature of each kg of fluid by some fixed amount say 1°C. If the increase in temperature can be controlled accurately and the heat can be measured then this would provide a way to measure mass flow rate. A simple thermal flow meter is shown in Fig. 4.16.

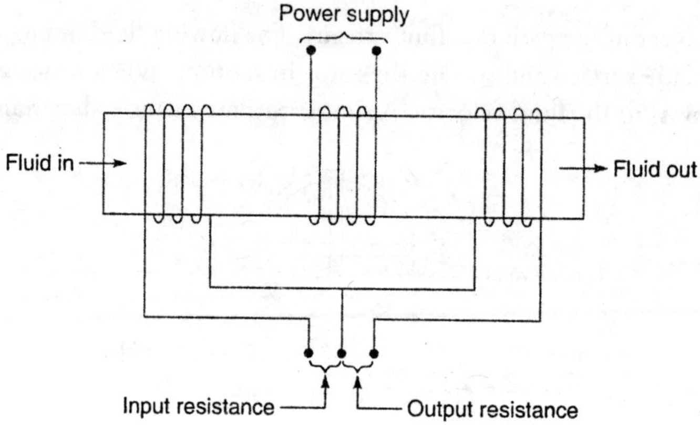

Fig. 4.16 Thermal flow meter.

The heat input required can be measured quite simply by using a wattmeter to indicate the instantaneous rate at which electric power is being applied to maintain the temperature increment.

The heat required is directly proportional to the rate at which the moving fluid is absorbing heat and thus directly proportional to the flow rate itself. The wattmeter can be calibrated in terms of flow rate.

Advantages :

(i) No obstruction is placed in the pipe and there are no wearing of parts.

(ii) Calibration for various gas or liquids depends only on heat capacity and are independent of density, viscosity and thermal conductivity.

(iii) Temperature rise or watt hour required is a linear function of mass flow rate.

Disadvantages :

(i) Location of heater directly in the stream.

(ii) Large power input required to measure high flow rate.

4.2.1.6 Radio Isotope Tracer Flow Meter

In this method a radio isotope is introduced into the system. This method gives the velocity of the fluid. In this method a gamma emitter such as C_0^{10} and Sb^{124} is injected quickly at a point close to the section of the pipe in which the velocity is to be determined. Two detector are placed in the stream at a known distance. A simple radio isotope tracer flow meter is shown in Fig. 4.17.

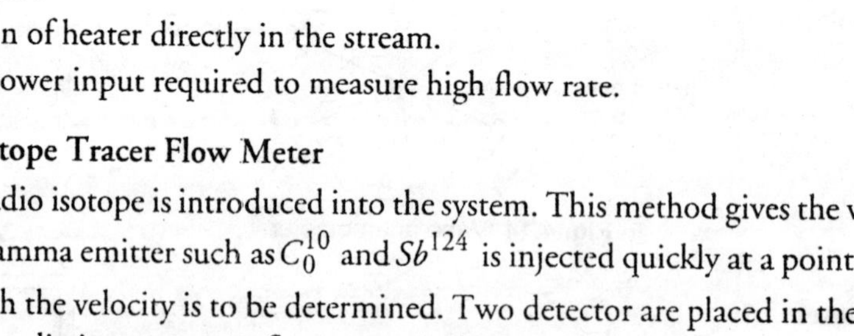

Fig. 4.17 Radio isotope tracer flow meter.

The velocity is calculated by finding the time required for the isotope to pass from first detector to second detector. Although this is a complex approach to measure the flow, is applicable where conventional approaches infact not possible.

4.2.1.7 Electro Magnetic Meter

This is similar to radio isotope method. The principle used here is that the electrical conductivity of salt solution is greater than that of pure water. The concentrated salt solution is injected to the flow at some point. The solution is carried forward with water at a speed equal to the velocity of water. A simple alum salt velocity meter is shown in Fig. 4.18.

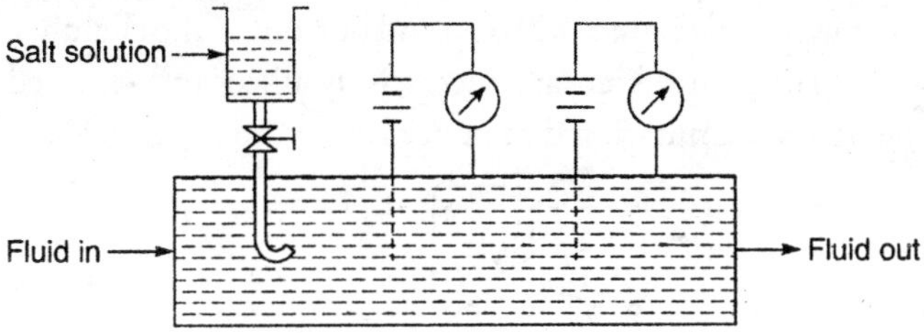

Fig. 4.18 Alum salt velocity meter.

Two conductivity cell is located in the flow stream at a fixed distance. When alum salt is injected into the flowing stream, the conductivity is recorded by the first instrument and then at a time interval by the second instrument. Once the time is determined by the instrument placed at a fixed distance for a constant value of thermal conductivity, the velocity can be calculated and hence the flow rate.

4.2.1.8 Alum Salt Velocity Meter

Its works on the principle of Faraday's law. We know that when electrical conductor moves through a magnetic field in a direction right angle to the conducting fluid moving, it forms an electro magnet. Its magnitude depends upon :

(i) The strength of the magnetic field.
(ii) The velocity of the conductive fluid flowing through the tube.
(iii) The distance between the electrode.

The induced emf is given by the equation

$$V = Blu, \text{ volts} \qquad \ldots(4.9)$$

where, V = Volts
B = Magnetic flux density, w/m^2
l = Length of the conductor, m
u = Velocity of the conductive material, m/s

Therefore, if the strength of the magnetic field and distance between the electrode held constant, the voltage produced will be entirely dependent on the flow velocity. This voltage is then amplified and connected to suitable indicating instrument which is calibrated in terms of velocity.

4.2.2 Quantity Meter

It is also called positive displacement meter. It measure the total quantity of fluid which has flown in a certain time interval. If the quantity of fluid measured in a certain time interval is known then the average rate of flow of fluid can be determined. The various quantity meter are as follows :

(i) Nutating disc meter
(ii) Rotating lobe meter
(iii) Reciprocating piston meter
(iv) Tilting trap meter

4.2.2.1 Nutating Disc Meter

This is used for measuring flow quantity of liquid and is widely used as domestic water meter.

It consists of an eccentrically mounted disc which mutates in a hemispherical chamber. The pecular mutating motion of the disc implies that the disc provide not only a rotation but also a sweeping motion. The disc maintain tangential contact with top and bottom of the chamber. The chamber is then sealed off into separate compartment which are successively filled and emptied. Each compartment holding a definite volume. A simple nutating disc meter is shown in Fig. 4.19.

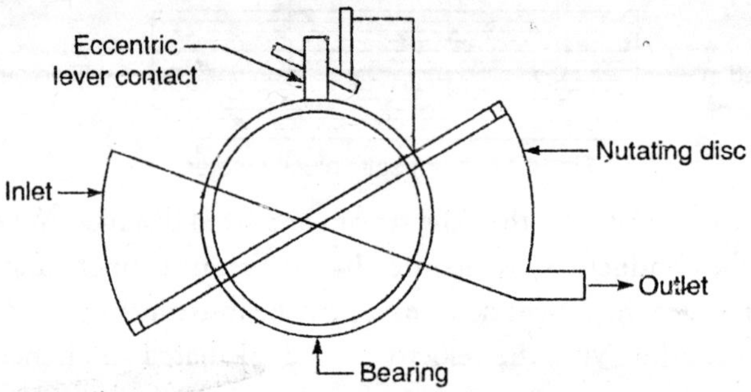

Fig. 4.19 Nutating disc meter.

The viscosity of the liquid takes care both sealing and lubrication. The compartments progress from suction to discharge side as the disc wobbles under liquid pressure from inlet to outlet. For the known capacity of the chamber and the number of disc oscillation which is counted by the rotation of eccentric lever, the total quantity of fluid flow can be metered.

The motion of disc can be made to drive a recording and indicating mechanism which canbe directly calibrated in terms of liquid discharged.

Characteristics :

(i) It has simple and rugged construction.
(ii) Low cost.
(iii) Low pressure drop.
(iv) Excellent accuracy.

4.2.2.2 Rotating Lobe Meter

The unit consists of two lobes which are mounted on separated parallel shaft. The lobes revolve in opposite direction in a close fitting chamber. The lobes are cycloidal or involute in form and ensure their correct mating. The swept volume between each impeller and the chamber wall is passed through the meter four times per revolution. The impeller rotation speed is proportional to the volume of fluid flow. A simple rotating lobe meter is shown in Fig. 4.20.

The meter is used primarily for metering gases. It has an advantage of small pressure drop and good accuracy particularly at high flow rate.

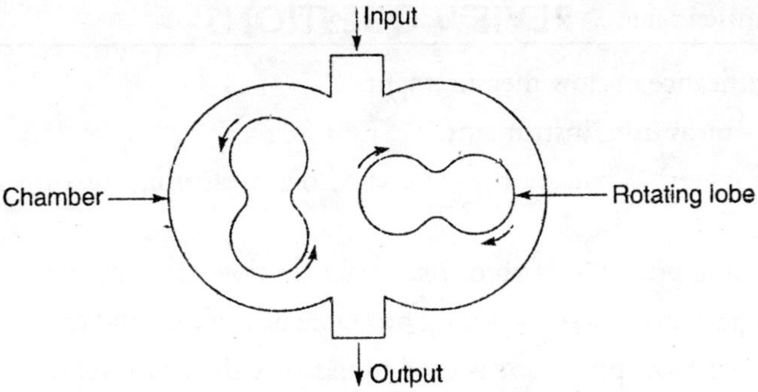

Fig. 4.20 Rotating lobe meter.

4.2.2.3 Reciprocating Piston Meter

The meter usually consists of a piston reciprocating inside a cylinder. Inlet and outlet valves with their opening and closing being controlled by a slide valve. The slide valve is actuated by the piston movement with the motion of the piston, the fluid is passed from the inlet to outlet alternately through each of the cylinder. The volume of the fluid flow is directly proportional to both the stroke length and piston speed. A simple reciprocating piston meter is shown in Fig. 4.21.

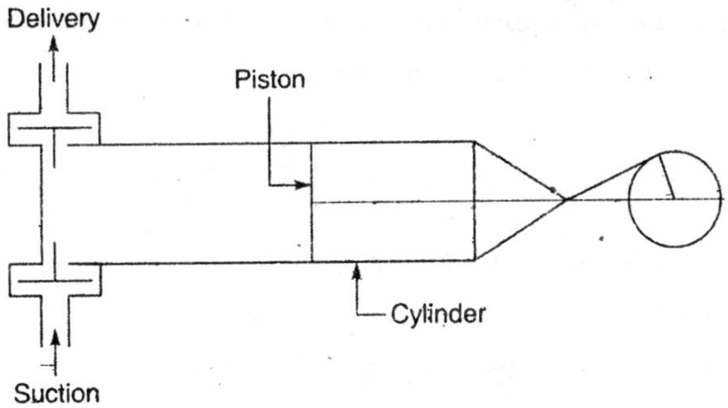

Fig. 4.21 Reciprocating piston meter.

The meter gives quite accurate measure of discharge in the range of 10 to 1000 ppm. The use is however limited to normal corrosive and low viscosity liquids.

4.2.2.4 Tilting Trap Meter

Tilting trap meter operate by gravity such that the trap is automatically upset when the centre of gravity passes over the stable equilibrium point. They are mechanically arranged for continuous flow by employing two tilting trap or by temporarily diverting the liquid to a holding chamber during the dumping operation. A simple tilting trap is shown in Fig. 4.22.

During each tilting a definite quantity of liquid flows out.

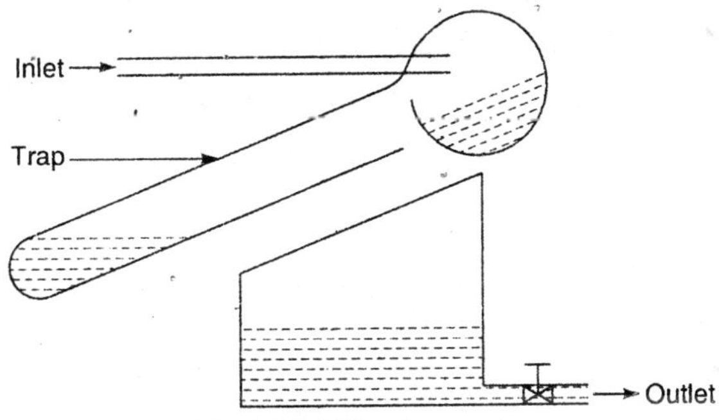

Fig. 4.22 Tilting trap meter.

REVIEW QUESTIONS

4.1 Explain the significance of flow measurement.

4.2 Classify the flow measuring instrument.

4.3 What is a head meter? What are the different flow measuring instruments comes under this category?

4.4 Explain the working principle of an orifice meter for flow measurement.

4.5 With a neat diagram explain the working principle of a venturimeter.

4.6 What is pitot tube? How pitot tube is used to measure the point velocity? Explain.

4.7 What is the principle involved in flow nozzle? Explain the working of a flow nozzle.

4.8 Explain the working of venturi, orifice meter, flow nozzle and pitot tube.

4.9 Write in detail on any one of the variable head flow meter.

4.10 Explain the functioning of deflection vane anemometer.

4.11 How are rotary vane meter used for measurement of fluid flow rate.

4.12 What is the principle of working of magnetic flow meter?

4.13 Explain the working of nutating disc meter.

4.14 Give the classification of flow measuring instrument. What are the rate and quantity meters?

4.15 Explain in detail with a neat sketch the construction and operation of one instrument used for flow measurement.

4.16 Discuss with neat sketch the working of mass flow meter indicating its specific advantages and disadvantages.

4.17 Discuss with neat sketch the working of electro magnetic flow meter indicating its specific advantages and disadvantages.

4.18 Write the working principle and field of operation of the following flow measuring devices.
   (i) Turbine type flow meter
   (ii) Positive displacement flow meter

4.19 Compare variable head and variable area meter in industrial flow metering.

4.20 What are the various type of orifice available and where they can be put to use?

4.21 Compare venturimeter and orifice meter in industrial practice.

4.22 What are the different manometer tappings available and at what situation they can be best used?

4.23 Write a notes on tilting trap meter.

□□□

Chapter 5

Level Measurements

5.1 INTRODUCTION

Liquid level refers to the position or height of a liquid surface above the datum line. The level measurements are made to ascertain the quantity of the liquid held within a container. Level effects both the pressure and rate of flow in and out of the container. It also indicate the quantity stored. Level measurement and control is an important function in variety of process.

5.2 CLASSIFICATION OF LEVEL MEASURING INSTRUMENTS

The liquid level measuring instruments may be broadly classified as :

(i) Direct method of level measurement
(ii) Indirect method of level measurement

5.2.1 Direct Method of Level Measurement

The simplest method of measuring liquid level directly is to use an ordinary sight glass or level gauge on the container. The level is also directly measured by a long stick or a thread and iron ball attached to it. This method is quite effective for clean coloured liquid if remote indication is not required. The various direct method of liquid level measurement are as follows :

(i) Bob and tape method
(ii) Sight glass method
(iii) Float operated mechanism
(iv) Ekstrom indicator method

5.2.1.1 Bob and Tape Method

The simplest of the direct method for liquid level measurement is the Bob and tape method. All the need is a Bob or weight suspended from a tape marked in feet or meters. The Bob is lowered to the bottom of the vessel containing the liquid and the level is determined by noting the point on the tape reached by the liquid. The actual reading is made after the tape is removed from the vessel. A simple Bob and tape method of level measurement is shown in Fig. 5.1.

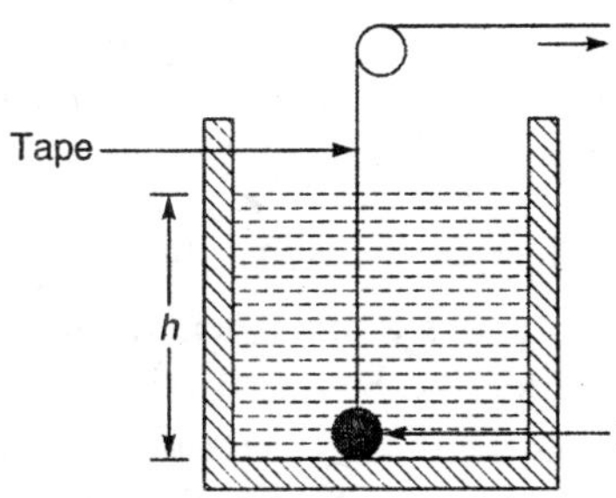

Fig. 5.1 Bob and tape.

The level can be measured in the similar manner by insisting a level rod or sight glass. A marked rod is inserted inside the tank to the bottom and the highest distance reached by the liquid on the rod is noted.

Precaution should be taken to note that the stick is not reacted with the liquid. This method is not suitable for continuous level measurement.

5.2.1.2 Sight Glass Method

A graduated sight glass tube mounted on the side of the vessel. As the level of liquid in the vessel changes, so does the level of liquid in the glass tube. Measurement is a simple matter of reading the position liquid level on the scale of the sight glass. Since the glass tube is fragile or brittle, they are often provided with metallic protecting cover around them. Further diameter of the glass tube be neither large enough to change the container liquid nor small enough to cause capillary action in the tube. A simple sight glass method of level measurement is shown in Fig. 5.2.

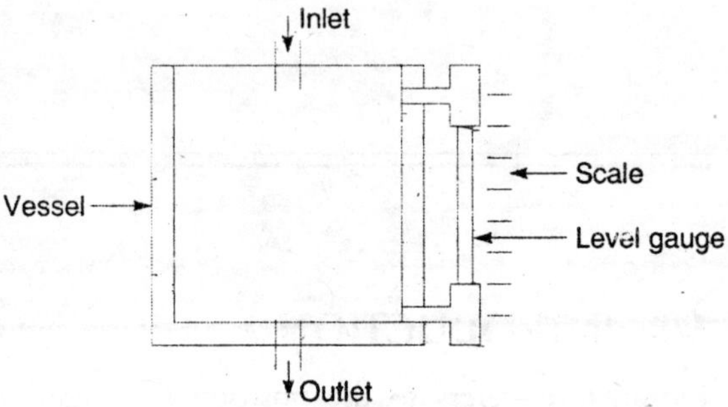

Fig. 5.2 Sight glass.

This is the simplest and continuous level indicators for clean and coloured liquids. But it can not be used with dirty, viscous and corrosive liquids.

5.2.1.3 Float Operated Mechanism

It consists of a primary device float, which because of the buoyancy effect follows the changing level of the liquid and a mechanism that transfer the float action to a pointer action.

Float is generally a hollow metal sphere, but cylindrical ceramic float and disc shaped float of synthetic material are also used. The indicating mechanism is attached with the level gauge. The float operated level gauge are of two different type :

(i) Float and pulley method

(ii) Float and shaft method

(i) Float and Pulley Method

In this type of mechanism float is usually attached to a cable, which wound around a pulley or drum and with other end of the cable a weight pointer is attached. The movement of float is thus transferred to the pointer which indicates the liquid level on an approximate scale. A simple float and pulley method of level measurement is shown in Fig. 5.3.

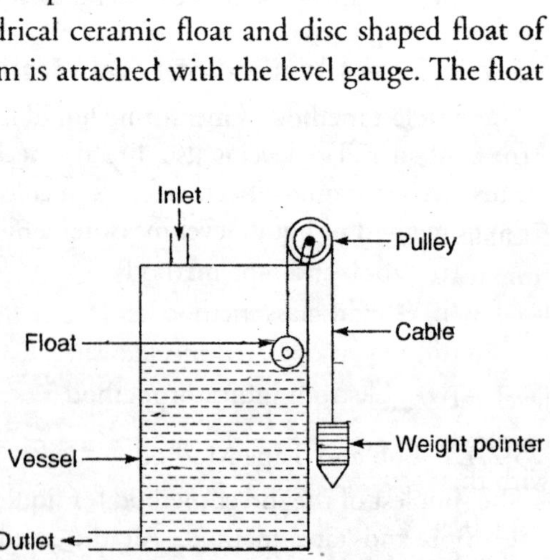

Fig. 5.3 Float and pulley mechanism.

(ii) Float and Shaft Method

In this arrangement float is attached to a shaft which transfer the motion of the float to an indicator. As the float rises from position *A* to *B* with increase in the liquid level, the float movement rotates the shaft which in term operate the pointer on scale. This is suitable for both open and closed vessel. A simple float and shaft method of level measurement is shown in Fig. 5.4.

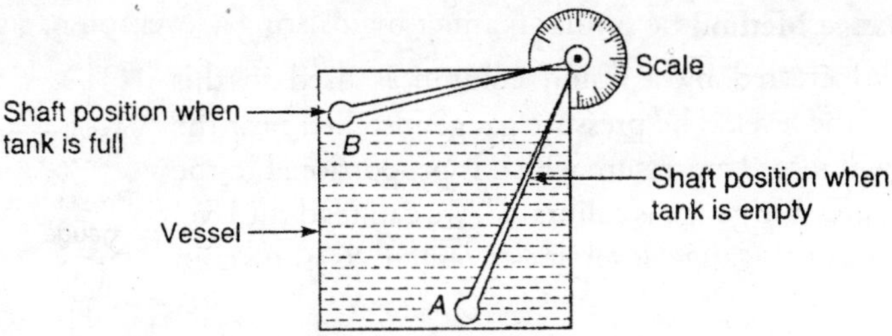

Fig. 5.4 Float and shaft mechanism.

There are many other arrangement in which movement of float is transformed into pointer movement on a calibrated scale so that level can be measured accurately.

5.2.1.4 Ekstrom Indicator Method

It is the development of sight glass with all the advantages and simplicity and there is no potential danger of breaking of the glasses. A simple Ekstrom level indicator is shown in Fig. 5.5.

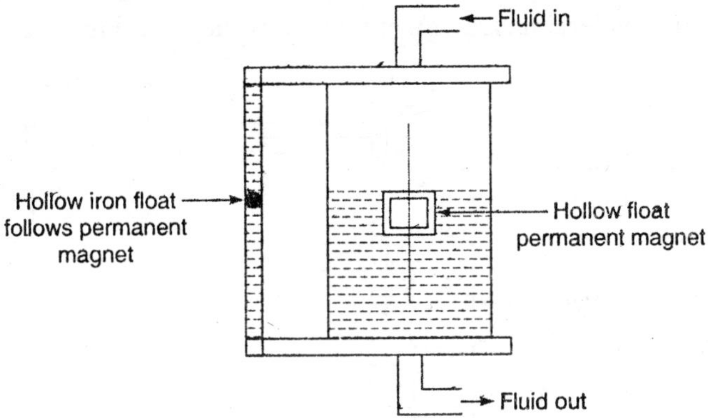

Fig. 5.5 Ekstrom indicator.

5.2.2 Indirect Method of Level Measurement

Indirect method of liquid level measurement generally employed where the direct method is not feasible. For example, consider a closed tank system in remote area where going of all the way, all the time for measuring level of the liquid in the tank is not comfortable. Therefore, in this case a indirect level measurement like bubbler air method is employed.

Consider an another case of tank containing corrosive and viscous liquid where direct method of level measurement is not feasible, the indirect level measurement like pressure gauge system can be employed. In this method no difficulties arises in level measurement but there are some disadvantage with this instrument. The various method of indirect level measurement are as follows :

(i) Pressure gauge method
(ii) Air trap method
(iii) Diaphragm method
(iv) Bubbler method
(v) Differential pressure method
(vi) Radio active method

5.2.2.1 Pressure Gauge Method

The hydrostatic head created by a liquid column is used in this method to determine the level. The pressure gauge installed near the bottom of the tank will record a pressure which is proportional to the height. Hence, if the pressure gauge is calibrated, we can read the level directly. A simple pressure gauge level measurement is shown in Fig. 5.6.

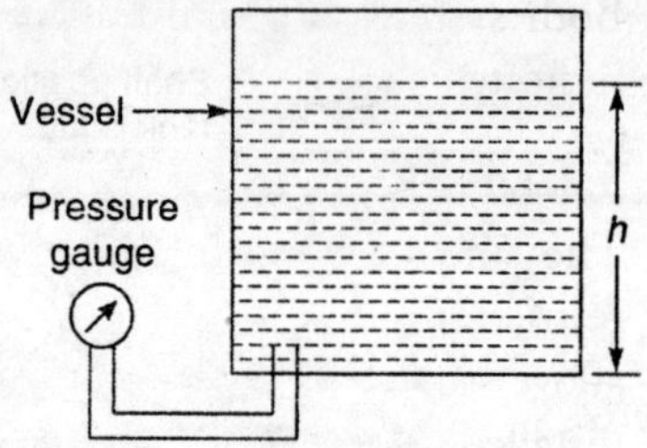

Fig. 5.6 Pressure gauge level measurement.

The gauge can serve as a pneumatic transmitter for remote indication, recording and control. For proper operation, the liquid neither corrode the gauge nor it should contain solid suspensions which would clog the gauge and can give erratic reading of the liquid level.

5.2.2.2 Air Trap Method

This is similar to the pressure gauge method. The pressure receiver operate from the pressure of the trapped air. A simple air trap level measurement method is shown in Fig. 5.7.

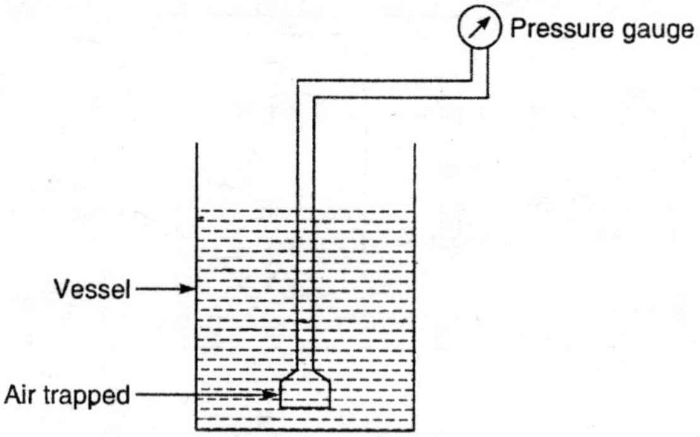

Fig. 5.7 Air trapped level measurement.

The main limitation of this type is the possibility of slow loss of air trapped in the system.

5.2.2.3 Diaphragm Method

Diaphragm method of level measurement is similar to air trap method. A simple diaphragm method of level measurement is shown in Fig. 5.8.

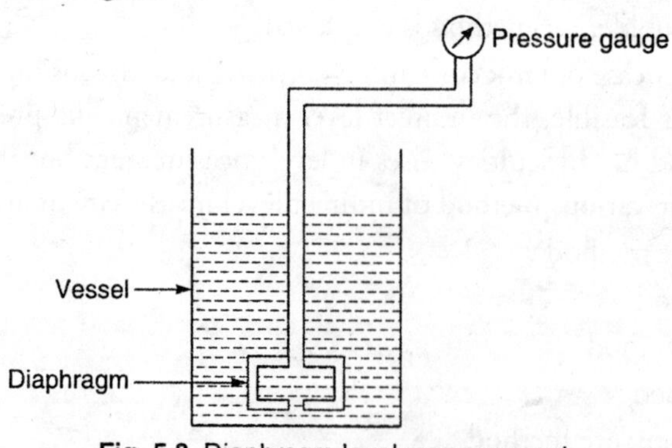

Fig. 5.8 Diaphragm level measurement.

For corrosive, thick, viscous and solid contaminated liquid diaphragm seals or air traps are used. Both systems are transmitting fluid mostly air between liquid and the gauge. The air pressure is indicated on the gauge scale in terms of liquid level.

5.2.2.4 Bubbler Method

The air pressure downstream of a flow restriction equals the hydrostatic pressure above the bubbler tube end. A pipe is installed vertically in the vessel with its open end at the zero level. The other end of the pipe is connected to a regular air supply and a pressure gauge. A simple bubbler method of level measurement is shown in Fig. 5.9.

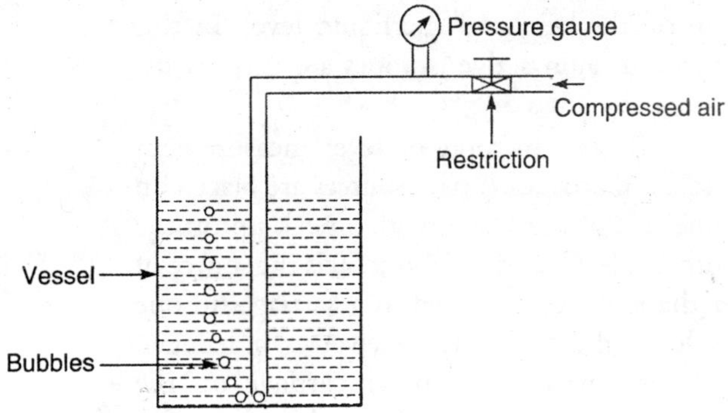

Fig. 5.9 Bubbler method of level measurement.

To make level measurement the air supply is adjusted so that the pressure is slightly higher than the pressure due to the height of the liquid. This is accomplished by regulating the air pressure until bubbles can be seen slowly leaving the open end of the tube. The pressure gauge then measure the air pressure needed to overcome the pressure of the liquid. The gauge is calibrated in terms of level and expressed as

$$h = P - P_a \qquad \text{...(5.1)}$$

where, h = Level of water, m
P = Applied pressure, N/m^2
P_a = Atmospheric pressure, N/m^2

5.2.2.5 Differential Pressure Method

This method is used to measure the level of closed vessel. When the liquid is in a pressure vessel, the liquid column pressure can not be used unless the vessel pressure is balanced. This is usually done using a differential pressure meter. A simple differential pressure meter is shown in Fig. 5.10.

The connections are made to the vessel at the top and bottom and also to the column of the differential pressure meter. The top connection is made to the low pressure column of the meter and the bottom connection to the high pressure column. In this the pressure in the column is balanced since it is fed to the both column of the meter. The difference in pressure detected by the meter is due to the

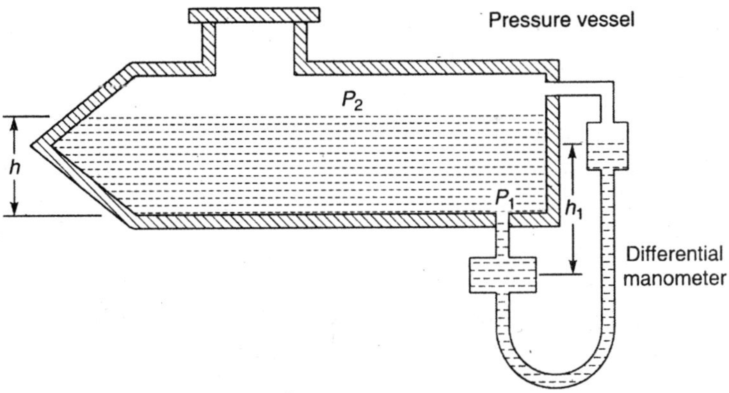

Fig. 5.10 Differential pressure method of level measurement.

changing level of liquid. Any of the differential pressure device can be used for this method of level measurement.

5.2.2.6 Radio Active Method

The radio active method is based on the principle of dependence of absorption of radio active radiation up on the height of the liquid level. In this method radio active isotopes are used as the source of emission.

In this method of level measurement one or more radio active sources are placed on one side of the vessel, with a peak up on the other side. Generally, about 60 is the source of radiation used is placed in a carriage on one side of the tank and receiver is a halogen counter placed in another carriage on the other side of the vessel. A simple radio active method of level measurement is shown in Fig. 5.11.

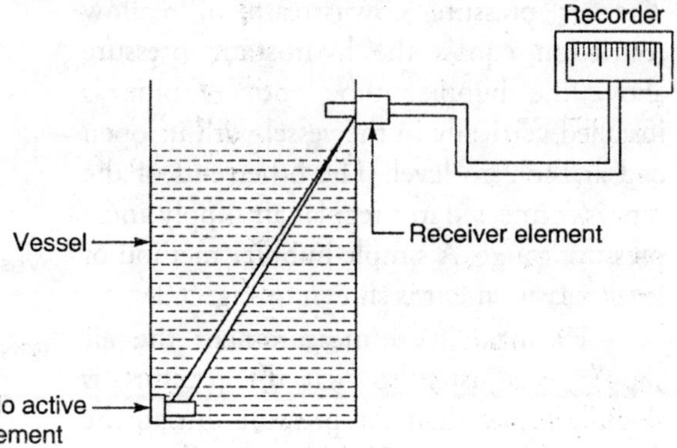

Fig. 5.11 Radio active method of level measurement.

As the level of the liquid changes it absorb more or less of radio active energy by the pick up carriage, which is connected to special electronic amplifier designed to produce enough electrical energy to actuate an electrical meter. The meter scale is calibrated to level unit.

REVIEW QUESTIONS

5.1 What is the significance of level measurement? Explain.

5.2 Classify the level measuring instrument.

5.3 Explain the working principle of float and pulley operated level indicating mechanism.

5.4 Explain the tank level gauge and float type level measuring gauge.

5.5 What are float switches?

5.6 How is radio active method used to measure the liquid level?

5.7 Explain the importance of liquid level measurement in process industry.

5.8 With the help of a neat sketch describe the working principle and functioning of an air bubbler type liquid level measuring instrument. What are its limitations?

5.9 Write short notes on level measuring instrument.

5.10 What are the possible sources of error in level measuring instruments? How will you remove them?

5.11 What is the basic principle of thermal level gauge? Where is it suitable?

5.12 With a neat diagram explain the working of a differential level gauge. Where is it suitable?

5.13 How do you measure liquid level if the liquid is corrosive?

5.14 How do you measure liquid level if liquid contain suspended solid particles? Suggest the method that can be used in this case. Also, briefly describe their principle of functioning.

5.15 With the help of a neat sketch explain the working principle and functioning of a pressure gauge method of level measurement. What are its limitations.

□□□

Chapter 6

Laplace Transform

6.1 INTRODUCTION

The Laplace transform is a mathematical tool which provide a systematic and relatively simple method of solving linear differential equations. Transforming a differential equation result in an algebraic equation with the complex variable 's' replacing time 't' as the independent variable. Solving this algebraic equation and performing the inverse transformation gives the solution of the original equation.

Another reason for the use of Laplace transform is that frequency can be obtained directly from the system transfer function.

The Laplace transform of a time function $f(t)$ is defined as function $f(t)$ multiplied by e^{-st} and then integrated from $t = 0$ to $t = \infty$. Such an operation may form a new function of variable 's'. Thus, by definition

$$f(s) = L\{f(t)\} = \int_0^{\infty} f(t).e^{-st}\, dt \qquad \text{...(6.1)}$$

where 'L' denotes the transformation process and 's' is a complex variable of the form $A + iB$. This process is restricted to function $f(t)$ only. In the complex variable 'i' is representative of $\sqrt{-1}$.

6.2 PROPERTIES OF LAPLACE TRANSFORM

Laplace transform has the following properties :

(i) The $f(s)$ contain no information about the behaviour of $f(t)$ for $t < 0$. Usually, the systems are defined for $t \geq 0$ and $f(t)$ is considered as '0' for $t < 0$.

(ii) The Laplace transform operator transform the function of variable 't' to a function variable 's'. Where, 't' is eliminated by integration.

(iii) The Laplace transform is linear.

Though a large number of other properties are enumerable, the above principal properties are sufficient for process control principles evaluation and application.

6.3 LAPLACE TRANSFORM OF SOME BASIC FUNCTION

The basic functions most commonly needed for the process control applications are as follows :

(i) Step function
(ii) Linear function
(iii) Exponential function
(iv) Sinusoidal function

6.3.1 Step Function

A function of the form

$$f(t) = 0 \quad : \quad t < 0$$
$$= A \quad : \quad t \geq 0 \qquad \ldots(6.2)$$

is called step function. Applying Laplace transform to step function, we get

$$L\{f(t)\} = \int_0^\infty f(t).e^{-st}\, dt$$
$$= \int_0^\infty A.e^{-st}\, dt$$
$$= A\left[\frac{e^{-st}}{-s}\right]_0^\infty$$
$$= A\left[0 - \frac{1}{(-s)}\right]$$
$$L\{A\} = \frac{A}{s}$$

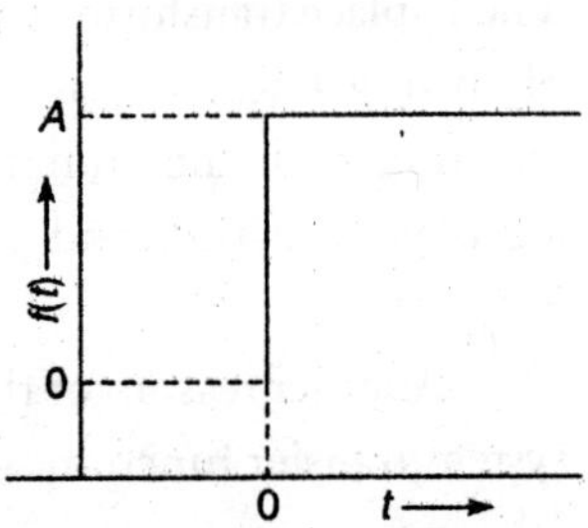

Fig. 6.1 Step function.

When, $A = 1$, $f(t) = 1$. This function is known as a unit step function and is generally defined as $u(t)$. A plot of $f(t)$ vs t for step function of size A is shown in Fig. 6.1.

6.3.2 Linear Function

A function of the form

$$f(t) = 0 \quad : \quad t < 0$$
$$= t \quad : \quad t \geq 0 \qquad \ldots(6.3)$$

is called linear function. Applying Laplace transform to linear function, we get

$$L\{f(t)\} = \int_0^\infty f(t).e^{-st}\, dt$$
$$= \int_0^\infty t.e^{-st}\, dt$$
$$= \left[t.\frac{e^{-st}}{-s}\right]_0^\infty - \int_0^\infty 1.\frac{e^{-st}}{-s}\, dt$$
$$= (0 - 0) + \frac{1}{s}\int_0^\infty e^{-st}\, dt$$

$$= \frac{1}{s}\left[\frac{e^{-st}}{-s}\right]_0^{\infty}$$

$$= \frac{1}{s}\left[0 - \frac{1}{-s}\right]$$

$$L\{t\} = \frac{1}{s^2}$$

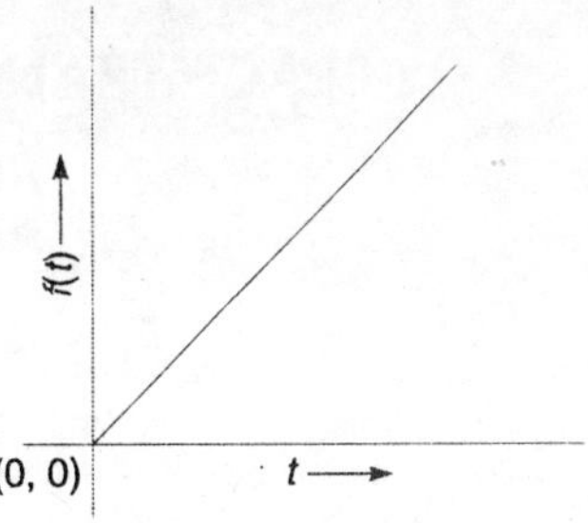

Fig. 6.2 Linear function.

A plot of $f(t)$ vs t for a linear function of unity size is shown in Fig. 6.2. This function is also called ramp function.

6.3.3 Exponential Function

A function of the form

$$\begin{aligned} f(t) &= 0 &&: t < 0 \\ &= e^{-at} &&: t \geq 0 \end{aligned} \qquad \ldots(6.4)$$

is called exponential function. Applying Laplace transform to exponential function, we get

$$L\{f(t)\} = \int_0^{\infty} f(t).e^{-st}\, dt$$

$$= \int_0^{\infty} e^{-at}.e^{-st}\, dt$$

$$= \int_0^{\infty} e^{-(s+a)t}\, dt$$

$$= \left[\frac{e^{-(s+a)t}}{-(s+a)}\right]_0^{\infty}$$

$$= 0 - \frac{1}{-(s+a)}$$

$$L\{e^{-at}\} = \frac{1}{(s+a)}$$

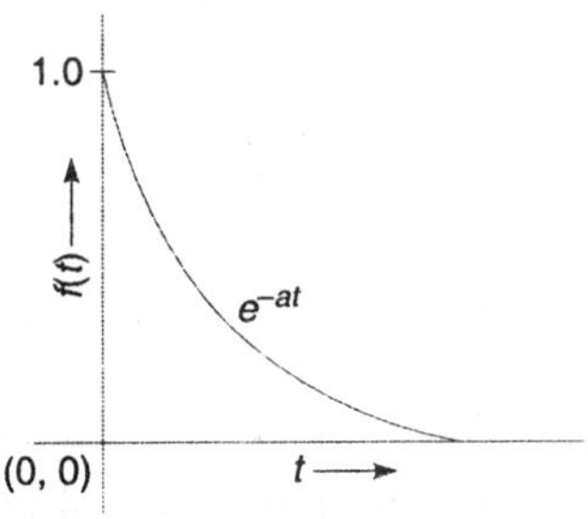

Fig. 6.3 Exponential function.

A plot of $f(t)$ vs t for an exponential function of unity size is shown in Fig. 6.3.

6.3.4 Sinusoidal Function

A function of the form

$$\begin{aligned} f(t) &= \sin kt &&: t \geq 0 \\ &= 0 &&: t < 0 \end{aligned} \qquad \ldots(6.5)$$

is called sinusoidal function. Applying Laplace transform to a sinusoidal function, we get

$$L\{f(t)\} = \int_0^{\infty} f(t).e^{-st}\, dt$$

$$= \int_0^{\infty} \sin kt.e^{-st}\, dt$$

$$= \left[\sin kt \,.\, \frac{e^{-st}}{-s} \right]_0^{\infty} - \int_0^{\infty} \cos kt \,.\, k \,.\, \frac{e^{-st}}{-s}\, dt$$

$$= (0-0) + \frac{k}{s} \int_0^{\infty} \cos kt \,.\, e^{-st}\, dt$$

$$L\{\sin kt\} = \frac{k}{s} \left\{ \left[\cos kt \,.\, \frac{e^{-st}}{-s} \right]_0^{\infty} - \int_0^{\infty} -\sin kt \,.\, k \,.\, \frac{e^{-st}}{-s}\, dt \right\}$$

$$L\{\sin kt\} = \frac{k}{s} \left\{ \left[0 - \frac{1}{-s} \right] - \int_0^{\infty} \sin kt \,.\, \frac{k}{s} \,.\, e^{-st}\, dt \right\}$$

$$L\{\sin kt\} = \frac{k}{s^2} - \frac{k^2}{s^2} \int_0^{\infty} \sin kt \,.\, e^{-st}\, dt$$

$$L\{\sin kt\} = \frac{k}{s^2} - \frac{k^2}{s^2} L\{\sin kt\}$$

$$L\{\sin kt\} \left(1 + \frac{k^2}{s^2} \right) = \frac{k}{s^2}$$

$$L\{\sin kt\} \left(\frac{s^2 + k^2}{s^2} \right) = \frac{k}{s^2}$$

$$L\{\sin kt\} = \frac{k}{s^2} \times \frac{s^2}{(s^2 + k^2)}$$

$$L\{\sin kt\} = \frac{k}{(s^2 + k^2)}$$

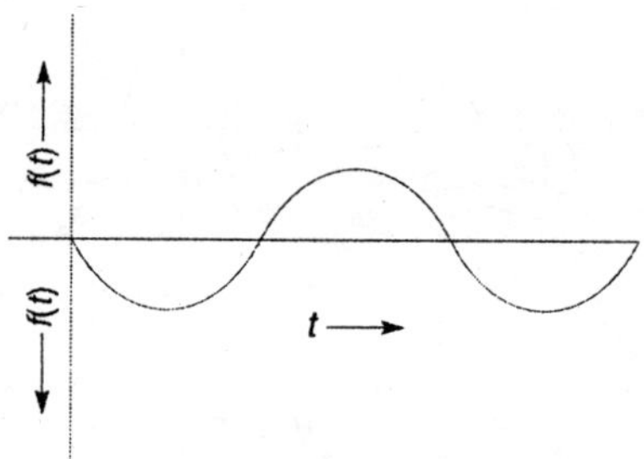

Fig. 6.4 Sinusoidal function.

A plot of $f(t)$ vs t for a sinusoidal function of unity size is shown in Fig. 6.4.
Laplace transforms of various functions are given in Table 6.1.

Table 6.1 : Laplace transforms.

S. No.	Time function $f(t)$	Laplace transform $f(s)$
1.	A	A/s
2.	At	A/s^2
3.	t^2	$2/s^3$
4.	t^n	$\frac{n!}{s^{n+1}}$
5.	e^{-at}	$\frac{1}{s+a}$

6.	$t^n . e^{-at}$	$\frac{n!}{(s+a)^{n+1}}$
7.	$\sin \omega t$	$\frac{\omega}{s^2+\omega^2}$
8.	$\cos \omega t$	$\frac{s}{s^2+\omega^2}$
9.	$e^{-at} \sin \omega t$	$\frac{\omega}{(s+a)^2+\omega^2}$
10.	$e^{-at} \cos \omega t$	$\frac{(s+a)}{(s+a)^2+\omega^2}$
11.	$\cosh (\omega t)$	$\frac{s}{(s^2-\omega^2)}$
12.	$\sinh (\omega t)$	$\frac{\omega}{(s^2-\omega^2)}$
13.	$e^{at} \sin \omega t$	$\frac{\omega}{(s-a)^2+\omega^2}$
14.	$e^{at} \cos \omega t$	$\frac{(s-a)}{(s-a)^2+\omega^2}$
15.	$e^{-at} \sinh (\omega t)$	$\frac{\omega}{(s+a)^2-\omega^2}$
16.	$e^{-at} \cosh (\omega t)$	$\frac{(s+a)}{(s+a)^2-\omega^2}$
17.	1	$1/s$

6.4 LAPLACE TRANSFORM OF DERIVATIVES

The Laplace transform of following derivatives are useful in process control :

(i) Laplace transform of first derivative

(ii) Laplace transform of second derivative

(iii) Laplace transform of third derivative

(iv) Laplace transform of *n*th derivative

6.4.1 Laplace Transform of First Derivative

Theorem : $$L\left\{\frac{df}{dt}\right\} = sf(s) - f(0)$$

where, $$f(s) = L\{f(t)\} = \int_0^{\infty} f(t) . e^{-st}\, dt$$

and $$f(0) = f(t) \quad \text{at} \quad t = 0$$

Proof : $$L\{f(t)\} = \int_0^\infty f(t).e^{-st}\,dt \qquad \text{...(6.1)}$$

From the given condition, we have

$$f(t) = \frac{df}{dt} \qquad \text{...(6.6)}$$

Substituting equation (6.6) in equation (6.1), we get

$$L\left\{\frac{df}{dt}\right\} = \int_0^\infty \frac{df}{dt}.e^{-st}\,dt$$

$$= [e^{-st}.f(t)]_0^\infty - \int_0^\infty (-s).e^{-st}.f(t).dt$$

$$= 0 - f(0) + s\int_0^\infty f(t).e^{-st}.dt$$

$$= -f(0) + sf(s)$$

$$L\left\{\frac{df}{dt}\right\} = sf(s) - f(0)$$

Proved

6.4.2 Laplace Transform of Second Derivative

Theorem : $$L\left\{\frac{d^2f}{dt^2}\right\} = s^2 f(s) - sf(0) - f'(0)$$

where, $$f(s) = L\{f(t)\} = \int_0^\infty f(t).e^{-st}\,dt$$

and $$f(0) = f(t) \quad \text{at} \quad t = 0$$

$$f'(0) = \frac{df}{dt} \quad \text{at} \quad t = 0$$

Proof : $$L\{f(t)\} = \int_0^\infty f(t).e^{-st}\,dt \qquad \text{...(6.1)}$$

From the given condition, we have

$$f(t) = \frac{d^2 f}{dt^2} \qquad \text{...(6.7)}$$

Substituting equation (6.7) in equation (6.1), we get

$$L\left\{\frac{d^2f}{dt^2}\right\} = L\left\{\frac{d}{dt}\left(\frac{df}{dt}\right)\right\}$$

$$= sL\left\{\frac{df}{dt}\right\} - \frac{df}{dt}$$

$$= s[sf(s) - f(0)] - f'(0)$$

$$= s^2 f(s) - sf(0) - f'(0)$$

Proved

6.4.3 Laplace Transform of Third Derivative

Theorem : $$L\left\{\frac{d^3 f}{dt^3}\right\} = s^3 f(s) - s^2 f(0) - sf'(0) - f''(0) \quad \text{at} \quad t = 0.$$

Proof : $$L\{f(t)\} = \int_0^{\infty} f(t).e^{-st}\, dt$$

From the given condition, we have

$$f(t) = \frac{d^3 f}{dt^3} \qquad \ldots(6.8)$$

Substituting equation (6.8) in equation (6.1), we get

$$L\left\{\frac{d^3 f}{dt^3}\right\} = L\left\{\frac{d}{dt}\left(\frac{d^2 f}{dt^2}\right)\right\}$$

$$= sL\left\{\frac{d^2 f}{dt^2}\right\} - \frac{d^2 f}{dt^2}$$

$$= s[s^2 f(s) - sf(0) - f'(0)] - f''(0)$$

$$= s^3 f(s) - s^2 f(0) - sf'(0) - f''(0)$$

Proved

6.4.4 Laplace Transform of *n*th Derivative

By close observation of Laplace transform of first derivative, second derivative and third derivative, we can generalized and write the Laplace transform of *n*th derivative as

$$L\left\{\frac{d^n f}{dt^n}\right\} = s^n f(s) - s^{n-1} f(0) - \ldots sf^{n-2}(0) - f^{n-1}(0)$$

In control engineering the functions and variables are defined such that its derivatives will be zero at initial conditions.

i.e., $$f(0) = f'(0) = f''(0) = 0$$

Therefore, $$L\left\{\frac{d^n f}{dt^n}\right\} = s^n f(s)$$

■ **Example 6.1** Find the Laplace transform of the function $f(t) = 1$.

Solution : By the definition of Laplace transform, we have

$$L\{f(t)\} = \int_0^{\infty} f(t).e^{-st}\, dt \qquad \ldots(6.1)$$

From the given data, we have

$$f(t) = 1 \qquad \ldots(6.9)$$

Substituting equation (6.9) in equation (6.1), we get

$$L\{1\} = \int_0^{\infty} 1.e^{-st}\, dt$$

$$= \left[\frac{e^{-st}}{-s}\right]_0^\infty$$

$$= 0 - \frac{1}{-s}$$

$$= \frac{1}{s}$$

■ **Example 6.2** Find the Laplace transform of 5.

Solution : By the definition of Laplace transform, we have

$$L\{f(t)\} = \int_0^\infty f(t).e^{-st}\, dt \qquad \ldots(6.1)$$

From the given data, we have

$$f(t) = 5 \qquad \ldots(6.10)$$

Substituting equation (6.10) in equation (6.1), we get

$$L\{5\} = \int_0^\infty 5.e^{-st}\, dt$$

$$= 5\left[\frac{e^{-st}}{-s}\right]_0^\infty$$

$$= 5\left[0 - \frac{1}{-s}\right]$$

$$= 5/s$$

■ **Example 6.3** Find the Laplace transform of

$$\frac{d^2x}{dt^2} + 5\frac{dx}{dt} + 6x = 1$$

Solution :

$$L\left\{\frac{d^2x}{dt^2}\right\} + 5L\left\{\frac{dx}{dt}\right\} + 6L\{x\} = L\{1\}$$

$$s^2x(s) + 5sx(s) + 6x(s) = \frac{1}{s}$$

$$x(s)\,[s^2 + 5s + 6] = \frac{1}{s}$$

$$x(s) = \frac{1}{(s)\,(s^2 + 5s + 6)}$$

■ **Example 6.4** Find the Laplace transform of

$$\frac{d^3x}{dt^3} + 4\frac{d^2x}{dt^2} + 5\frac{dx}{dt} + 2x = 2$$

Solution : $$L\left\{\frac{d^3x}{dt^3}\right\}+4\left\{\frac{d^2x}{dt^2}\right\}+5L\left\{\frac{dx}{dt}\right\}+2L\{x\}=L\{2\}$$

$$s^3x(s)+4s^2x(s)+5s\,x(s)+3x(s)=\frac{2}{s}$$

$$x(s)\,[s^3+4s^2+5s+2]=\frac{2}{s}$$

$$x(s)=\frac{2}{(s)\,(s^3+4s^2+5s+2)}$$

■ **Example 6.5** Find the Laplace transform of $f(t)$ defined as

$$f(t)=\frac{t}{T} \quad ; \quad \text{when } 0<t<T$$
$$=1 \quad ; \quad \text{when } t>T$$

Solution : By the definition of Laplace transform, we have

$$L\{f(t)\}=\int_0^\infty f(t).e^{-st}\,dt \qquad \ldots(6.1)$$

$$=\int_0^T \frac{t}{T}.e^{-st}\,dt+\int_T^\infty 1.e^{-st}\,dt$$

$$=\frac{1}{T}\int_0^T t.e^{-st}\,dt+\int_T^\infty e^{-st}\,dt$$

$$=\frac{1}{T}\left\{\left[t.\frac{e^{-st}}{-s}\right]_0^T-\int_0^T\frac{1.e^{-st}}{-s}\,dt\right\}+\left[\frac{e^{-st}}{-s}\right]_T^\infty$$

$$=\frac{1}{T}\left\{\left[\frac{Te^{-st}}{-s}-0\right]+\frac{1}{s}\left[\frac{e^{-st}}{-s}\right]_0^T\right\}+\left[0-\frac{e^{-st}}{-s}\right]$$

$$=-\frac{e^{-st}}{s}+\frac{1}{Ts}\left[\frac{e^{-st}}{-s}-\frac{1}{-s}\right]-\frac{e^{-st}}{-s}$$

$$=-\frac{e^{-sT}}{s}-\frac{e^{-sT}}{Ts^2}+\frac{1}{Ts^2}+\frac{e^{-sT}}{s}$$

$$=-\frac{e^{-sT}}{Ts^2}+\frac{1}{Ts^2}$$

$$=\frac{1-e^{-Ts}}{Ts^2}$$

■ **Example 6.6** Find the Laplace transform of $f(t)$ defined as :

$$f(t)=\text{sint} \quad : \quad 0<t<\pi$$
$$=0 \quad ; \quad t>\pi$$

Solution : By the definition of Laplace transform, we have

$$L\{f(t)\} = \int_0^{\infty} f(t).e^{-st}\, dt \qquad \text{...(6.1)}$$

$$= \int_0^{\pi} \sin t.e^{-st}\, dt + \int_{\pi}^{\infty} 0.e^{-st}\, dt$$

$$= \left[\frac{\sin t.e^{-st}}{-s}\right]_0^{\pi} - \int_0^{\pi} \frac{\cos t.e^{-st}}{-s}\, dt + 0$$

$$= (0-0) + \frac{1}{s}\int_0^{\pi} \cos t.e^{-st}\, dt$$

$$= \frac{1}{s}\left\{\left[\frac{\cos t.e^{-st}}{-s}\right]_0^{\pi} - \int_0^{\pi} \frac{-\sin t.e^{-st}}{-s}\, dt\right\}$$

$$= \frac{1}{s}\left\{\left[\frac{-e^{-\pi s}}{-s} - \frac{1}{-s}\right] - \frac{1}{s}\int_0^{\pi} \sin t.e^{-st}\, dt\right\}$$

$$L\{f(t)\} = \frac{e^{-\pi s}}{s^2} + \frac{1}{s^2} - \frac{1}{s^2} L\{f(t)\}$$

$$L\{f(t)\}\left(1 + \frac{1}{s^2}\right) = \frac{1+e^{-\pi s}}{s^2}$$

$$L\{f(t)\} = \frac{1+e^{-\pi s}}{s^2} \times \frac{s^2}{s^2+1}$$

$$L\{f(t)\} = \frac{1+e^{-\pi s}}{s^2+1}$$

■ **Example 6.7** Find the Laplace transform of the given function

$$\frac{dx}{dt} + x = 1, \quad \text{given that } x(0) = 0$$

Solution :

$$\frac{dx}{dt} + x = 1$$

$$L\left\{\frac{dx}{dt}\right\} + L\{x\} = L\{1\}$$

$$sx(s) - x(0) + x(s) = \frac{1}{s}$$

From the given data, we have

$$x(0) = 0$$

$$\Rightarrow \qquad sx(s) + x(s) = \frac{1}{s}$$

$$\Rightarrow \qquad x(s)\,(s+1) = \frac{1}{s}$$

$$\Rightarrow \qquad x(s) = \frac{1}{(s)\,(s+1)}$$

■ **Example 6.8** Find the Laplace transform of the given function

$$\frac{d^2x}{dt^2} + 3\frac{dx}{dt} + x = 1, \quad \text{given that } x(0) = x'(0) = 0.$$

Solution :

$$\frac{d^2x}{dt^2} + 3\frac{dx}{dt} + x = 1$$

$$s^2 x(s) - sx(0) - x'(0) + 3\,[sx(s) - x(s)] + x(s) = \frac{1}{s}$$

From the given data, we have

$$x(0) = x'(0) = 0$$

$$\Rightarrow \qquad s^2 x(s) + 3x\,x(s) + x(s) = \frac{1}{s}$$

$$\Rightarrow \qquad x(s)\,[s^2 + 3s + 1] = \frac{1}{s}$$

$$\Rightarrow \qquad x(s) = \frac{1}{(s)\,[s^2 + 3s + 1]}$$

■ **Example 6.9** Find $L\{e^{-at} \sin kt\}$.

Solution : By the definition of Laplace transform, we have

$$L\{f(t)\} = \int_0^\infty f(t)\,.\,e^{-st}\,dt \qquad \ldots(6.1)$$

From the given data, we have

$$f(t) = e^{-at}\,.\sin kt \qquad \ldots(6.11)$$

Substituting equation (6.11) in equation (6.1), we get

$$L\{e^{-at}\,.\sin kt\} = \int_0^\infty e^{-at}\,.\sin kt\,.\,e^{-st}\,dt$$

$$= \int_0^\infty \sin kt\,.\,e^{-(s+a)t}\,dt$$

$$= \left[\frac{\sin kt\,.\,e^{-(s+a)t}}{-(s+a)}\right]_0^\infty - \int_0^\infty \frac{\cos kt\,.\,k\,.\,e^{-(s+a)t}}{-(s+a)}\,dt$$

$$= (0-0) + \frac{k}{(s+a)} \int_0^\infty \cos kt\,.\,e^{-st}\,.\,dt$$

$$= \frac{k}{s+a}\left\{\left[\frac{\cos kt\,.\,e^{-(s+a)t}}{-(s+a)}\right]_0^\infty - \int_0^\infty \frac{-\sin kt\,.\,k\,.\,e^{-(s+a)t}}{-(s+a)}\,dt\right\}$$

$$= \frac{k}{s+a}\left\{\left[0 - \frac{1}{-(s+a)}\right] - \frac{k}{(s+a)}\int_0^{\infty} \sin kt \,.\, e^{-(s+a)t}\, dt\right\}$$

$$= \frac{k}{(s+a)^2} - \frac{k^2}{(s+a)^2}\int_0^{\infty} \sin kt \,.\, e^{-(s+a)t}\, dt$$

$$L\{e^{-at} \,.\sin kt\} = \frac{k}{(s+a)^2} - \frac{k^2}{(s+a)^2} L\{e^{-at} \,.\sin kt\}$$

$$L\{e^{-at} \,.\sin kt\} = \frac{k}{(s+a)^2} \times \frac{(s+a)^2}{(s+a)^2 + k^2}$$

$$L\{e^{-at} \,.\sin kt\} = \frac{k}{(s+a)^2 + k^2}$$

Example 6.10 Find $L\{e^{-at} \sinh kt\}$.

Solution : By the definition of Laplace transform, we have

$$L\{f(t)\} = \int_0^{\infty} f(t) \,.\, e^{-st}\, dt \qquad ...(6.1)$$

From the given data, we have

$$f(t) = e^{-at} \sinh kt \qquad ...(6.12)$$

Substituting equation (6.12) in equation (6.1), we get

$$L\{e^{-at} \sinh kt\} = \int_0^{\infty} e^{-at} \,.\sinh kt \,.\, e^{-st}\, dt$$

$$= \int_0^{\infty} \sinh kt \,.\, e^{-(s+a)t}\, dt$$

$$= \left[\frac{\sinh kt \,.\, e^{-(s+a)t}}{-(s+a)}\right]_0^{\infty} - \int_0^{\infty} \frac{\cosh kt \,.\, k \,.\, e^{-(s+a)t}}{-(s+a)}\, dt$$

$$= (0-0) + \frac{k}{(s+a)}\int_0^{\infty} \cosh kt \,.\, e^{-(s+a)t}\, dt$$

$$= \frac{k}{(s+a)}\left\{\left[\frac{\cosh kt \,.\, e^{-(s+a)t}}{-(s+a)}\right]_0^{\infty} - \int_0^{\infty} \frac{\sinh kt \,.\, k \,.\, e^{-(s+a)t}}{-(s+a)}\, dt\right\}$$

$$= \frac{k}{(s+a)}\left\{\left[0 - \frac{1}{-(s+a)}\right] + \frac{k}{(s+a)}\int_0^{\infty} \sinh kt \,.\, e^{-(s+a)t} \,.\, dt\right\}$$

$$= \frac{k}{(s+a)^2} + \frac{k^2}{(s+a)^2}\int_0^{\infty} \sinh kt \,.\, e^{-(s+a)t} \,.\, dt$$

$$L\{e^{-at}.\sinh kt\} = \frac{k}{(s+a)^2} + \frac{k^2}{(s+a)^2} L\{e^{-at}.\sinh kt\}$$

$$L\{e^{-at}.\sinh kt\} = \frac{k}{(s+a)^2} \times \frac{(s+a)^2}{(s+a)^2 - k^2}$$

$$L\{e^{-at}.\sinh kt\} = \frac{k}{(s+a)^2 - k^2}$$

6.5 LAPLACE INVERSE

If Laplace transform for $f(t)$ is defined as $f(s)$ according to the equation

$$f(s) = L\{f(t)\}$$

Then, Laplace inverse, $L^{-1}\{f(s)\} = f(t)$

■ **Example 6.11** Find $f(t)$ whose transform $f(s) = \frac{1}{s} + \frac{2}{s^2}$.

Solution :

$$f(s) = \frac{1}{s} + \frac{2}{s^2}$$

$$L^{-1}\{f(s)\} = L^{-1}\left\{\frac{1}{s} + \frac{2}{s^2}\right\}$$

$$f(t) = L^{-1}\left\{\frac{1}{s}\right\} + L^{-1}\left\{\frac{2}{s^2}\right\}$$

$$f(t) = 1 + 2t$$

Example 6.12 Solve the equation, $\frac{dx}{dt} + x = 1$.

Solution :

$$\frac{dx}{dt} + x = 1$$

$$L\left\{\frac{dx}{dt}\right\} + L\{x\} = L\{1\}$$

$$s\,x(s) + x(s) = \frac{1}{s}$$

$$x(s)\,[s+1] = \frac{1}{s}$$

$$x(s) = \frac{1}{(s)\,(s)+1}$$

$$x(s) = \frac{1}{(s)\,(s+1)} = \frac{A}{s} + \frac{B}{(s+1)}$$

$$\Rightarrow \qquad 1 = A\,(s+1) + B(s)$$

Put, $s = 0$; $A = 1$

Put, $s = -1$; $B = -1$

$$\Rightarrow \qquad x(s) = \frac{1}{s} - \frac{1}{s+1} \qquad \text{...(6.13)}$$

Taking Laplace inverse of equation (6.13), we get

$$x(t) = 1 - e^{-t}$$

■ **Example 6.13** Solve : $\dfrac{d^2x}{dt^2} + 2\dfrac{dx}{dt} + x = 1.$

Solution :

$$\frac{d^2x}{dt^2} + 2\frac{dx}{dt} + x = 1$$

$$L\left\{\frac{d^2x}{dt^2}\right\} + 2L\left\{\frac{dx}{dt}\right\} + L\{x\} = L\{1\}$$

$$s^2\,x(s) + 2s\,x(s) + x(s) = \frac{1}{s}$$

$$x(s)\,[s^2 + 2s + 1] = \frac{1}{s}$$

$$x(s) = \frac{1}{(s)\,(s^2 + 2s + 1)}$$

$$x(s) = \frac{1}{(s)\,(s+1)^2}$$

$$x(s) = \frac{1}{(s)\,(s+1)^2} = \frac{A}{s} + \frac{B}{(s+1)} + \frac{C}{(s+1)^2}$$

$$\Rightarrow \qquad 1 = A\,(s+1)^2 + B(s)\,(s+1) + C(s) \qquad \text{...(6.14)}$$

Put, $s = 0$; $A = 1$

Put, $s = -1$; ; $C = -1$

Comparing the coefficient of s^2 of equation (6.14), we get

$$0 = A + B$$

$$0 = 1 + B$$

$$B = -1$$

$$\Rightarrow \qquad x(s) = \frac{1}{s} - \frac{1}{(s+1)} - \frac{1}{(s+1)^2} \qquad \text{...(6.15)}$$

Taking Laplace inverse of equation (6.15), we get

$$x(t) = 1 - e^{-t} - t\,.\,e^{-t}$$

■ **Example 6.14** Solve : $\frac{d^2x}{dt^2}+4\frac{dx}{dt}+2x=2$, given that $x(0)=x'(0)=0$.

Solution :

$$\frac{d^2x}{dt^2}+4\frac{dx}{dt}+4x=2$$

$$s^2x(s)-s\,x(0)-x'(0)+4[s\,x(s)-x(0)]+4x(s)=\frac{2}{s}$$

$$s^2x(s)+4s\,x(s)+4x(s)=\frac{2}{s}$$

$$x(s)[s^2+4s+4]=\frac{2}{s}$$

$$x(s)=\frac{2}{(s)(s^2+4s+4)}$$

$$x(s)=\frac{2}{(s)(s+2)^2}$$

$$x(s)=\frac{2}{(s)(s+2)^2}=\frac{A}{s}+\frac{B}{(s+2)}+\frac{C}{(s+2)^2}$$

$$\Rightarrow \quad 2=A(s+2)^2+B(s)(s+2)+C(s) \qquad \ldots(6.16)$$

Put, $s=0$; $A=1/2$

Put, $s=-2$; $C=-1$

Comparing the coefficient of s^2 of equation (6.16), we get

$$0=A+B$$

$$0=1/2+B$$

$$B=-1/2$$

$$\Rightarrow \quad x(s)=\frac{1}{2s}-\frac{1}{2}e^{-2t}-t.e^{-2t}$$

■ **Example 6.15** Find the Laplace inverse of $1/s$.

Solution :

$$x(s)=1/s$$

$$L^{-1}\{x(s)\}=L^{-1}\left\{\frac{1}{s}\right\}$$

$$x(t)=1$$

■ **Example 6.16** Find the Laplace inverse of $\frac{3}{s}$.

Solution :

$$x(s)=\frac{3}{s}$$

$$L^{-1}\{x(s)\}=L^{-1}\left\{\frac{3}{s}\right\}$$

$$x(t)=3$$

■ Example 6.17 Solve : $\frac{d^3x}{dt^3} + 4\frac{d^2x}{dt^2} + 5\frac{dx}{dt} + 2x = 2.$

Solution :

$$s^3x(s) + 4s^2x(s) + 5s\,x(s) + 2x(s) = \frac{2}{s}$$

$$x(s)\,[s^3 + 4s^2 + 5s + 2] = \frac{2}{s}$$

$$x(s) = \frac{2}{(s)\,(s^3 + 4s^2 + 5s + 2)}$$

Factorisation :

$$s^3 + 4s^2 + 5s + 2 = 0 \qquad \ldots(6.17)$$

Put, $s = -1$

$$(-1)^3 + 4(-1)^2 + 5(-1) + 2 = 0$$

$$-1 + 4 - 5 + 2 = 0$$

$$0 = 0$$

Therefore, $(s + 1)$ is a factor.
Dividing equation (6.17) by $(s + 1)$, we get

$$s^2(s+1) + 3s(s+1) + 2(s+1) = 0$$

$$(s+1)\,(s^2 + 3s + 2) = 0$$

Middle term break :

$$s^2 + 3s + 2 = 0$$

$$s^2 + s + 2s + 2 = 0$$

$$s(s+1) + 2(s+1) = 0$$

$$(s+1)\,(s+2) = 0$$

Therefore, $(s+1)\,(s+1)\,(s+2)$ are the factors.

$$\Rightarrow \qquad x(s) = \frac{2}{(s)\,(s+1)^2\,(s+2)}$$

$$x(s) = \frac{2}{(s)\,(s+1)^2\,(s+2)} = \frac{A}{s} + \frac{B}{(s+1)} + \frac{C}{(s+1)^2} + \frac{D}{(s+2)}$$

$$\Rightarrow \qquad 2 = A(s+1)^2(s+2) + B(s)(s+1)(s+2) + C(s)(s+2) + D(s)(s+1)^2$$

Put, $s = 0$; $A = 1$
Put, $s = -1$; $C = -2$
Put, $s = -2$; $D = -1$

Comparing the coefficient of s^3, we get

$$0 = A + B + D$$

$$0 = 1 + B - 1$$

$$0 = 0$$

$$\Rightarrow \qquad x(s) = \frac{1}{s} - \frac{2}{(s+1)^2} - \frac{1}{(s+2)}$$

$$\Rightarrow \qquad x(t) = 1 - 2t\,.\,e^{-t} - e^{-2t}$$

■ **Example 6.18** Solve : $4\dfrac{d^2y}{dt^2} + 4\dfrac{dy}{dt} + y = 2t,\ y'(0) = y(0) = 0.$

Solution : $4[s^2 y(s) - sy(0) - y'(0)] + 4[sy(s) - y(0)] + y(s) = \dfrac{2}{s^2}$

$$4s^2 y(s) + 4s\,y(s) + y(s) = \frac{2}{s^2}$$

$$y(s)[4s^2 + 4s + 1] = \frac{2}{s^2}$$

$$y(s) = \frac{2}{s^2(4s^2 + 4s + 1)}$$

$$y(s) = \frac{2}{s^2(2s+1)^2}$$

$$\Rightarrow \qquad y(s) = \frac{2}{s^2(2s+1)^2} = \frac{A}{s} + \frac{B}{s^2} + \frac{C}{(2s+1)} + \frac{D}{(2s+1)^2}$$

$$\Rightarrow \qquad 2 = A(s)(2s+1)^2 + B(2s+1)^2 + C(s^2)(2s+1) + D(s^2) \qquad \ldots(6.1$$

Put, $s = 0$; $B = 2$

Put, $s = -1/2$; $D = 8$

Comparing the coefficient of s of equation (6.18), we get

$$0 = A + 4B$$
$$0 = A + 4 \times 2$$
$$A = -8$$

Comparing the coefficient of s^2 of equation (6.18), we get

$$0 = 4A + 4B + C + D$$
$$0 = 4 \times (-8) + 4 \times (2) + C + 8$$
$$0 = -32 + 8 + C + 8$$
$$C = 16$$

$$\Rightarrow \qquad y(s) = -\frac{8}{s} + \frac{2}{s^2} + \frac{C}{(2s+1)} + \frac{8}{(2s+1)^2}$$

$$\Rightarrow \qquad y(s) = -\frac{8}{s} + \frac{2}{s^2} + \frac{16}{2(s+1/2)} + \frac{8}{[2(s+1/2)]^2}$$

$$\Rightarrow \qquad y(s) = -\frac{8}{s} + \frac{2}{s^2} + \frac{8}{(s+1/2)} + \frac{2}{(s+1/2)^2}$$

$$\Rightarrow \qquad y(t) = -8 + 2t + 8e^{-t/2} + 2\,.\,t\,.\,e^{-t/2}$$

■ **Example 6.19** Find the Laplace inverse of $\frac{1}{s}e^{-1/s}$.

Solution : From the given data, we have

$$\frac{1}{s}e^{-1/s} = \frac{1}{s}\left\{1 - \frac{1}{s^1} + \frac{1}{2!s^2} - \frac{1}{3!s^3} + \ldots\right\}$$

$$\frac{1}{s}e^{-1/s} = \frac{1}{s} - \frac{1}{s^2} + \frac{1}{2!s^3} - \frac{1}{3!s^4} + \ldots$$

$$L^{-1}\left\{\frac{1}{s}e^{-1/s}\right\} = L^{-1}\left\{\frac{1}{s} - \frac{1}{s^2} + \frac{1}{2!s^3} - \frac{1}{3!s^4} + \ldots\right\}$$

$$= 1 - t + \frac{t^2}{2!} - \frac{t^3}{3!} + \ldots$$

6.6 INITIAL VALUE THEOREM

If $f(s)$ is the Laplace transform of $f(t)$, then

$$\underset{t \to 0}{Lt}\ f(t) = \underset{s \to \infty}{Lt}\ sf(s)$$

The conditions of this theorem are not so stringent because of the function of interest to us, the order of integration and limiting process need not interchanged to establish the result.

Proof : From the Laplace transform of first derivative, we have

$$L\left\{\frac{df}{dt}\right\} = \int_0^{\infty} \frac{df}{dt}.e^{-st}\ dt = sf(s) - f(0) \qquad \ldots(6.19)$$

Taking limit as $s \to \infty$ on both side of equation (6.19), we get

$$\underset{s \to \infty}{Lt}\left[\int_0^{\infty} \frac{df}{dt}.e^{-st}\ dt\right] = \underset{s \to \infty}{Lt}\ [sf(s) - f(0)]$$

$$0 = \underset{s \to \infty}{Lt}\ [sf(s) - f(0)]$$

$$f(0) = \underset{s \to \infty}{Lt}\ sf(s)$$

$$\underset{t \to 0}{Lt}\ f(t) = \underset{s \to \infty}{Lt}\ sf(s)$$

Hence proved.

6.7 FINAL VALUE THEOREM

If $f(s)$ is the Laplace transform of $f(t)$, then

$$\underset{t \to \infty}{Lt}\ f(t) = \underset{s \to 0}{Lt}\ sf(s)$$

provided that $sf(s)$ does not become infinite for any value of 's' satisfying $\text{Re}(s) \geq 0$. If this condition does not hold good, $f(t)$ does not approach a limit as $t \to \infty$. In the practical application of this theorem, the limit of $f(t)$ that is found by use of the theorem is correct only if $f(t)$ is bounded as 't' approaches infinity.

Proof : From the Laplace transform of first derivative, we have

$$L\left\{\frac{df}{dt}\right\} = \int_0^{\infty} \frac{df}{dt}\, e^{-st}\, dt = sf(s) - f(0) \qquad \ldots(6.19)$$

Taking limit as $s \to 0$ on both side of equation (6.19), we get

$$\underset{s \to 0}{Lt}\left[\int_0^{\infty} \frac{df}{dt}.e^{-st}\, dt\right] = \underset{s \to 0}{Lt}\,[sf(s) - f(0)]$$

$$[f(t)]_0^{\infty} = \underset{s \to 0}{Lt}\,[sf(s) - f(0)]$$

$$f(\infty) - f(0) = \underset{s \to 0}{Lt}\, sf(s) - f(0)$$

$$f(\infty) = \underset{s \to 0}{Lt}\, sf(s)$$

$$\underset{t \to \infty}{Lt}\, f(t) = \underset{s \to 0}{Lt}\, sf(s)$$

Hence proved.

■ **Example 6.20** Find the initial and final value for the function $x(t)$ for which

$$x(s) = \frac{1}{(s)(s^3 + 3s^2 + 3s + 2)}$$

Solution : From the initial value theorem, we have

$$\underset{t \to 0}{Lt}\, x(t) = \underset{s \to \infty}{Lt}\, s\,x(s)$$

$$= \underset{s \to \infty}{Lt}\, s \times \frac{1}{(s)(s^3 + 3s^2 + 3s + 2)}$$

$$= \underset{s \to \infty}{Lt}\, \frac{1}{(s^3 + 3s^2 + 3s + 2)}$$

$$= \underset{s \to \infty}{Lt}\, \frac{1}{s^3\left(1 + \frac{3}{s} + \frac{3}{s^2} + \frac{2}{s^3}\right)}$$

$$= \frac{1}{\infty} = 0$$

From the final value theorem, we have

$$\underset{t \to \infty}{Lt}\, x(t) = \underset{s \to 0}{Lt}\, s\,x(s)$$

$$= \underset{s \to 0}{Lt}\, s \times \frac{1}{(s)(s^3 + 3s^2 + 3s + 2)}$$

$$= \underset{s \to 0}{Lt}\, \frac{1}{(s^3 + 3s^2 + 3s + 2)}$$

$$= \frac{1}{2}$$

Example 6.21 Find the initial and final value of the function $x(t)$ whose transform is given as

$$x(s) = \frac{s^4 - 6s^2 + 9s - 8}{(s)(s-1)(s-2)(s+1)(s+2)}$$

Solution : From the initial value theorem, we have

$$\underset{t \to 0}{Lt}\ x(t) = \underset{s \to \infty}{Lt}\ s\,x(s)$$

$$= \underset{s \to \infty}{Lt}\ s \times \left[\frac{s^4 - 6s^2 + 9s - 8}{(s)(s-1)(s-2)(s+1)(s+2)}\right]$$

$$= \underset{s \to \infty}{Lt}\left[\frac{s^4 - 6s^2 + 9s - 8}{(s-1)(s-2)(s+1)(s+2)}\right]$$

$$= \underset{s \to \infty}{Lt}\ \frac{s^4\left[1 - \frac{6}{s^2} + \frac{9}{s^3} - \frac{8}{s^4}\right]}{s^4\left(1-\frac{1}{s}\right)\left(1-\frac{2}{s}\right)\left(1+\frac{1}{s}\right)\left(1+\frac{2}{s}\right)}$$

$$= \underset{s \to \infty}{Lt}\ \frac{\left[1 - \frac{6}{s^2} + \frac{9}{s^3} - \frac{8}{s^4}\right]}{\left(1-\frac{1}{s}\right)\left(1-\frac{2}{s}\right)\left(1+\frac{1}{s}\right)\left(1+\frac{2}{s}\right)}$$

$$= 1$$

From the final value theorem, we have

$$\underset{t \to \infty}{Lt}\ x(t) = \underset{s \to 0}{Lt}\ s\,x(s)$$

$$= \underset{s \to 0}{Lt}\ s \times \frac{(s^4 - 6s^2 + 9s - 8)}{(s)(s-1)(s-2)(s+1)(s+2)}$$

$$= \underset{s \to 0}{Lt}\ \frac{(s^4 - 6s^2 + 9s - 8)}{(s-1)(s-2)(s+1)(s+2)}$$

$$= \infty$$

In the above problem $s\,x(s)$ become infinity for real value of $s = 1$ or 2. Therefore, final value theorem is not applicable and $x(t)$ will not approach a limit.

6.8 TRANSLATION OF TRANSFORM

If $L\{f(t)\} = f(s)$. then

$L\{e^{-at} f(t)\} = f(s+a)$.

In other words, the variable in the translation 's' is translated by 'a'.

Proof :

$$L\{e^{-at} f(t)\} = \int_0^{\infty} f(t) \,.\, e^{-(s+a)t} \,.\, dt$$

$$= f(s+a)$$

Translation function :

If $$L\{f(t)\} = f(s)$$

then, $$L\{f(t-t_0)\} = e^{-st_0} f(s)$$

Provided that $f(t) = 0$ for $t < 0$.

Before proving this theorem, it may be desirable to clarify the relationship between $f(t-t_0)$ and $f(t)$. This is done for an arbitrary function $f(t)$ as shown in Fig. 6.5, where it can be seen that $f(t-t_0)$ is simply translated horizontally from $f(t)$ through a distance t_0.

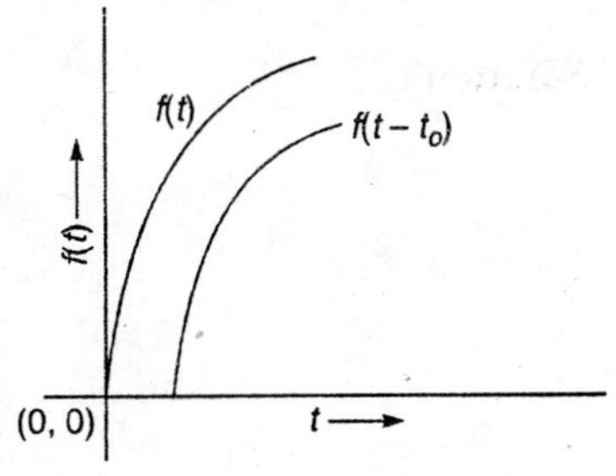

Fig. 6.5 Translation of transform.

$$L\{f(t-t_0)\} = \int_0^{\infty} f(t-t_0)\,.\,e^{-st}\,dt$$

$$= e^{-st_0} \int_{t_0}^{\infty} f(t-t_0)\,.\,e^{-s(t-t_0)}\,.\,d(t-t_0)$$

$$= e^{-st_0} f(s)$$

Since, $f(t) = 0$ for $t < 0$, the lower limit of this integral may be replaced by zero. Since $(t-t_0)$ is now dummy variable of integration. The integral may be recognized as the Laplace transform of $f(t)$. Thus, the theorem is proved.

This result is also useful in investing transforms. It follows that, if $f(t)$ is the inverse transform of $f(s)$, then the inverse transform of $e^{-st_0} f(s)$ is the function

$$g(t) = \begin{cases} 0 & ; \quad t < t_0 \\ f(t-t_0) & ; \quad t > t_0 \end{cases}$$

6.9 TRANSFORM OF INTEGRAL

If $L\{f(t)\} = f(s)$, then

$$L\left\{\int_0^{\infty} f(t)\,dt\right\} = \frac{f(s)}{s}$$

This is an important theorem. It is closely related to the theorem of differentiation. The operations of differentiation and integration are inverse of each other when applied to the time function.

Therefore, $$\frac{d}{dt}\left[\int_0^t f(t)\,dt\right] = \int_0^t \frac{df}{dt}\,.\,dt = f(t)$$

It is expected that these operations when applied to the transform will also be inverse. Thus assuming the theorem to be valid, the above equation in the transformed variable 's' become

$$s \times \frac{f(s)}{s} = \frac{1}{s} \times sf(s) = f(s)$$

The proof follows from a straight forward integration by parts.

$$f(s) = \int_0^{\infty} f(t)\,.\,e^{-st}\,dt$$

$$= e^{-st}\,.\int_0^{\infty} f(t)\,dt - \int_0^{\infty}\left[(-s)\,.\,e^{-st}\int_0^{\infty} f(t)\,dt\right]dt$$

$$= e^{-st}\,.\int_0^{\infty} f(t)\,dt + \int_0^{\infty}\left[(s)\,e^{-st}\int_0^{\infty} f(t)\,dt\right]dt$$

■ **Example 6.22** Find the Laplace transform of

$$f(t) = \begin{cases} 0 & ; \quad t < 0 \\ 1/h & ; \quad 0 < t < h \\ 0 & ; \quad t > h \end{cases}$$

Solution :

$$L\{f(t)\} = \int_0^{\infty} f(t) \, . \, e^{-st} \, dt$$

$$f(t) = \frac{1}{h}[u(t) - u(t-h)]$$

$$f(s) = \frac{1}{h}\left[\frac{1}{s} - \frac{e^{-sh}}{s}\right]$$

■ **Example 6.23** Solve the following equation for $x(t)$

$$\frac{dx}{dt} = \int_0^t x(t) \, dt - t \quad ; \quad x(0) = 3$$

Solution : From the given equation, we have

$$\frac{dx}{dt} = \int_0^t x(t) \, dt - t \qquad \ldots(6.20)$$

Taking Laplace transform of both side of equation (6.20) and making use of the previous theorem transform of integral, we get

$$s\,x(s) - x(0) = \frac{x(s)}{s} - \frac{1}{s^2}$$

$$s\,x(s) - 3 = \frac{x(s)}{s} - \frac{1}{s^2}$$

$$s\,x(s) - \frac{x(s)}{s} = 3 - \frac{1}{s^2}$$

$$x(s)\left[s - \frac{1}{s}\right] = \frac{(3s^2 - 1)}{s^2}$$

$$x(s) = \frac{(3s^2 - 1)}{s^2} \times \frac{s}{(s^2 - 1)}$$

$$x(s) = \frac{(3s^2 - 1)}{(s)\,(s^2 - 1)}$$

$$x(s) = \frac{(3s^2 - 1)}{(s)\,(s-1)\,(s+1)} = \frac{A}{s} + \frac{B}{(s-1)} + \frac{C}{(s+1)}$$

$$\frac{(3s^2 - 1)}{(s)\,(s-1)\,(s+1)} = \frac{A(s-1)\,(s+1) + B(s)\,(s+1) + C(s)\,(s-1)}{(s)\,(s-1)\,(s+1)}$$

$$\Rightarrow \qquad (3s^2 - 1) = A(s-1)\,(s+1) + B(s)\,(s+1) + C(s)\,(s-1)$$

Put, $s = 0$; $A = 1$

Put, $s = 1$; $B = 1$

Put, $s = -1$; $C = 1$

$$\Rightarrow \qquad x(s) = \frac{1}{s} + \frac{1}{(s-1)} + \frac{1}{s+1}$$

$$\Rightarrow \qquad x(t) = 1 + e^{t} + e^{-t}$$

6.10 COMPLEX VARIABLE

Number of the form $z = x + iy$, where 'x' and 'y' are real number, $i^2 = -1$ is called complex number. If both x and y or one of them is a variable then z is a complex variable. The number $z = x + iy$ is representable in the plane as shown in the Fig. 6.6.

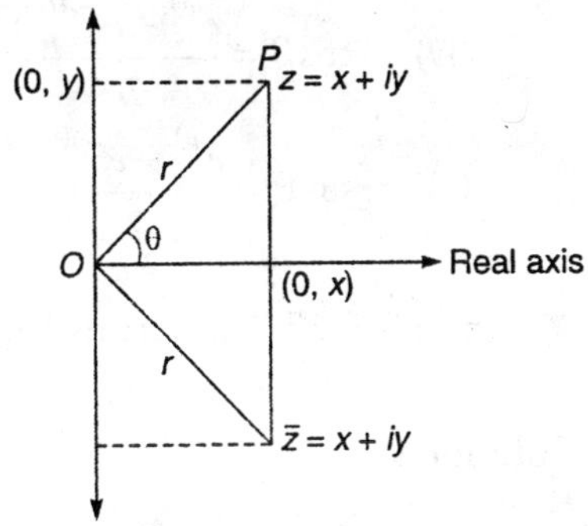

Fig. 6.6 Complex plane.

The following definitions and terminology are used in complex variables :

(i) Distance $OP = r =$ Modulus of 'z' written in the form $|z|$, where

$$|z| = \sqrt{x^2 + y^2}$$

(ii) 'x' is the real part of z.

(iii) 'y' is the imaginary part of z.

(iv) The angle θ, $0 \leq \theta \leq 2\pi$, measured counter clockwise from the positive x-axis to OP is the argument of z.

$$\theta = \text{arc sin } y/r$$
$$\theta = \text{arc cos } x/r$$

(v) The r, θ are the polar coordinate of z.

(vi) $z = x - iy$ is the complex conjugate of z.

■ **Example 6.24** Express the complex quantity $\frac{1+2i}{3-i}$ in the form $z = x + iy$.

Solution :

$$\frac{1+2i}{3-i} = \frac{(1+2i)(3+i)}{(3-i)(3+i)}$$
$$= \frac{3+6i+i-2}{9-i^2}$$
$$= \frac{1+7i}{9+1}$$
$$= \frac{1}{10} + i\frac{7}{10}$$

6.11 POLAR CO-ORDINATE

The number of the form $x = r\cos\theta$; $y = r\sin\theta$, so that $z = (x + iy) = r(\cos\theta + i\sin\theta)$ is called the polar co-ordinate of the complex number. The following definitions and terminology are used in polar co-ordinate :

(i) $\cos\theta + i\sin\theta = e^{i\theta}$

$$x + iy = z$$

(ii) $r(\cos\theta + i\sin\theta) = r\,e^{i\theta} = z$

(iii) $r(\cos\theta - i\sin\theta) = r\,e^{-i\theta} = \bar{z}$

(iv) $\cos\theta = \dfrac{e^{i\theta} + e^{-i\theta}}{2}$

(v) $\sin\theta = \dfrac{e^{i\theta} - e^{-i\theta}}{2}$

■ **Example 6.25** Convert the equation $G(j\omega) = \dfrac{1}{1 + j\omega\tau}$ in $z = x + iy$ form.

Solution :

$$G(j\omega) = \frac{1}{1 + j\omega\tau}$$

$$G(j\omega) = \frac{1}{1 + j\omega\tau} \times \frac{1 - j\omega\tau}{1 - j\omega\tau}$$

$$G(j\omega) = \frac{1 - j\omega\tau}{1 - j^2\omega^2\tau^2}$$

$$G(j\omega) = \frac{1 - j\omega\tau}{1 + \omega^2\tau^2}$$

$$G(j\omega) = \frac{1}{1 + \omega^2\tau^2} + \frac{j(-\omega\tau)}{1 + \omega^2\tau^2}$$

REVIEW QUESTIONS

6.1 Find the Laplace transform of

(i) $\cos kt$ (ii) $\cosh kt$

(iii) $e^{-at}\cos kt$ (iv) $e^{at}\cosh kt$

6.2 Find the Laplace transform of third derivative.

6.3 Solve : $\dfrac{d^3x}{dt^3} + 6\dfrac{d^2x}{dt^2} + 11\dfrac{dx}{dt} + 6x = 2$

6.4 Solve : $\dfrac{d^3x}{dt^3} + 5\dfrac{d^2x}{dt^2} + 8\dfrac{dx}{dt} + 4x = e^{-t}$

6.5 Solve : $\dfrac{d^3x}{dt^3} - 6\dfrac{d^2x}{dt^2} + 11\dfrac{dx}{dt} - 6x = 1$

6.6 Find the initial value and final value for the function $x(t)$ for which

$$x(s)=\frac{(s^2+2s+3)}{(s^3+2s^2+2s+1)}$$

6.7 Find the initial value and final value for the function $x(t)$ for which

$$x(s)=\frac{(3s+1)}{(s^4+2s^3+3s^2+4s)}$$

6.8 Find the Laplace transform of

$$f(t)=\begin{cases}0 & ; \quad t<0\\ 10 & ; \quad 0<t<1\\ 0 & ; \quad t>1\end{cases}$$

6.9 Express the complex quantity $\frac{1+2i}{2+3i}$ in the form $z=x+iy$.

□□□

Chapter **7**

Basic Concept of Process Control and First Order System

7.1 INTRODUCTION

Process control is an engineering subject that deals with architect plan mechanism and algorithm for maintaining the output of a specific process within a desired range. For example, consider the maintenance of temperature of a chemical reactor. It may be controlled to obtain a constant product output. Process control is extensively used in industry and enable mass production of consistent products from continuously operated processes such as oil refinery, cement manufacturing unit and many others. Process control enable us to operate a complex process from a central control room.

7.2 BASIC CONCEPT OF PROCESS CONTROL

Basic concept of process control involves following important terms.

7.2.1 Process

A process is one in which a system undergoes series of exchanges in a state. For example, reaction carried out in a reactor.

7.2.2 Controlled Variable

It is the variable whose value is to be maintained at a desired level with the help of controller. Ex : The output temperature of liquid.

7.2.3 Process Variable

It is the current status of a process under control. For an example, temperature of a reactor.

7.2.4 Load Variable

When something added to a process in such a way that it bring the change in controlled variable is called load variable.

7.2.5 Manipulated Variable

It is one of the load variable which is adjusted to compensate for any disturbance in the system.

7.2.6 Set Point

It is the desired value of the controlled variable.

7.2.7 Error

The difference between the set value and the measured value of the controlled variable is called error.

7.2.8 Controlled Medium

The process energy or material which is to be adjusted to a definite value is called controlled medium.

7.2.9 Controlled Agent

The process energy or material which is manipulated to hold the controlled medium at its desired value is called controlled agent.

7.2.10 Measuring Element

An instrument used to measure the controlled variable and pass the information to comparator is called measuring element.

7.2.11 Comparator

The device which compare measured and desired value of the controlled variable and passes the information about the difference if any, to the controller is called comparator.

7.2.12 Controller

The device which convert the output signal from the comparator to a signal which adjust the final control element is called controller.

7.2.13 Final Control Element

The device which manipulate the controlled agent according to the signal from the controller is called final control element.

7.2.14 Open Loop Control System

A system in which no comparison is made between the actual value and the desired value of process variable is called open loop control system.

7.2.15 Closed Loop Control System

A control system in which comparison is made between the actual value and the desired value of process variable and then passes the information to the controller in such a way that it will reduce the deviation from its standard value is called closed loop control system.

7.2.16 Offset

A sustained deviation of the controlled variable from the set point is called offset.

7.2.17 Dynamic Response

The behaviour of a system output in response to a changing input as a function of time is called dynamic response.

7.2.18 Feed Forward Control System

A control action which does not require the feed back information regarding the controlled variable is called feed forward control system. It is also known as open loop control system.

In feed forward control system, no comparison is made between the input and output of the system. Hence, the error can not be estimated. As a result of this, there is no control over the output of the system. The generalised feed forward or open loop control system can be schematically represented as shown in Fig. 7.1.

Fig. 7.1 Feed forward control.

For example, consider a room heating system. It consists of heating element and the room. The objective is to control the actual room temperature. The input to the system is the desired temperature setting and the output of the system is the actual temperature. As there is no controller, no comparison is made between the output and the input. Hence, there is no control over the output of the system. An open loop system is unreliable and inaccurate. Proper judgement and experience is required to control the system.

7.2.19 Feed Back Control System

A control action which required the feed back information regarding the controlled variable is called feed back control system.

In other words a feed back control system compares what we are getting (output) with what we are asked for (input) and uses the difference between the two to drive the output into close correspondence with the input. It is this comparison that makes a feed back control system so effective. The generalised feed back or closed loop control system can be schematically represented as shown in Fig. 7.2.

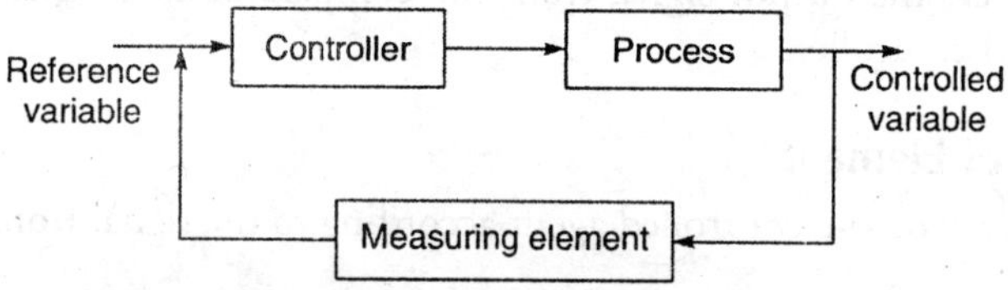

Fig. 7.2 Feed back control.

Driving an automobile is a classic example of feed back control system. In this case, the driver is controller, the steering wheel with its mechanism is the controlling element and the front two wheels are the controlled element. The input to the system is the idea or a safe position of the automobile on the road. The output of the system is the actual position of the front two wheels on the road. The driver constantly observes the actual position of the front two wheels and compares with actual position with his idea of safe position and uses the difference between the two to drive the output into close correspondence with the input by activating the corrective action proportional to the error. In this process of driving, the driver acts as a controller initiating a corrective action and thus maintaining the error.

A close loop control system is more reliable and accurate than open loop system. The major advantages of a closed loop system is that it can adopt itself to the variation in the environmental conditions. Less experience and unskilled person can easily operate a closed loop system and it requires no judgement to operate the system. Using simple components which are inexpensive a perfect closed loop system can be obtained.

7.3 BASIC COMPONENT OF CONTROL SYSTEM

The basic component of control system are as follows :

(i) Controller
(ii) Final control element
(iii) Process
(iv) Measuring element

7.3.1 Controller

The automatic controller including its measuring means determines the value of the controlled variable, determines the deviation and produces the concentration necessary to maintain the smallest possible deviation.

7.3.2 Final Control Element

The final control element is the mechanism which alter the value of the manipulated variable in response to the output signal from the automatic control device.

The final control element often consists of two parts :

(i) An actuator which is used to translate the output signal of the automatic controller into a position of a member exerting large power.
(ii) A device to adjust the value of the manipulated variable.

7.4 BLOCK DIAGRAM OF CONTROL SYSTEM

Any control system can be represented by a simple block diagram as shown in Fig. 7.3.

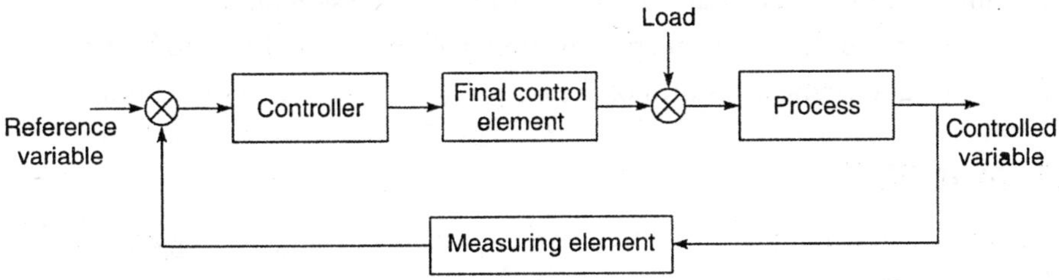

Fig. 7.3 Block diagram of control system.

7.4.1 Summing Junction

In a block diagram when two or more variable are to be added or subtracted before being operated by the block a summing junction is used.

A summing junction is a circle with a cross mark inside it and the symbol + or – just outside the circle along with the variables to represent whether that particular variable is to be added or subtracted. A typical representation of summing junction is shown in Fig. 7.4.

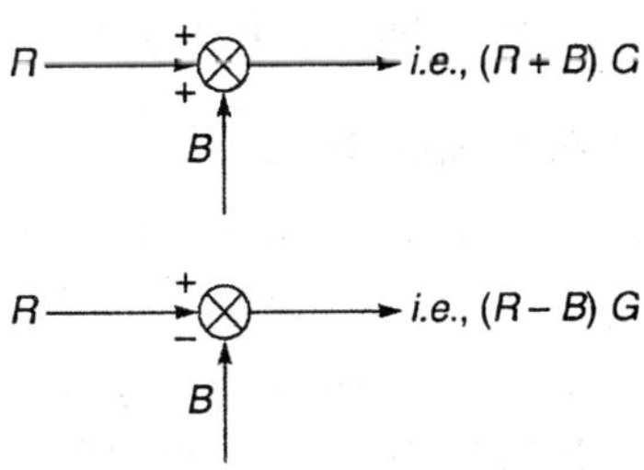

Fig. 7.4 Summing junction.

7.5 TRANSFER FUNCTION

Transfer function is the ratio of the transform of the response in deviation form to the transform of the input in deviation form. It can be represented by simple block diagram as shown in Fig. 7.5.

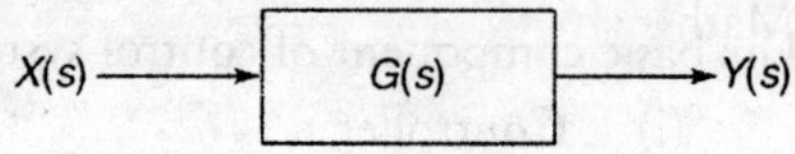

Fig. 7.5 Representation of transfer function.

$$G(s) = \frac{Y(s)}{X(s)}$$

where, $G(s)$ = Transfer function

$Y(x)$ = Transform of output

$X(s)$ = Transform of input

The transfer function gives the dynamic behaviour of the system. We can find the change in output of the system for any change in input if we know the transfer function.

If $x(t)$ is the change in input, taking Laplace transform of the input and substituting in the transfer function, we can get an equation for Laplace transform of the output and we get the response, *i.e.,* the change in output by taking Laplace inverse.

$$\Rightarrow \qquad G(s) = \frac{Y(s)}{X(s)}$$

$$\Rightarrow \qquad Y(s) = G(s).X(s)$$

$$\Rightarrow \qquad Y(t) = L^{-1}\,[G(s).X(s)]$$

7.5.1 Procedure to Obtain the Transfer Function

The step by step procedure to obtain the transfer function of a system are as follows :

(i) Obtain the differential (characteristic) equation of the system.

(ii) Transform the differential equation assuming all initial condition to be zero.

(iii) Form the ratio of the transform of the output to the transform of the input.

7.6 SYSTEM INPUT

The common types of system input encountered in control system engineering are as follows :

(i) Step input

(ii) Ramp input

(iii) Impulse input

(iv) Sinusoidal input

7.6.1 Step Input

Mathematically the step input of magnitude '*A*' can be expressed as

$$x(t) = A\,u(t)$$

where, $u(t)$ is the unit step function. The graphical representation of step input function is shown in Fig. 7.6.

It is so called, because of the shape of the plot. When the time t = 0, the input is suddenly increased to a value and remains constant for any value of 't'.

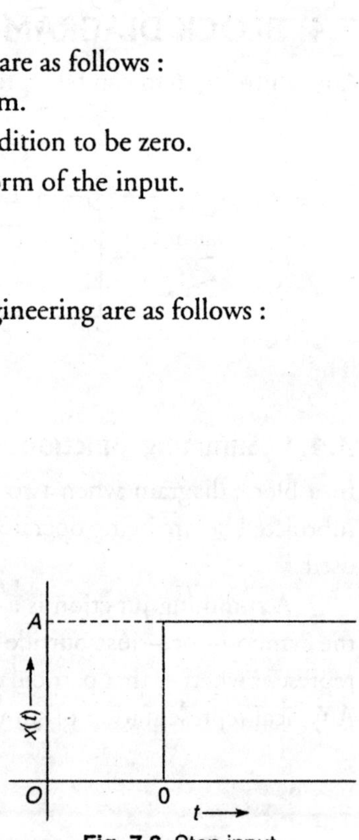

Fig. 7.6 Step input.

7.6.2 Ramp Input

Mathematically the ramp input of magnitude 'A' can be expressed as

$$x(t) = At$$

The graphical representation of ramp input function is shown in Fig. 7.7. At time $t = 0$, the input is A. As 't' increases the input gradually increases, which is proportional to the time.

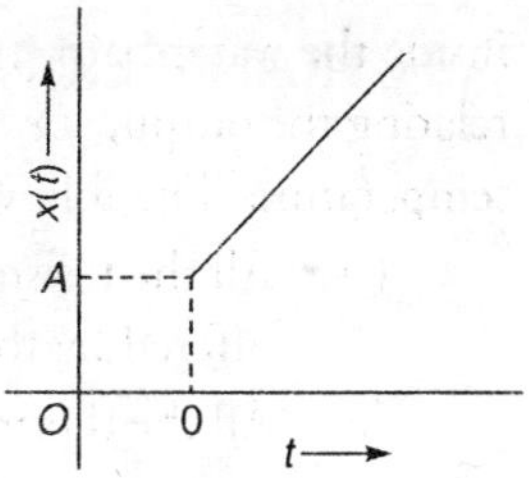

Fig. 7.7 Ramp input.

7.6.3 Impulse Input

Mathematically the impulse input of magnitude 'A' can be expressed as

$$x(t) = A\,\partial(t)$$

where, $\partial(t)$ is unit impulse function. The graphical representation of this function is shown in Fig. 7.8.

If the time duration 't' is sufficiently small the forcing function gives a response which closely resembles the response of a true impulse.

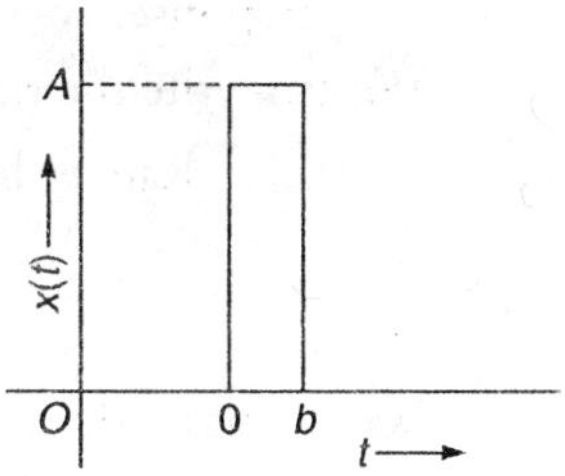

Fig. 7.8 Impulse input.

7.6.4 Sinusoidal Input

Mathematically the sinusoidal input of magnitude 'A' can be represented as

$$x(t) = \begin{cases} 0 & ; \quad t < 0 \\ A \sin \omega t & ; \quad t \geq 0 \end{cases}$$

where, A is the amplitude and ω is the radian frequency which is related to the cyclic frequency 'f' by the equation

$$\omega = 2\pi f$$

The graphical representation of sinusoidal function is shown in Fig. 7.9. As time $t = 0$, input is zero. As 't' increases the input follows a sinusoidal wave form.

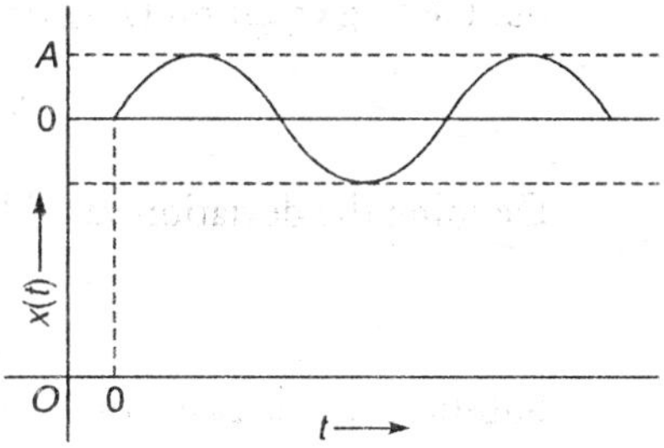

Fig. 7.9 Sinusoidal input.

7.7 FIRST ORDER SYSTEM

A system which can be represented by a linear first order differential equation is called first order system. The example of first order system are :

(i) Thermometer
(ii) Liquid level system
(iii) Mixing in a tank
(iv) Continuous stirred tank reactor
(v) Resistance capacitance circuit

7.7.1 Thermometer

Let us consider a mercury thermometer placed inside a water bath as shown in Fig. 7.10.

The bath temperature is constant and it is represented by 'x'. The thermometer reading is variable and it is represented by 'y'. As the thermometer placed

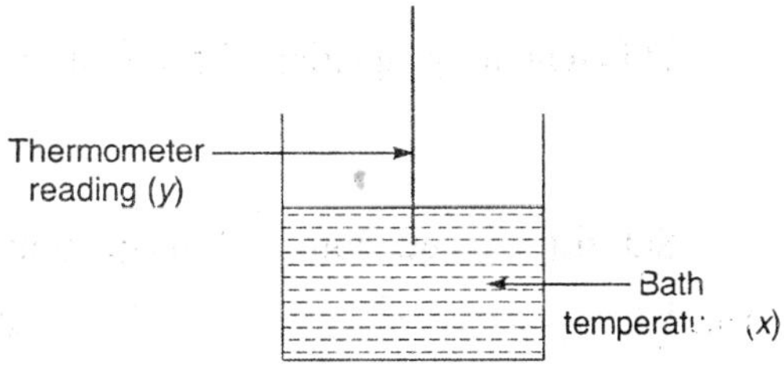

Fig. 7.10 Mercury thermometer.

inside the water bath 'y' increases till steady state is reached. We have to develop the transfer function relating the output, *i.e.*, change in thermometer reading and the input, *i.e.*, change in the surrounding temperature. The following assumptions are made for the development of transfer function :

(i) All the resistance to heat transfer resides in the film surrounding the bulb, *i.e.*, the resistance offered by the glass and mercury is neglected.

(ii) All the thermal capacity is in mercury and further we assume that at any instant the mercury assumes a uniform temperature throughout.

(iii) The glass wall containing the mercury does not expand or contract during the transient response.

We can write the transient heat energy balance for the thermometer as :

Rate of heat input – Rate of heat output = Rate of heat accumulation

$$hA(x - y) - 0 = mC_p \frac{dy}{dt} \quad \text{...(7.1)}$$

We can write the steady state heat energy balance for the thermometer as

$$hA(x_s - y_s) - 0 = 0 \quad \text{...(7.2)}$$

where, x_s = Steady state bath temperature

y_s = Steady state thermometer reading

Subtracting equation (7.2) from equation (7.1), we get

$$hA[(x - x_s) - (y - y_s)] = mC_p \frac{dy}{dt} \quad \text{...(7.3)}$$

Defining the deviation variable as

$$X = (x - x_s) \quad \text{...(7.4)}$$

$$Y = (y - y_s) \quad \text{...(7.5)}$$

Substituting the equation (7.4) and (7.5) in equation (7.3), we get

$$hA[X - Y] = mC_p \frac{dy}{dt}$$

$$X - Y = \frac{mC_p}{hA} \frac{dy}{dt} \quad \text{...(7.6)}$$

Let us assume, $$\tau = \frac{mC_p}{hA}$$

$$X - Y = \tau \frac{dy}{dt} \quad \text{...(7.7)}$$

From the deviation variable, we have

$$Y = y - y_s \quad \text{...(7.5)}$$

Differentiating equation (7.5), w.r.t. 't' we get

$$\frac{dY}{dt} = \frac{dy}{dt} \quad \text{...(7.8)}$$

Substituting equation (7.8) in equation (7.7), we get

$$X - Y = \tau \frac{dY}{dt} \quad \text{...(7.9)}$$

Taking Laplace transform of equation (7.9), we get

$$X(s) - Y(s) = \tau s\, Y(s)$$

$$X(s) = Y(s)[\tau s + 1]$$

$$\frac{Y(s)}{X(s)} = \frac{1}{\tau s + 1} \qquad ...(7.10)$$

where, $\tau = \dfrac{mC_p}{hA}$

τ = Time constant of thermometer, s

C_p = Specific heat of mercury, $\dfrac{\text{J}}{(\text{kg})\,(°\text{C})}$

h = Heat transfer coefficient of fluid media, $\dfrac{\text{W}}{(\text{m}^2)\,(°\text{C})}$

A = Surface area of thermometer bulb, m^2

m = Mass of mercury, kg

The common type of system response are as follows :

(i) Step response
(ii) Ramp response
(iii) Impulse response
(iv) Sinusoidal response

7.7.1.1 Step Response

It is the response of the system for a step change in input.

From th transfer function of first order system, we have

$$\frac{Y(s)}{X(s)} = \frac{1}{\tau s + 1} \qquad ...(7.10)$$

If the step change of magnitude 'A' is introduced to the system, we have

$$X(t) = A$$

$$\Rightarrow \qquad X(s) = \frac{A}{s} \qquad ...(7.11)$$

Substituting equation (7.11) in equation (7.10), we get

$$Y(s) = \frac{A}{(s)\,(\tau s + 1)}$$

$$Y(s) = \frac{A}{(s)\,(\tau s + 1)} = \frac{C_1}{(s)} + \frac{C_2}{(\tau s + 1)}$$

$$\frac{A}{(s)\,(\tau s + 1)} = \frac{C_1(\tau s + 1) + C_2(s)}{(s)\,(\tau s + 1)}$$

$$\Rightarrow \qquad A = C_1(\tau s + 1) + C_2(s)$$

Put, $s = 0 \quad ; \quad C_1 = A$

Put, $s = -\dfrac{1}{\tau} \quad ; \quad C_2 = -A\tau$

$$\Rightarrow \quad Y(s) = \frac{A}{(s)} - \frac{A\tau}{(\tau s + 1)}$$

$$\Rightarrow \quad Y(s) = \frac{A}{(s)} - \frac{A\tau}{\tau(s + 1/\tau)}$$

$$\Rightarrow \quad Y(s) = \frac{A}{(s)} - \frac{A\tau}{(s + 1/\tau)}$$

$$\Rightarrow \quad Y(t) = A - A\,e^{-t/\tau}$$

$$\Rightarrow \quad Y(t) = A[1 - e^{-t/\tau}] \quad \text{...(7.12)}$$

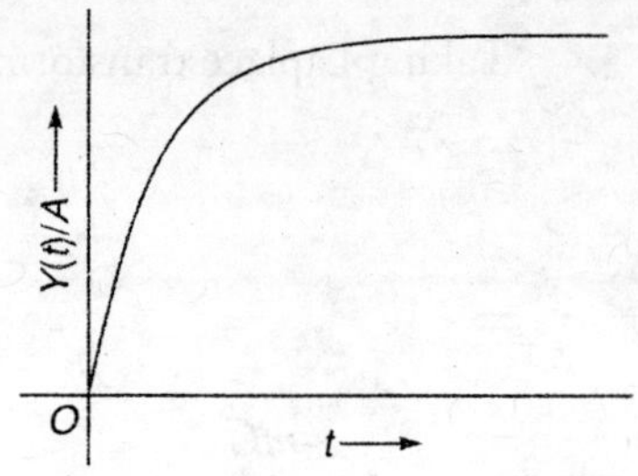

Fig. 7.11 Step response.

This is the response of first order system for a step input of magnitude '*A*'. It can be seen from the equation (7.12) that response of a system for step input of magnitude '*A*' is the multiplication of unit step response of the system and its magnitude '*A*'. The step response of a system is shown in Fig. 7.11.

7.7.1.2 Ramp Response

It is the response of the system for a ramp change in in input.

From the transfer function of first order system, we have

$$\frac{Y(s)}{X(s)} = \frac{1}{\tau s + 1} \quad \text{...(7.10)}$$

If the ramp change of magnitude '*A*' is introduced to the system, we have

$$X(t) = At$$

$$\Rightarrow \quad X(s) = \frac{A}{s^2} \quad \text{...(7.13)}$$

Substituting equation (7.13) in equation (7.10), we get

$$Y(s) = \frac{A}{s^2(\tau s + 1)}$$

$$Y(s) = \frac{A}{s^2(\tau s + 1)} = \frac{C_1}{s} + \frac{C_2}{s^2} + \frac{C_3}{(\tau s + 1)}$$

$$\frac{A}{s^2(\tau s + 1)} = \frac{C_1}{s} + \frac{C_2}{s^2} + \frac{C_3}{(\tau s + 1)}$$

$$\frac{A}{s^2(\tau s + 1)} = \frac{C_1(s)(\tau s + 1) + C_2(\tau s + 1) + C_3(s^2)}{s^2(\tau s + 1)}$$

$$A = C_1(s)(\tau s + 1) + C_2(\tau s + 1) + C_3(s^2)$$

Put, $s = 0$; $C_2 = A$

Put, $s = -1/\tau$; $C_3 = A\tau^2$

Comparing the coefficient of s^2, we get

$$0 = C_1 + C_3 \quad ; \quad C_1 = -A\tau^2$$

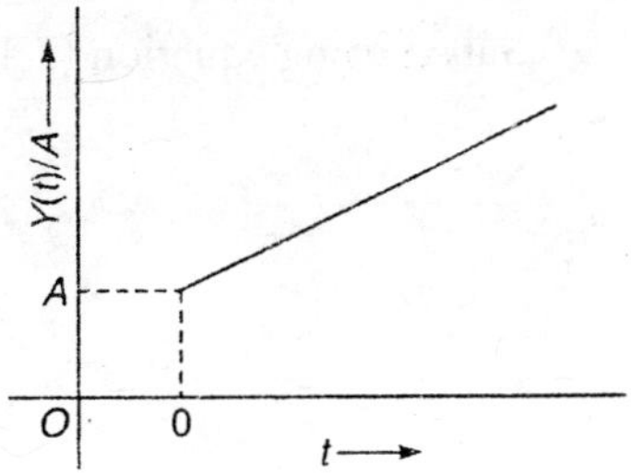

Fig. 7.12 Ramp response.

$$\Rightarrow \quad Y(s) = -\frac{A\tau^2}{s} + \frac{A}{s^2} + \frac{A\tau^2}{(\tau s + 1)}$$

$$\Rightarrow \quad Y(s) = -\frac{A\tau^2}{s} + \frac{A}{s^2} + \frac{A\tau^2}{\tau(s + 1/\tau)}$$

$$\Rightarrow \quad Y(s) = -\frac{A\tau^2}{s} + \frac{A}{s^2} + \frac{A\tau}{(s + 1/\tau)}$$

$$\Rightarrow \quad Y(t) = -A\tau^2 + At + A\tau e^{-t/\tau}$$

$$\Rightarrow \quad Y(t) = A[-\tau^2 + t + \tau e^{-t/\tau}] \qquad \ldots(7.14)$$

This is the response of first order system for a ramp input of magnitude 'A'. The ramp response of a system is shown in Fig. 7.12.

7.7.1.3 Impulse Response

It is the response of the system for the impulse change in input.

From the transfer function of first order system, we have

$$\frac{Y(s)}{X(s)} = \frac{1}{\tau s + 1} \qquad \ldots(7.10)$$

If the impulse change of magnitude 'A' is introduced to the system, we have

$$X(t) = A\,\delta(t)$$

$$\Rightarrow \quad X(s) = A \qquad \ldots(7.15)$$

Substituting equation (7.15) in equation (7.10), we get

$$Y(s) = \frac{A}{(\tau s + 1)}$$

$$\Rightarrow \quad Y(s) = \frac{A}{\tau(s + 1/\tau)}$$

$$\Rightarrow \quad Y(t) = \frac{A}{\tau} . e^{-t/\tau} \qquad \ldots(7.16)$$

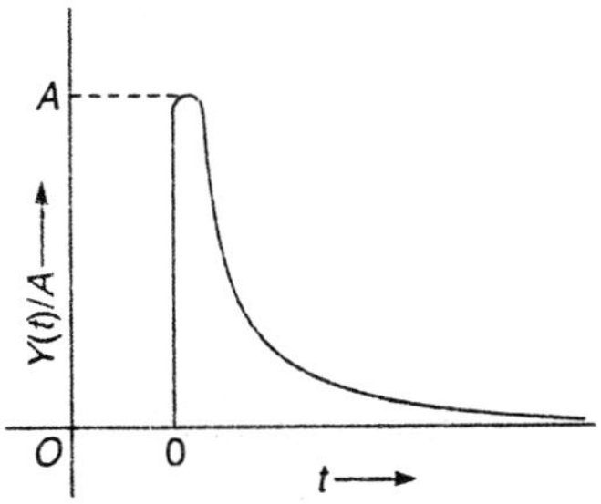

Fig. 7.13 Impulse response.

This is the response of first order system for a impulse input of magnitude 'A'. The impulse response of a system is shown in Fig. 7.13.

7.7.1.4 Sinusoidal Response

It is the response of the system for a sinusoidal change in input.

From the transfer function of first order system, we have

$$\frac{Y(s)}{X(s)} = \frac{1}{(\tau s + 1)} \qquad \ldots(7.10)$$

If the sinusoidal change of magnitude 'A' is introduced to the system, we have

$$X(t) = A \sin \omega t$$

$$\Rightarrow \quad X(s) = \frac{A\omega}{s^2 + \omega^2} \qquad \ldots(7.17)$$

Substituting equation (7.17) in equation (7.10), we get

$$Y(s) = \frac{A\omega}{(\tau s + 1)(s^2 + \omega^2)}$$

$$Y(s) = \frac{A\omega}{\tau(s + 1/\tau)(s^2 + \omega^2)}$$

$$Y(s) = \frac{(A\omega/\tau)}{(s + 1/\tau)(s^2 + \omega^2)} = \frac{C_1}{(s + 1/\tau)} + \frac{C_2 s + C_3}{(s^2 + \omega^2)} \quad \text{...(7.18)}$$

$$\frac{A\omega/\tau}{(s + 1/\tau)(s^2 + \omega^2)} = \frac{C_1}{(s + 1/\tau)} + \frac{C_2 s + C_3}{(s^2 + \omega^2)}$$

$$\frac{A\omega/\tau}{(s + 1/\tau)(s^2 + \omega^2)} = \frac{C_1(s^2 + \omega^2) + (C_2 s + C_3)(s + 1/\tau)}{(s + 1/\tau)(s^2 + \omega^2)}$$

$$\frac{A\omega}{\tau} = C_1(s^2 + \omega^2) + (C_2 s + C_3)(s + 1/\tau) \quad \text{...(7.19)}$$

Comparing the coefficient of 's' of equation (7.19), we get

$$C_3 + \frac{C_2}{\tau} = 0 \quad \text{...(7.20)}$$

Comparing the coefficient of 's^2' of equation (7.19), we get

$$C_2 + C_1 = 0 \quad \text{...(7.21)}$$

Comparing the constant terms of equation (7.19), we get

$$\frac{A\omega}{\tau} = C_1\omega^2 + \frac{C_3}{\tau} \quad \text{...(7.22)}$$

$$\Rightarrow \quad \frac{A\omega}{\tau} = \frac{C_1\omega^2\tau + C_3}{\tau}$$

$$\Rightarrow \quad C_3 = A\omega - C_1\omega^2\tau \quad \text{...(7.23)}$$

From equation (7.21), we have

$$C_2 + C_1 = 0$$

$$\Rightarrow \quad C_1 = -C_2 \quad \text{...(7.24)}$$

Substituting equation (7.24) in equation (7.23), we get

$$C_3 = A\omega + C_2\omega^2\tau \quad \text{...(7.25)}$$

From equation (7.20), we have

$$C_2 = -C_3\tau \quad \text{...(7.26)}$$

Substituting equation (7.26) in equation (7.25), we get

$$C_3 = A\omega - C_3\omega^2\tau^2$$

$$\Rightarrow \quad C_3(1 + \omega^t\tau^2) = A\omega$$

$$\Rightarrow \qquad C_3 = \frac{A\omega}{(1+\omega^2\tau^2)} \qquad \text{...(7.27)}$$

$$\Rightarrow \qquad C_2 = \frac{-A\omega\tau}{(1+\omega^2\tau^2)} \qquad \text{...(7.28)}$$

$$\Rightarrow \qquad C_1 = \frac{A\omega\tau}{(1+\omega^2\tau^2)} \qquad \text{...(7.29)}$$

Substituting equation (7.27), (7.28) and (7.29) in equation (7.18), we get

$$\Rightarrow \qquad Y(s) = \frac{A\omega\tau}{(1+\omega^2\tau^2)} \times \frac{1}{(s+1/\tau)} + \left[\frac{\left(\dfrac{-A\omega\tau}{(1+\omega^2\tau^2)}\right)s + \dfrac{A\omega}{(1+\omega^2\tau^2)}}{(s^2+\omega^2)}\right]$$

$$\Rightarrow \qquad Y(s) = \frac{A\omega\tau}{(1+\omega^2\tau^2)} \times \frac{1}{(s+1/\tau)} + \left(\frac{-A\omega\tau}{1+\omega^2\tau^2}\right)\left(\frac{s}{s^2+\omega^2}\right) + \left(\frac{A}{1+\omega^2\tau^2}\right)\left(\frac{\omega}{s^2+\omega^2}\right)$$

$$\Rightarrow \qquad Y(t) = \left(\frac{A\omega\tau}{1+\omega^2\tau^2}\right)e^{-t/\tau} + \left(\frac{-A\omega\tau}{1+\omega^2\tau^2}\right)\cos\omega t + \left(\frac{A}{1+\omega^2\tau^2}\right)\sin\omega t \qquad \text{...(7.30)}$$

Equation (7.30) can be further simplified by using the trigonometric relation

$$P\cos\theta + q\sin\theta = R\sin(\theta+\phi) \qquad \text{...(7.31)}$$

where,

$$R = \sqrt{P^2+q^2} \qquad \text{...(7.32)}$$

$$\phi = \tan^{-1}(P/q) \qquad \text{...(7.33)}$$

Therefore,

$$Y(t) = \left(\frac{A\omega\tau}{1+\omega^2\tau^2}\right)e^{-t/\tau} + P\cos\omega t + q\sin\omega t \qquad \text{...(7.34)}$$

$$\Rightarrow \qquad Y(t) = \left(\frac{A\omega\tau}{1+\omega^2\tau^2}\right)e^{-t/\tau} + R\sin(\omega t+\phi) \qquad \text{...(7.35)}$$

Comparing equation (7.30) and (7.34), we get

$$P = \frac{-A\omega\tau}{1+\omega^2\tau^2} \qquad \text{...(7.36)}$$

$$q = \frac{A}{1+\omega^2\tau^2} \qquad \text{...(7.37)}$$

$$\Rightarrow \qquad R = \sqrt{\left(\frac{-A\omega\tau}{1+\omega^2\tau^2}\right)^2 + \left(\frac{A}{1+\omega^2\tau^2}\right)^2}$$

$$\Rightarrow \quad R = \sqrt{\frac{A^2\omega^2\tau^2}{(1+\omega^2\tau^2)^2} + \frac{A^2}{(1+\omega^2\tau^2)^2}}$$

$$\Rightarrow \quad R = \sqrt{\frac{A^2(1+\omega^2\tau^2)}{(1+\omega^2\tau^2)^2}}$$

$$\Rightarrow \quad R = \frac{A}{\sqrt{1+\omega^2\tau^2}} \qquad \ldots(7.38)$$

Also, $$\phi = \tan^{-1}\left[\frac{-A\omega\tau/(1+\omega^2\tau^2)}{A/(1+\omega^2\tau^2)}\right]$$

$$\Rightarrow \quad \phi = \tan^{-1}(-\omega\tau)$$

$$\Rightarrow \quad \phi = -\tan^{-1}(\omega\tau) \qquad \ldots(7.39)$$

Substituting equation (7.38) and (7.39) in equation (7.38), we get

$$Y(t) = \left(\frac{A\omega\tau}{1+\omega^2\tau^2}\right)e^{-t/\tau} + \frac{A}{\sqrt{1+\omega^2\tau^2}}\sin[\omega t - \tan^{-1}(\omega\tau)] \qquad \ldots(7.40)$$

This is the response of first order system for a sinusoidal input of magnitude 'A'. The sinusoidal response of a system is shown in Fig. 7.14.

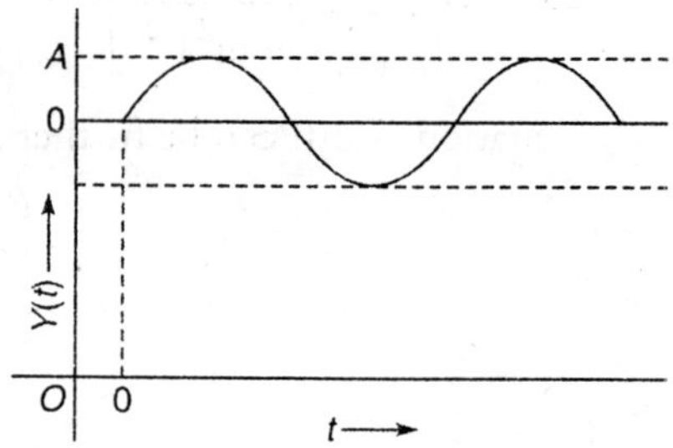

Fig. 7.14 Sinusoidal response.

The sinusoidal response is also known as frequency response. The ultimate response of sinusoidal function of the same frequency we get by eliminating the first part from equation (7.40).

Therefore,

$$Y(t) = \frac{A}{\sqrt{1+\omega^2\tau^2}}\sin[\omega t - \tan^{-1}(\omega\tau)] \qquad \ldots(7.41)$$

Frequency response analysis is an important branch of control engineering. For example, it is required to study the dynamic behaviour of control system in design.

■ **Example 7.1** A mercury thermometer of first order characteristics takes 5 seconds to indicate 70% of the final temperature. Calculate the time constant of the thermometer.

Solution : From the step response of first order system, we have

$$Y(t) = A[1 - e^{-t/\tau}] \qquad \ldots(7.12)$$

Ultimate value of $$Y(t) = \underset{t\to\infty}{Lt}\; Y(t)$$

$$= \underset{t\to\infty}{Lt}\; A[1 - e^{-t/\tau}] = A$$

Data given : $t = 5$ sec ; $Y(t) = 0.7A$

$$\Rightarrow \quad 0.7A = A[1-e^{-5/\tau}]$$

$$\Rightarrow \quad 0.7 = [1-e^{-5/\tau}]$$

$$\Rightarrow \quad e^{-5/\tau} = 1-0.7$$

$$\Rightarrow \quad e^{-5/\tau} = 0.3$$

$$\Rightarrow \quad \tau = 4.15 \text{ sec}$$

Example 7.2 A mercury thermometer of first order characteristics having a time constant of 4.5 sec. Calculate the time needed to indicate 75% of the final temperature.

Solution : From the step response of first order system, we have

$$Y(t) = A[1-e^{-t/\tau}] \quad \ldots(7.12)$$

Ultimate value of

$$Y(t) = \underset{t \to \infty}{Lt} A[1-e^{-t/\tau}] = A$$

Data given :

$$t = 4.5 \text{ sec} \quad ; \quad Y(t) = 0.75A$$

$$\Rightarrow \quad 0.75A = A[1-e^{-t/4.5}]$$

$$\Rightarrow \quad 0.75 = [1-e^{-t/4.5}]$$

$$\Rightarrow \quad e^{-t/4.5} = 1-0.75$$

$$\Rightarrow \quad e^{-t/4.5} = 0.25$$

$$\Rightarrow \quad \tau = 6.24 \text{ sec}$$

Example 7.3 A thermometer having a time constant of 0.1 minute is at a steady state temperature of 30°C. At time $t = 0$, the thermometer is placed in a temperature bath maintained at 40°C. Determine the time needed for the thermometer to read 38°C.

Solution : Data given : $x_s = y_s = 30°C$

$x = 40°C$

$y = 38°C$

From the deviation variable, we have

$$X(t) = x - x_s \quad \ldots(7.4)$$

$$X(t) = 40 - 30$$

$$X(t) = 10$$

It is the magnitude of step change and represented by the symbol 'A'.

Therefore,

$$A = 10 \quad \ldots(7.42)$$

From the deviation variable, we have

$$Y(t) = y - y_s \quad \ldots(7.5)$$

$$Y(t) = 38 - 30$$

$$Y(t) = 8 \quad \ldots(7.43)$$

From the step response of first order system, we have

$$Y(t) = A[1-e^{-t/\tau}] \quad \ldots(7.12)$$

Substituting equation (7.42) and (7.43) in equation (7.12), we get

$$8 = 10[1 - e^{-t/0.1}]$$

$$0.8 = 1 - e^{-t/0.1}$$

$$e^{-t/0.1} = 1 - 0.8$$

$$e^{-t/0.1} = 0.2$$

$$t = 0.161 \text{ min}$$

■ **Example 7.4** A thermometer having a time constant of 0.1 min is placed in a temperature bath. After the thermometer comes to equilibrium with bath, the temperature of the bath is increased linearly with time at a rate of 1°C per min. What is the difference between the indicated temperature and the bath temperature at t = 0.1 min after the change in temperature begin.

Solution : From the transfer function of first order system, we have

$$\frac{Y(s)}{X(s)} = \frac{1}{\tau s + 1} \quad \text{...(7.10)}$$

Data given :

$$\tau = 0.1 \text{ min}$$

$$X(t) = t \quad \text{Linear (ramp) input}$$

$$\Rightarrow \quad X(s) = \frac{1}{s^2} \quad \text{...(7.44)}$$

Substituting equation (7.44) in equation (7.10), we get

$$Y(s) = \frac{1}{s^2(0.1s + 1)}$$

$$\Rightarrow \quad Y(s) = \frac{10}{s^2(s + 10)}$$

$$\Rightarrow \quad Y(s) = \frac{10}{s^2(s + 10)} = \frac{C_1}{s} + \frac{C_2}{s^2} + \frac{C_3}{(s + 10)} \quad \text{...(7.45)}$$

$$\Rightarrow \quad \frac{10}{s^2(s + 10)} = \frac{C_1(s)(s + 10) + C_2(s + 10) + C_3(s^2)}{s^2(s + 10)}$$

$$\Rightarrow \quad 10 = C_1(s)(s + 10) + C_2(s + 10) + C_3(s^2) \quad \text{...(7.46)}$$

Put, $s = 0$; $C_2 = 1$

Put, $s = -10$; $C_3 = 0.1$

Comparing the coefficient of 's^2' of equation (7.46), we get

$$0 = C_1 + C_3$$

$$\Rightarrow \quad 0 = C_1 + 0.1$$

$$\Rightarrow \quad C_1 = -0.1$$

Substituting the value of C_1, C_2 and C_3 in equation (7.45), we get

$$Y(s) = -\frac{0.1}{s} + \frac{1}{s^2} + \frac{0.1}{(s+10)}$$

$$\Rightarrow \qquad Y(t) = -0.1 + t + 0.1e^{-10t}$$

$$\text{Bath temp} - \text{Indicated temp} = X(t) - Y(t)$$

$$= t - (-0.1 + t + 0.1e^{-10t})$$

$$= t + 0.1 - t - 0.1e^{-10t}$$

$$= 0.1 - 0.ie^{-10t}$$

$$= 0.1[1 - e^{-10t}]$$

$$= 0.1[1 - e^{-10 \times 0.1}]$$

$$= 0.063°\text{C}$$

■ **Example 7.5** A mercury thermometer bulb is 12 mm long and its diameter is 3 mm. If the glass envelope is very thin, calculate the time constant of thermometer placed in water flowing with a velocity of 3 m/s and at a temperature of 40°C.

Solution : Data given : $d = 3$ m

$$L = 12 \text{ mm}$$

$$m = V \times \rho = \left(\frac{\pi d^2}{4} \times L\right) \times \rho$$

$$= \frac{\pi(3 \times 10^{-3})^2}{4} \times (12 \times 10^{-3}) \times 13600$$

$$= 1.15 \times 10^{-3} \text{ kg}$$

$$A = \pi dL = \pi \times (3 \times 10^{-3}) \times (12 \times 10^{-3})$$

$$= 1.13 \times 10^{-4} \text{ m}^2$$

Data from the Perry hand book :

$$h_w = 2500, \text{ W/m}^2 \text{°C}$$

$$C_{pm} - 502, \text{ J/kg°C}$$

From the theoretical equation of time constant of thermometer, we have

$$\tau = \frac{mC_p}{hA}$$

$$= \frac{(1.15 \times 10^{-3}) \times 502}{2500 \times 1.13 \times 10^{-4}}$$

$$= 2.04 \text{ sec}$$

■ **Example 7.6** A thermometer is kept inside a constant temperature bath at 70°C. This is suddenly transferred into another bath kept at 60°C at $t = 0$. The following readings are recorded.

Time (min)	Temperature (°C)
0	70
1	68
2	66
4	64
5	63
8	62
10	61.5
12	61

Calculate the time constant of thermometer by graphical method.

Solution : Data given : $x_s = y_s = 70°\text{C}$

$x = 60°\text{C}$

From the deviation variable, we have

$$X(t) = x - x_s = 60 - 70 = -10$$

It is the magnitude of step change and represented by the symbol 'A'.

Therefore, $A = -10$

From the deviation variable, we have

$$Y(t) = y - y_s = 70 - 70 = 0$$

From the step response of first order system, we have

$$Y(t) = A[1 - e^{-t/\tau}] \quad \text{...(7.12)}$$

$$\Rightarrow \quad \frac{Y(t)}{A} = [1 - e^{-t/\tau}]$$

Table 7.1 : Calculation of $Y(t)/A$

t (min)	Temperature (°C) (y)	A	$Y(t)$	$Y(t)/A$
0	70	– 10	0	0
1	68	– 10	– 2	0.2
2	66	– 10	– 4	0.4
4	64	– 10	– 6	0.6
6	63	– 10	– 7	0.7
8	62	– 10	– 8	0.8
10	61.5	– 10	– 8.5	0.85
12	61	– 10	– 9	0.9

From the graph time constant of thermometer, $\tau = 4.5$ min.

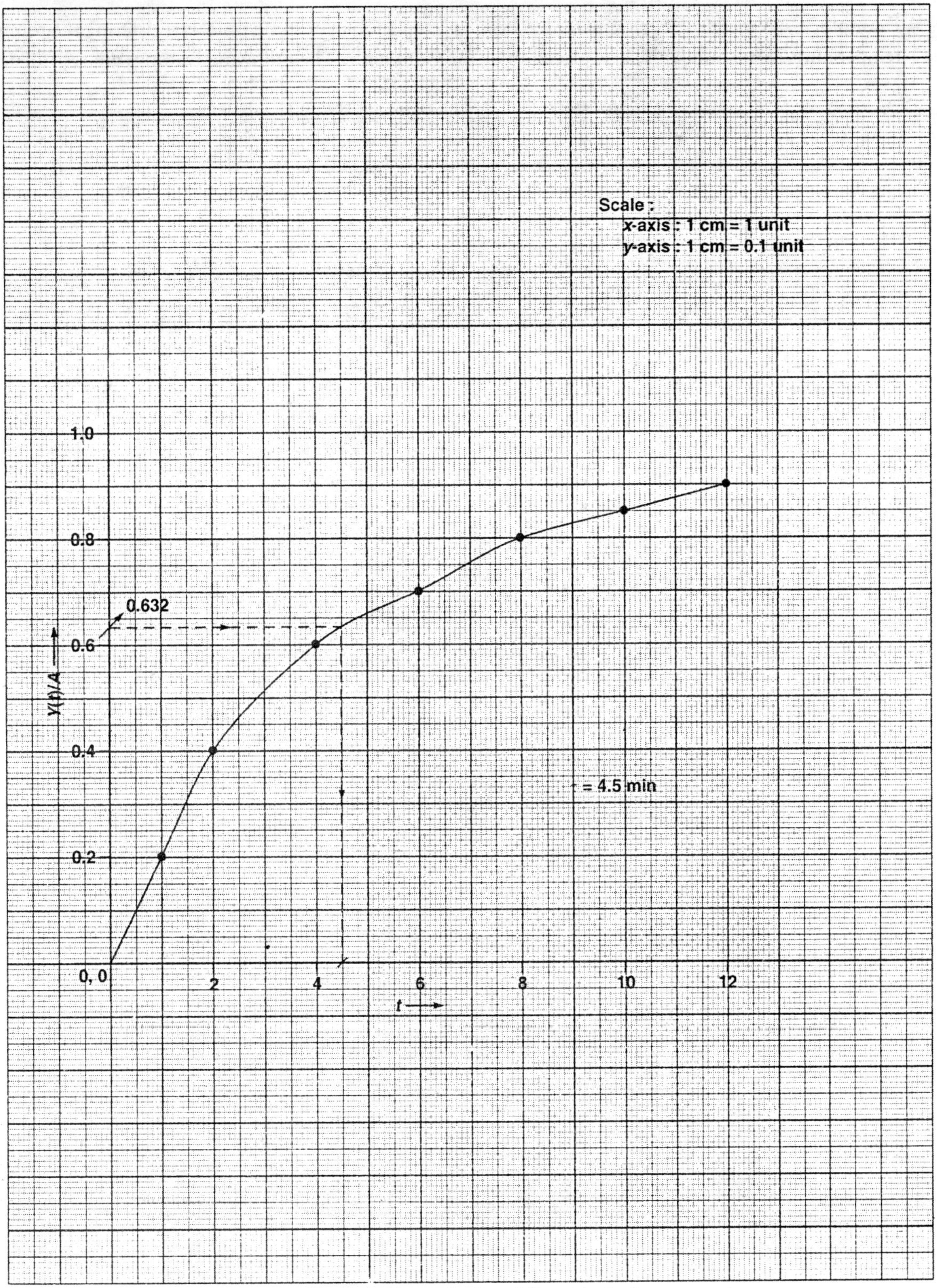

Fig. 7.15 Determination of time constant.

■ **Example 7.7** A mercury thermometer has been kept on a table for some time, is registering the room temperature 25°C. Suddenly, it is placed in a 210°C oil bath. The following readings are recorded.

Time (sec)	Thermometer reading (°C)
0	25
1	42
2	60
5	98
8	120
10	140
15	165
25	200

Calculate the time constant of thermometer by graphical method.

Solution : Data given : $x_s = y_s = 25°C$

$$x = 210°C$$

From the deviation variable, we have

$$X(t) = x - x_s = 210 - 25 = 185$$

It is the magnitude of step change and represented by the symbol '*A*'.

Therefore, $A = 185$

From the deviation variable, we have

$$Y(t) = y - y_s = 25 - 25 = 0$$

Table 7.2 : Calculation of $Y(t)/A$

t (s)	Temperature (y) (°C)	A	$Y(t)$	$Y(t)/A$
0	25	185	0	0
1	42	185	17	0.09
2	60	185	35	0.19
5	98	185	73	0.39
8	120	185	95	0.51
10	140	185	115	0.62
15	165	185	140	0.76
25	200	185	175	0.95

From the graph time constant of thermometer, $\tau = 10.5$ sec.

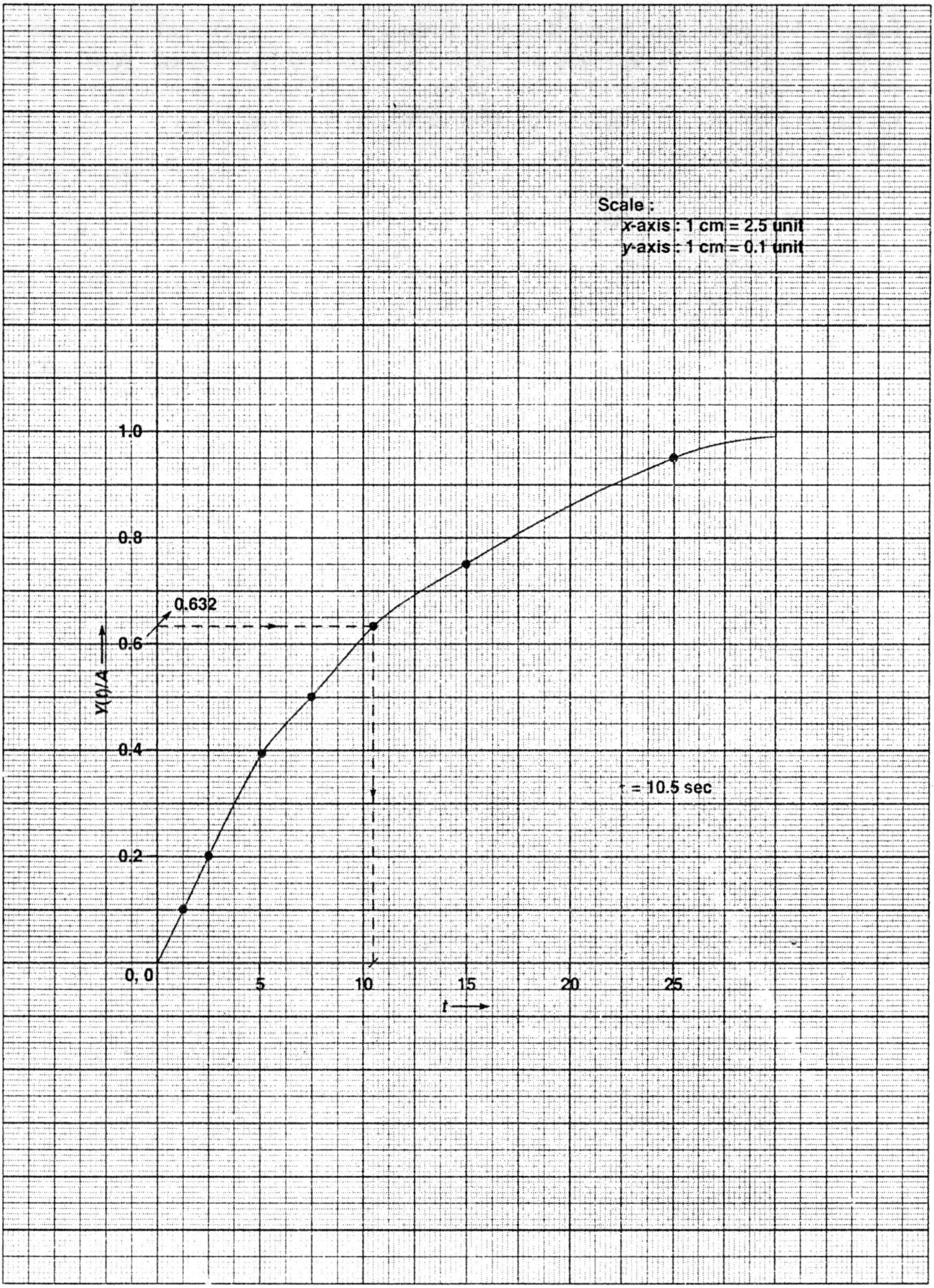

Fig. 7.16 Determination of time constant.

■ **Example 7.8** A thermometer having a first order dynamic with a time constant of 1 min is placed in a temperature bath at 100°C. After the thermometer reaches steady state, it is suddenly placed in a bath at 110°C at time $t = 0$ and left for 1 min after which it is immediately returned to the bath at 100°C. Draw a sketch showing the variation of thermometer reading with time.

Solution : Data given : $x(s) = y(s) = 100°\text{C}$

$$\tau = 1\,\text{min}$$

$$x = \begin{cases} 100 & ; \quad t<0 \\ 110 & ; \quad 0<t<1 \\ 100 & ; \quad t>1 \end{cases}$$

$$X(t) = x - x_s$$

$$X(t) = \begin{cases} 0 & ; \quad t<0 \\ 10 & ; \quad 0<t<1 \\ 0 & ; \quad t>1 \end{cases}$$

$$\Rightarrow \qquad X(t) = 10\,[u(t) - u(t-1)]$$

$$\Rightarrow \qquad X(s) = 10\left[\frac{1}{s} - \frac{e^{-s}}{s}\right]$$

$$\Rightarrow \qquad X(s) = 10\left[\frac{1-e^{-s}}{s}\right] \qquad \text{...(7.47)}$$

From the transfer function of first order system, we have

$$\frac{Y(s)}{X(s)} = \frac{1}{\tau s + 1} \qquad \text{...(7.10)}$$

Substituting equation (7.47) in equation (7.10), we get

$$Y(s) = \frac{1}{(s+1)} \times \frac{10\,[1-e^{-s}]}{s}$$

$$Y(s) = \frac{10\,(1-e^{-s})}{(s)\,(s+1)}$$

$$Y(s) = \frac{10\,(1-e^{-s})}{(s)\,(s+1)} = \frac{C_1}{s} + \frac{C_2}{(s+1)}$$

$$\Rightarrow \qquad \frac{10\,(1-e^{-s})}{(s)\,(s+1)} = \frac{C_1(s+1) + C_2(s)}{(s)\,(s+1)}$$

$$\Rightarrow \qquad 10\,(1-e^{-s}) = C_1(s+1) + C_2(s)$$

Put, $\quad s = 0 \quad ; \quad C_1 = 0$

Put, $\quad s = -1 \quad ; \quad C_2 = 17.18$

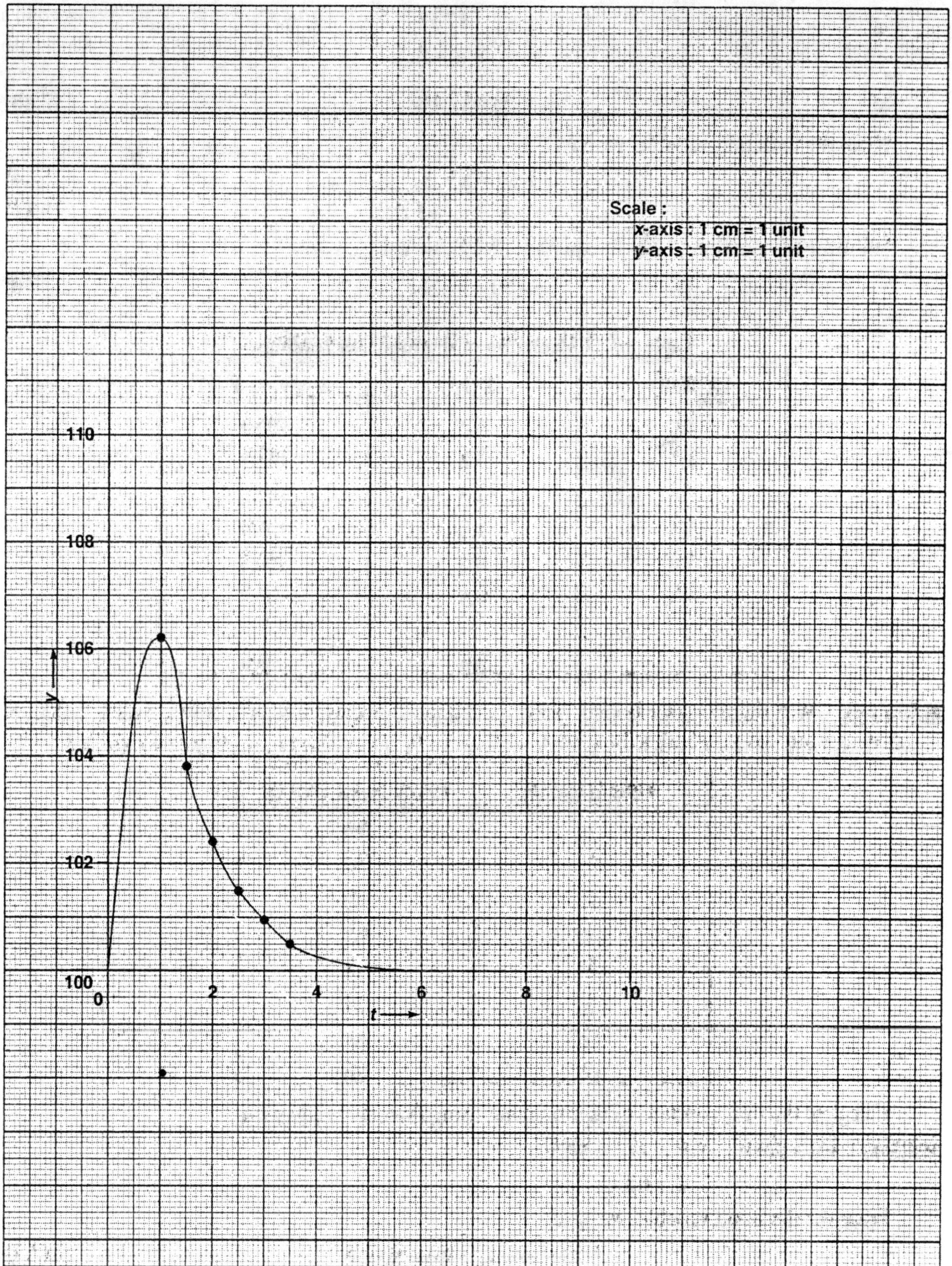

Fig. 7.17 Variation of thermometer reading with time.

$$\Rightarrow \qquad Y(s) = \frac{17.18}{(s+1)}$$

$$\Rightarrow \qquad Y(t) = 17.18e^{-t} \qquad \text{...(7.48)}$$

From the deviation variable, we have

$$Y(t) = y - y_s \qquad \text{...(7.4)}$$

$$y = y_s + Y(t) \qquad \text{...(7.49)}$$

Substituting equation (7.48) in equation (7.49), we get

$$y = 100 + 17.18e^{-t} \qquad \text{...(7.50)}$$

Table 7.3 : Calculation of thermometer readings.

t (min)	$Y(t)$	$y = y_s + Y(t)$
1.0	6.3	106.3
1.5	3.8	103.8
2.0	2.3	102.3
2.5	1.4	101.4
3.0	0.8	100.8
3.5	0.5	100.5
⋮	⋮	⋮
∞	0.0	100.0

■ **Example 7.9** A thermometer is kept inside a constant temperature bath at 80°C. This is suddenly transferred into another bath kept at 60°C at time $t = 0$. The following readings are recorded.

Time (min) (t)	Temperature (°C) (y)
0	80
1	72
2	69
4	65
5	63
8	62
10	61
12	60

Calculate the time constant of thermometer using theoretical equation.

Solution : Data given : $x(s) = y(s) = 80°C$

$$x = 60°C$$

From the deviation variable, we have

$$Y(t) = y - y_s = 80 - 80 = 0 \qquad \text{...(7.4)}$$

From the step response of first order system, we have

$$Y(t) = A[1-e^{-t/\tau}] \quad \text{...(7.12)}$$

$$\frac{Y(t)}{A} = 1 - e^{-t/\tau}$$

$$e^{-t/\tau} = \left[1 - \frac{Y(t)}{A}\right]$$

$$-t/\tau = \ln\left[1 - \frac{Y(t)}{A}\right]$$

$$\tau = -\frac{t}{\ln\left[1 - \frac{Y(t)}{A}\right]} \quad \text{...(7.51)}$$

Table 7.4 : Calculation of time constant.

t (min)	$Y(t)$	A	$Y(t)/A$	$\ln\left[1-\frac{Y(t)}{A}\right]$	Time constant (min.) τ
0	0	– 20	0.00	0.00	0.0
1	– 8	– 20	0.40	– 0.51	1.96
2	– 11	– 20	0.55	– 0.80	2.50
4	– 15	– 20	0.75	– 1.39	2.88
5	– 17	– 20	0.85	– 1.90	2.63
8	– 18	– 20	0.90	– 2.30	3.48
10	– 19	– 20	0.95	– 2.99	3.34

The average time constant of thermometer by theoretical equation is 2.39 min.

■ **Example 7.10** A mercury thermometer having a time constant of 0.1 min is placed in a temperature bath at 100°C and is allowed to come to equilibrium with bath. At time $t = 0$ the temperature of the bath begins to very sinusoidally about its average temperature of 100°C with an amplitude of 2°C. If the frequency of oscillation is $10/\pi$ cycles/min. Plot the ultimate response of the thermometer against time. What is the phase lag?

Solution : Data given : $x_s = y_s = 100°C$

$$A = 2°C$$

$$f = 10/\pi \text{ cycles/min}$$

$$\omega = 2\pi f = 2\pi \times \frac{10}{\pi} = 20 \text{ radian/min}$$

From the sinusoidal response of first order system, we have

$$Y(t) = \frac{A}{\sqrt{1+\omega^2\tau^2}} \sin[\omega t - \tan^{-1}(\omega\tau)] \quad \text{...(7.41)}$$

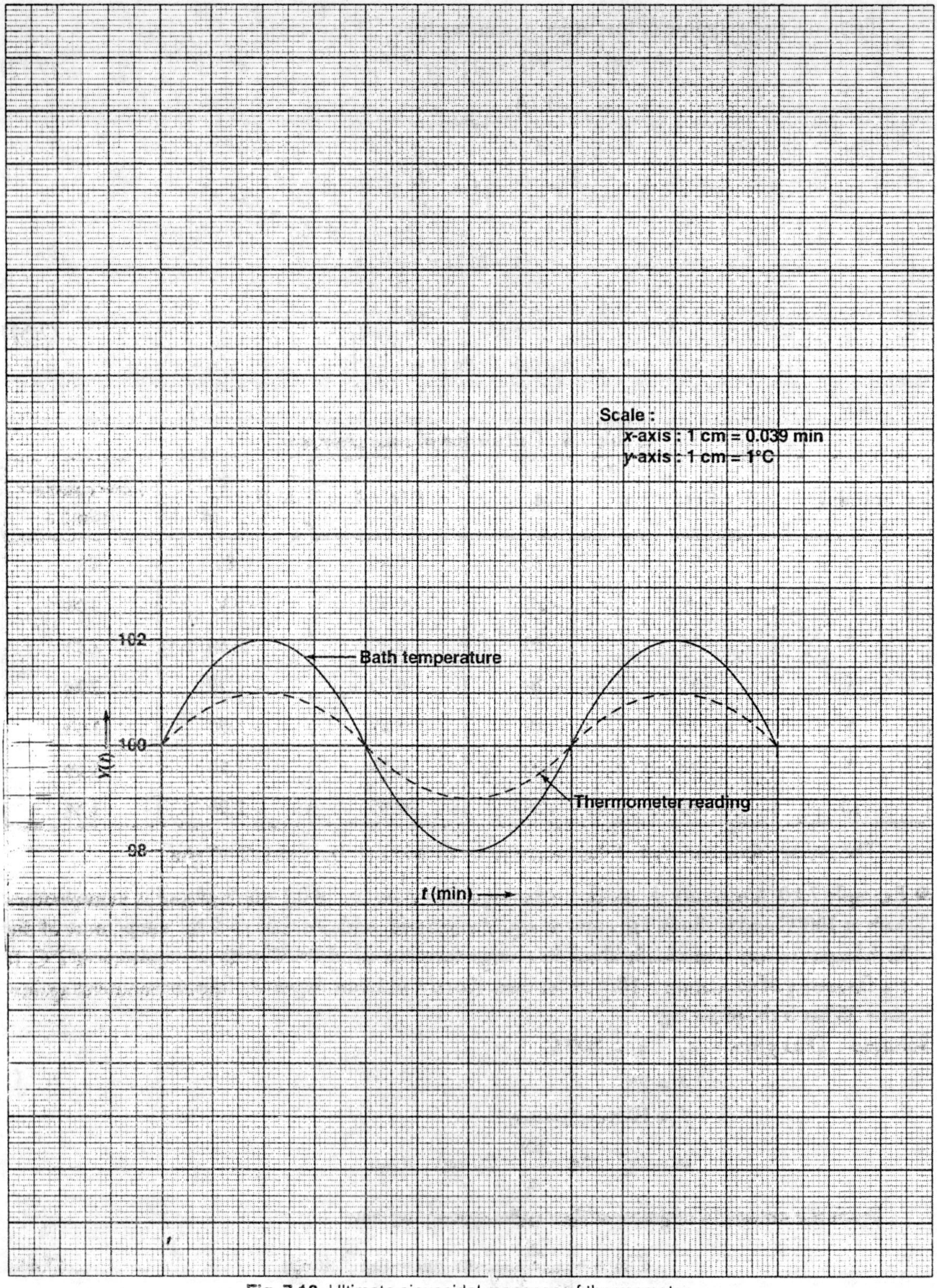

Fig. 7.18 Ultimate sinusoidal response of thermometer.

$$\Rightarrow \qquad Y(t) = \frac{2}{\sqrt{1+2^2}} \sin\,[20t - \tan^{-1}(20 \times 0.1)]$$

$$\Rightarrow \qquad Y(t) = 0.894 \sin\,[20t - 1.107] \qquad \text{...(7.52)}$$

From the deviation variable, we have

$$Y(t) = y - y_s \qquad \text{...(7.4)}$$

$$\Rightarrow \qquad y = y_s + Y(t)$$

$$\Rightarrow \qquad y = 100 + 0.894 \sin\,[20t - 1.107] \qquad \text{...(7.53)}$$

Phase lag, $\phi = \tan^{-1}(\omega\tau) = \tan^{-1}(20 \times 0.1)$

$$= \tan^{-1}(2) = 1.107 \text{ radians}$$

■ **Example 7.11** A mercury thermometer is registering equilibrium room temperature at 25°C. Then, it is placed in the 205°C oil bath for a length of time less than 1 sec and quickly removed from the bath and re-exposed to the 25°C ambient conditions. It may be estimated that the heat transfer co-efficient to the thermometer in air is 1/15th of that in oil bath. If 10 sec after the thermometer is removed from the bath its read 37°C, estimate the length of time that the thermometer was in the bath.

Solution : Data given : $x_s = y_s = 25°\text{C}$

$$x = 205°\text{C}$$

$$y = 37°\text{C}$$

$$h_{\text{air}} = \frac{1}{15} h_{\text{oil}}$$

From the deviation variable, we have

$$Y(t) = y - y_s = 37 - 25 = 12 \qquad \text{...(7.4)}$$

From the deviation variable, we have

$$X(t) = x - x_s = 205 - 25 = 180 \qquad \text{...(7.3)}$$

It is the magnitude of the step response and represented by the symbol 'A'.

Therefore, $A = 180$

From the transfer function of first order system, we have

$$\frac{Y(s)}{X(s)} = \frac{1}{\tau s + 1}$$

$$Y(s) = \frac{180}{(s)\,(\tau s + 1)}$$

$$Y(s) = \frac{180}{(s)\,\tau\,(s + 1/\tau)} = \frac{C_1}{s} + \frac{C_2}{(s + 1/\tau)}$$

$$\frac{180/\tau}{(s)\,(s + 1/\tau)} = \frac{C_1(s + 1/\tau) + C_2(s)}{(s)\,(s + 1/\tau)}$$

$$\frac{180}{\tau} = C_1(s + 1/\tau) + C_2(s)$$

Put, $s = 0$; $C_1 = 180$

Put, $s = -1/\tau$; $C_2 = -180$

$$\Rightarrow \quad Y(s) = \frac{180}{(s)} - \frac{180}{(s+1/\tau)}$$

$$\Rightarrow \quad Y(t) = 180[1 - e^{-t/\tau}] \qquad \ldots(7.54)$$

$$\Rightarrow \quad 12 = 180[1 - e^{-10/\tau}]$$

$$\Rightarrow \quad 0.067 = (1 - e^{-10/\tau})$$

$$\Rightarrow \quad e^{-10/\tau} = 1 - 0.067$$

$$\Rightarrow \quad -10/\tau = 0.933$$

$$\Rightarrow \quad -10/\tau = -0.069$$

$$\Rightarrow \quad \tau_{air} = 145 \text{ sec}$$

From the given data, we have

$$\tau_{oil} = \frac{1}{15}\tau_{air} = \frac{1}{15} \times (145) = 9.66 \text{ sec}$$

From the equation (7.54), we have

$$Y(t) = 180[1 - e^{-t/\tau}] \qquad \ldots(7.54)$$

$$\Rightarrow \quad 12 = 180[1 - e^{-t/9.66}]$$

$$\Rightarrow \quad 0.067 = 1 - e^{-t/9.66}$$

$$e^{-t/9.66} = 1 - 0.067$$

$$e^{-t/9.66} = 0.983$$

$$-\tau/9.66 = -0.069$$

$$t = 0.67 \text{ sec}$$

The length of time that the thermometer was in the oil bath is 0.67 sec.

7.7.2 Liquid Level System

Let us consider a tank of uniform cross-sectional area 'A' where liquid is flowing into the tank at the flow rate of $q(t)$ and flowing out at the flow rate of $q_0(t)$.

Let 'h' be the corresponding level in the tank. If there is any change in the inlet flow rate, it results in the change in level and also the exit flow rate. A typical liquid level system is shown in Fig. 7.19.

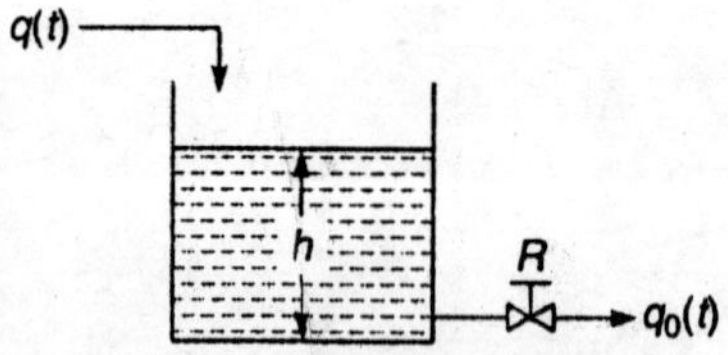

Fig. 7.19 Liquid level system.

Let us consider that there is a linear resistance 'R' in the exit line such that the level and exit flow rate are related by the equation

$$q_0 = \frac{h}{R} \qquad \ldots(7.55)$$

We can write the transient material balance around the tank as

Mass of liquid in – Mass of liquid out = Mass of liquid accumulation

$$\rho q - \rho q_0 = \frac{d}{dt}(\rho A h) \qquad ...(7.56)$$

$$\Rightarrow \qquad \rho q - \rho q_0 = \rho A \frac{dh}{dt}$$

$$\Rightarrow \qquad q - q_0 = A \frac{dh}{dt} \qquad ...(7.57)$$

We can also write the steady state material balance around the tank as

$$q_s - q_{os} = 0 \qquad ...(7.58)$$

Subtracting equation (7.58) from equation (7.57), we get

$$(q - q_s) - (q_o - q_{os}) = A \frac{dh}{dt} \qquad ...(7.59)$$

Defining the deviation variable as

$$Q = q - q_s \qquad ...(7.60)$$

$$Q_o = q_o - q_{os} \qquad ...(7.61)$$

$$H = h - h_s \qquad ...(7.62)$$

Substituting equation (7.60) and (7.61) in equation (7.59), we get

$$Q - Q_s = A \frac{dh}{dt} \qquad ...(7.63)$$

Differentiating equation (7.62) w.r.t. 't', we get

$$\frac{dH}{dt} = \frac{dh}{dt} \qquad ...(7.64)$$

Substituting equation (7.64) in equation (7.63), we get

$$Q - Q_o = A \frac{dH}{dt} \qquad ...(7.65)$$

From the linear resistance through the valve at transient state, we have

$$q_0 = \frac{h}{R} \qquad ...(7.55)$$

At steady state, we have

$$q_{os} = \frac{h_s}{R} \qquad ...(7.66)$$

Subtracting equation (7.66) from equation (7.55), we get

$$q_o - q_{os} = \frac{(h - h_s)}{R} \qquad ...(7.67)$$

Substituting equation (7.61) and (7.62) in equation (7.67), we get

$$Q_o = \frac{H}{R} \qquad ...(7.68)$$

Substituting equation (7.68) in equation (7.65), we get

$$Q - \frac{H}{R} = A \frac{dH}{dt} \qquad ...(7.69)$$

Taking Laplace transform of equation (7.69), we get

$$Q(s) - \frac{H(s)}{R} = As\,H(s)$$

$$\Rightarrow \quad Q(s) = As\,H(s) + \frac{H(s)}{R}$$

$$\Rightarrow \quad Q(s) = \frac{ARs\,H(s) + H(s)}{R}$$

$$\Rightarrow \quad Q(s) = \frac{[ARs + 1]\,H(s)}{R}$$

$$\Rightarrow \quad \frac{H(s)}{Q(s)} = \frac{R}{ARs + 1}$$

$$\Rightarrow \quad \frac{H(s)}{Q(s)} = \frac{R}{\tau s + 1} \qquad \ldots(7.70)$$

where, $\tau = AR$

τ = Time constant, sec.

A = Cross-sectional area, m^2

R = Resistance, s/m^2

Taking Laplace transform of equation (7.68), we get

$$Q_o(s) = \frac{H(s)}{R}$$

$$\Rightarrow \quad R\,Q_o(s) = H(s) \qquad \ldots(7.71)$$

Substituting equation (7.71) in equation (7.70), we get

$$\frac{R\,Q_o(s)}{Q(s)} = \frac{R}{\tau s + 1}$$

$$\Rightarrow \quad \frac{Q_o(s)}{Q(s)} = \frac{1}{\tau s + 1} \qquad \ldots(7.72)$$

■ **Example 7.12** Derive the transfer function $H(s)/Q(s)$ for the liquid level system shown in Fig. 7.20. When,

(i) The tank level operates around the steady state value of $h_s = 1\,\text{m}$.

(ii) The tank level operates around the steady state value of $h_s = 3\,\text{m}$.

The pump removes water at a constant rate of 10 m^3/min. This rate is independent of head. The cross-sectional area of the tank is 2 m^2 and R is 0.5 min/m^2.

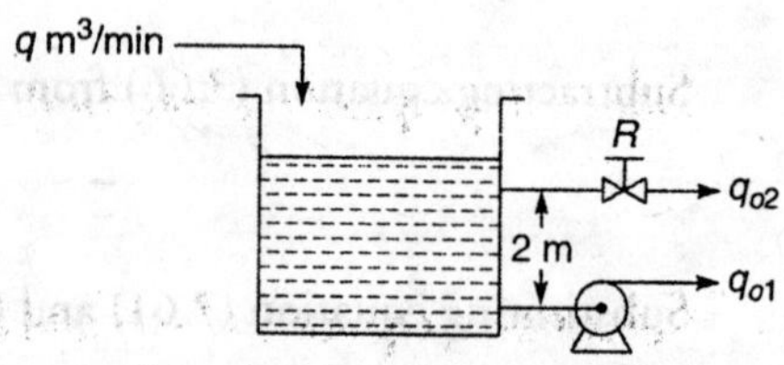

Fig. 7.20 Liquid level system.

Solution :

(i) The tank level operates around the steady state value of $h_s = 1\,\text{m}$.

The pump remove water at a constant rate of 10 m^3/min.

From the given data, we have

$$q_o = q_{os} \quad ...(7.73)$$

$$\Rightarrow \quad q - 10 = 2\frac{dh}{dt} \quad ...(7.74)$$

From the steady state material balance, we have

$$q_s - q_{os} = 0$$

$$\Rightarrow \quad q_s - 10 = 0 \quad ...(7.75)$$

Subtracting equation (7.75) from equation (7.74), we get

$$(q - q_s) - (10 - 10) = 2\frac{dh}{dt}$$

$$\Rightarrow \quad (q - q_s) = 2\frac{dh}{dt} \quad ...(7.76)$$

Defining the deviation variables

$$Q = (q - q_s)$$

$$H = (h - h_s)$$

$$\Rightarrow \quad Q = 2\frac{dH}{dt} \quad ...(7.77)$$

Taking Laplace transform of equation (7.77), we get

$$Q(s) = 2s\, H(s)$$

$$\Rightarrow \quad \frac{H(s)}{Q(s)} = \frac{1}{2s}$$

(ii) The tank level operates around the steady state value of $h_s = 3$ m.
From the given data, we have

$$q_o = q_{o1} + q_{o2}$$

$$\Rightarrow \quad q_o = 10 + \frac{h-2}{R}$$

$$\Rightarrow \quad q_o = 10 + \frac{h-2}{0.5}$$

$$\Rightarrow \quad q_o = 10 + 2(h-2) \quad ...(7.78)$$

From the transient state material balance, we have

$$q - q_o = A\frac{dh}{dt}$$

$$q - [10 + 2(h-2)] = 2\frac{dh}{dt} \quad ...(7.79)$$

From the steady state material balance, we have

$$q_s - [10 + 2(h_s - 2)] = 0 \quad ...(7.80)$$

Subtracting equation (7.80) from equation (7.79), we get

$$(q - q_s) - 2(h - h_s) = 2\frac{dh}{dt}$$

Defining deviation variables

$$Q = q - q_s$$
$$H = h - h_s$$
$$\Rightarrow \qquad Q - 2H = 2\frac{dH}{dt} \qquad \text{...(7.81)}$$

Taking Laplace transform of equation (7.81), we get

$$Q(s) - 2H(s) = 2s\,H(s)$$
$$\Rightarrow \qquad Q(s) = 2s\,H(s) + 2H(s)$$
$$\Rightarrow \qquad Q(s) = 2(s+1)\,H(s)$$
$$\Rightarrow \qquad \frac{H(s)}{Q(s)} = \frac{1}{2(s+1)}$$

■ **Example** 7.13 A tank having a cross-sectional area 2 m^2 is operating at steady state with an inlet flow rate of 2 m^3/min. The flow head character is shown in Fig. 7.21.

(i) Find the transfer function $\frac{H(s)}{Q(s)}$

(ii) If the flow to the tank is increases from 2 to 2.2 m^3/min according to step change. Calculate the level '*h*', 2 min after the change occurs.

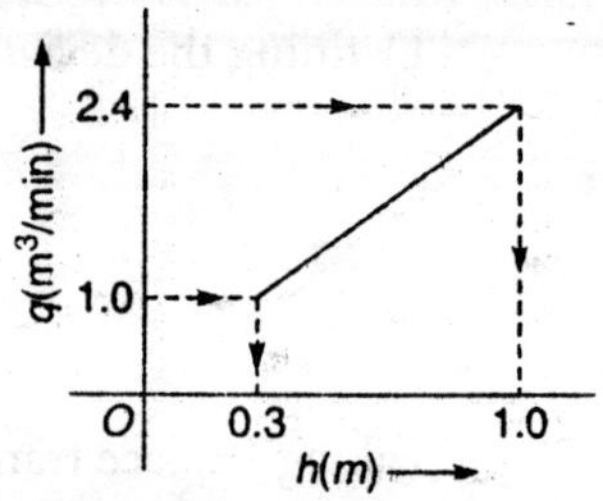

Fig. 7.21 Flow head character.

Solution :

(i) **Transfer Function :** From the Fig. 7.21 it is observed that the flow head character is linear. It can be represented by the equation

$$q_o = mh + C \qquad \text{...(7.82)}$$

where,

q_o = Exit flow rate, m^3/min

m = Constant

h = Level of liquid in the tank, m

C = Constant

Substituting the corresponding data in equation (7.82), we get

$$1 = 0.3\,m + C \qquad \text{...(7.83)}$$
$$2.4 = m + C \qquad \text{...(7.84)}$$

Subtracting equation (7.83) from equation (7.84), we get

$$1.4 = 0.7\,m$$
$$\Rightarrow \qquad m = 2$$
$$\Rightarrow \qquad C = 0.4$$
$$\Rightarrow \qquad q_o = 2h + 0.4 \qquad \text{...(7.85)}$$

From the transient state material balance, we have

$$q - q_o = A\frac{dh}{dt}$$

$$\Rightarrow \quad q-(2h+0.4)=2\frac{dh}{dt} \quad ...(7.86)$$

From the steady state material balance, we have

$$q_s-(2h_s+0.4)=0 \quad ...(7.87)$$

Subtracting equation (7.87) from equation (7.86), we get

$$(q-q_s)-2(h-h_s)=2\frac{dh}{dt} \quad ...(7.88)$$

Defining deviation variables

$$Q=q-q_s$$
$$H=h-h_s$$

$$\Rightarrow \quad Q-2H=2\frac{dH}{dt} \quad ...(7.89)$$

Taking Laplace transform of equation (7.89), we get

$$Q(s)-2H(s)=2s\,H(s)$$

$$\Rightarrow \quad Q(s)=2s\,H(s)+2H(s)$$

$$\Rightarrow \quad Q(s)=2[s+1]\,H(s)$$

$$\Rightarrow \quad \frac{H(s)}{Q(s)}=\frac{1}{2(s+1)} \quad ...(7.90)$$

(ii) Flow to the tank is increases from 2 to 2.2 m^3/min.

Data given :

$$Q(t)=2.2-2.0$$
$$Q(t)=0.2$$

$$\Rightarrow \quad Q(s)=\frac{0.2}{s} \quad ...(7.91)$$

From equation (7.90), we have

$$\frac{H(s)}{Q(s)}=\frac{1}{2(s+1)} \quad ...(7.90)$$

Substituting equation (7.91) in equation (7.90), we get

$$H(s)=\frac{0.2}{2(s)(s+1)}$$

$$H(s)=\frac{0.1}{(s)(s+1)}$$

$$H(s)=\frac{0.1}{(s)(s+1)}=\frac{C_1}{s}+\frac{C_2}{(s+1)}$$

$$\Rightarrow \quad \frac{0.1}{(s)(s+1)}=\frac{C_1(s+1)+C_2(s)}{(s)(s+1)}$$

$$\Rightarrow \quad 0.1=C_1(s+1)+C_2(s)$$

Put, $s=0$; $C_1=0.1$

Put, $s=-1$; $C_2=-0.1$

$$\Rightarrow \qquad H(s) = \frac{0.1}{s} - \frac{0.1}{(s+1)}$$

$$\Rightarrow \qquad H(t) = 0.1 - 0.1e^{-t}$$

$$\Rightarrow \qquad H(t) = 0.1(1 - e^{-t}) \qquad \text{...(7.92)}$$

From the data given, we have

$$q_{os} = 2, \text{m}^3/\text{min}$$

From the equation (7.85), we have

$$q_o = 2h + 0.4 \qquad \text{...(7.85)}$$

At steady state condition equation (7.85) can be written as

$$q_{os} = 2h_s + 0.4$$

$$\Rightarrow \qquad 2 = 2h_s + 0.4$$

$$\Rightarrow \qquad h_s = 0.9$$

From the deviation variable, we have

$$H(t) = h - h_s$$

$$h = H(t) + h_s \qquad \text{...(7.93)}$$

Substituting equation (7.92) in equation (7.93), we get

$$h = 0.1(1 - e^{-t}) + 0.8$$

$$h = 0.1(1 - e^{-2}) + 0.8 \qquad t = 2 \text{ min}$$

$$h = 0.886 \text{ m}$$

■ **Example 7.14** A tank having a time constant of 1 min and a valve at the exit stream with a valve resistance of 0.1 min/m^2. At time $t = 0$ the flow is suddenly increased to 100 m^3/min by adding 9 m^3 of water to the tank uniformly over a period of 0.1 min. Find an expression for the liquid level as a function of time. Also, find what would be the change in liquid level from its steady state value at $t = 1$ min.

Solution :

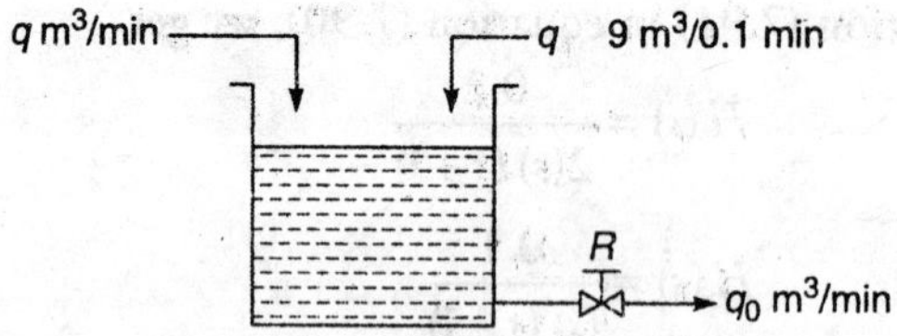

Fig. 7.22 Liquid level in a tank.

From the given data, we have

$$q = \frac{9}{0.1}$$

$$q = 90 \text{ m}^3/\text{min}$$

$$\tau = 1 \text{ min}$$

$$R = 0.1 \text{ min/m}^2$$

From the deviation variable, we have

$$Q(t) = q - q_s$$

$$Q(t) = \begin{cases} 0 & ; \quad t < 0 \\ 90 & ; \quad 0 < t < 0.1 \\ 0 & ; \quad t > 0.1 \end{cases}$$

$$\Rightarrow \quad Q(t) = 90\,[u(t) - u(t-0.1)]$$

$$\Rightarrow \quad Q(s) = 90\left[\frac{1}{s} - \frac{e^{-0.1s}}{s}\right]$$

$$Q(s) = 90\left[\frac{1-e^{-0.1s}}{s}\right] \qquad \ldots(7.94)$$

From equation (7.70), we have

$$\frac{H(s)}{Q(s)} = \frac{R}{\tau s + 1} \qquad \ldots(7.70)$$

Substituting equation (7.94) in equation (7.70), we get

$$H(s) = \frac{90\,[1-e^{-0.1s}]}{(s)} \times \frac{0.1}{(s+1)}$$

$$H(s) = \frac{9\,(1-e^{-0.1s})}{(s)\,(s+1)}$$

$$\Rightarrow \quad H(s) = \frac{9\,(1-e^{-0.1s})}{(s)\,(s+1)} = \frac{C_1}{(s)} + \frac{C_2}{(s+1)}$$

$$\Rightarrow \quad \frac{9\,(1-e^{-0.1s})}{(s)\,(s+1)} = \frac{C_1(s+1) + C_2(s)}{(s)\,(s+1)}$$

$$\Rightarrow \quad 9\,(1-e^{-0.1s}) = C_1(s+1) + C_2(s)$$

Put, $s = 0 \quad ; \quad C_1 = 0$

Put, $s = -1 \quad ; \quad C_2 = 0.946$

$$\Rightarrow \quad H(s) = \frac{0.946}{(s+1)}$$

$$\Rightarrow \quad H(t) = 0.946e^{-t} \qquad \ldots(7.45)$$

From the deviation variable, we have

$$H(t) = h - h_s$$

$$\Rightarrow \quad h = H(t) + h_s$$

$$\Rightarrow \quad h = 0.946e^{-t} + h_s$$

From equation (7.95), we have

$$H(t) = 0.946e^{-t}$$

$$H(t) = 0.946 \times e^{-1} \qquad \text{for } t = 1 \text{ min}$$

$$H(t) = 0.946 \times (0.367)$$

$$H(t) = 0.348 \text{ m}$$

■ **Example 7.15** In a process control laboratory experiment is carried out on single tank first order system and the following readings are recorded.

Inclined tube reading (cm)	Volumetric flow rate (lpm)
9	5
12	9
15	12
18	15
21	18
24	21

Calculate the time constant of the tank.

Data : Length of the tank, $l = 44$ cm

Breadth of the tank, $b = 28$ cm

Inclined tube angle, $\theta = 34°$

Solution :

Table 7.5 : Data for calculation of time constant.

Actual Level (h) (m)	Volumetric flow rate (q) (m^3/min)
0.048	0.005
0.064	0.009
0.079	0.012
0.095	0.015
0.111	0.018
0.127	0.021

From the graph, $R = \dfrac{1}{\text{slope}} = \dfrac{1}{0.2} = 5 \text{ min/m}^2$

$\Rightarrow \quad \tau = AR$

$\Rightarrow \quad \tau = (44 \times 10^{-2}) \times (28 \times 10^{-2}) \times 5$

$\Rightarrow \quad \tau = 0.62 \text{ min}$

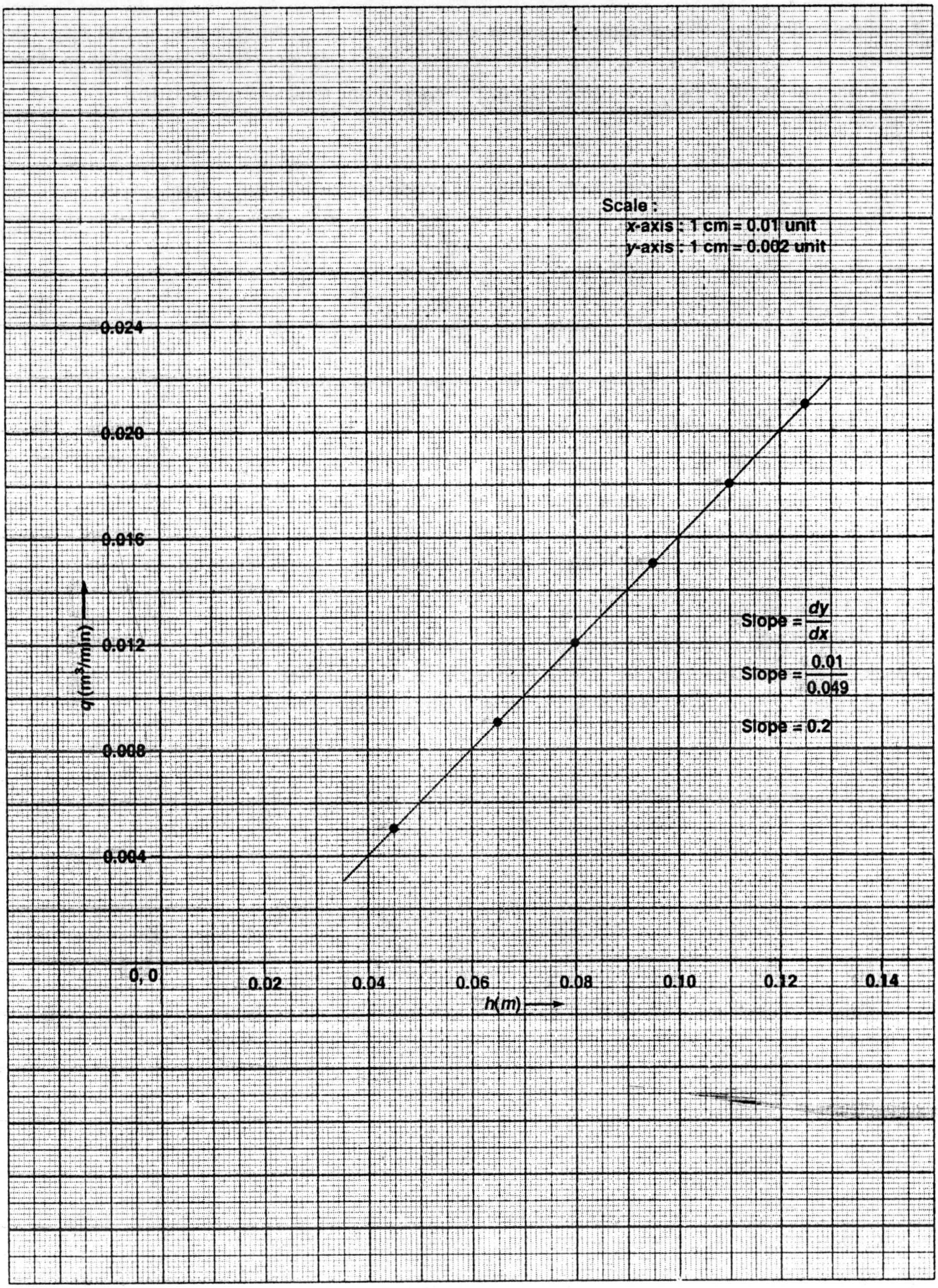

Fig. 7.23 Determination of time constant.

■ **Example 7.16** Develop a formula for finding the time constant of a liquid level system, when the average operating level is h_o and the resistance R is linear. The tank has 3 vertical walls and one slopes with an angle α from the vertical wall. The distance separating the parallel wall is 1 m.

Solution :

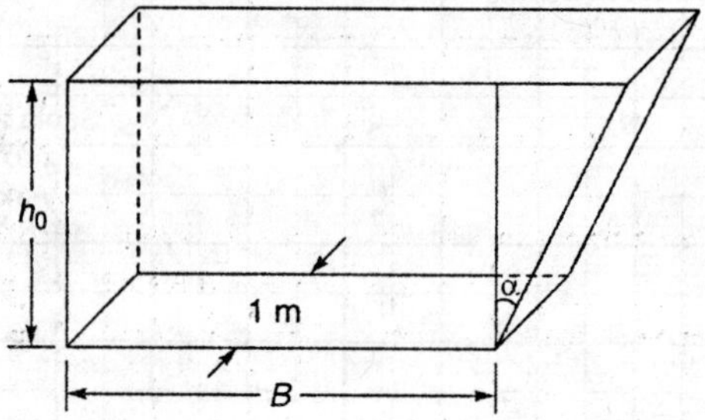

Fig. 7.24 A tank with one inclined vertical wall.

Time constant of a liquid level in a tank is given by

$$\tau = AR$$

$$\Rightarrow \qquad \tau = \left[\frac{h_o \times B \times 1 + h_o\,(h_o\ \tan\alpha)/2}{h_o}\right] \times R$$

$$\tau = [B + (h_o\ \tan\alpha)/2] \times R$$

7.7.2.1 Non-Linear Liquid Level System

Some of the physical systems are not linear. This type of system can not be represented by a linear differential equation. In such a case, it is difficult to take Laplace transform of the equation representing the system and it is also difficult to represent the system in terms of transfer function. In such a case, it should be linearized before taking the Laplace transform of the equation and the technique is called linearization.

For example, consider a liquid level system shown in Fig. 7.25.

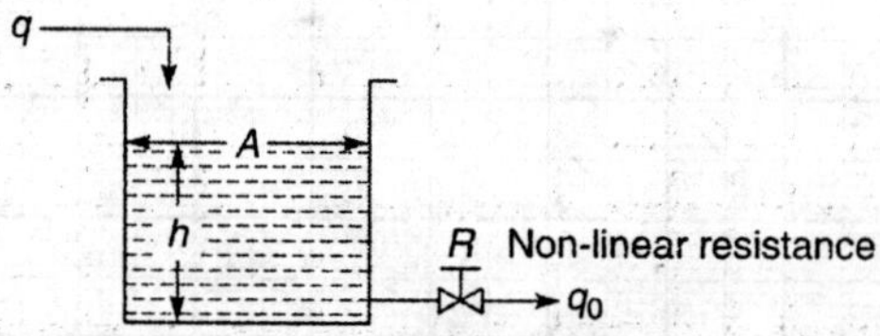

Fig. 7.25 Liquid level system with non-linear resistance.

There is a non-linear resistance in the exit line such that the level and flow is related by a non-linear relationship.

$$q_o = Ch^{1/2} \qquad \text{...(7.96)}$$

where, q_o = Exit flow, m^3/s

C = Constant

h = Level of liquid, m

Let 'A' be the cross-sectional area of the tank and 'q' be the inlet flow rate. Assuming the density of the liquid is constant, we can write the transient state material balance as

$$q - q_o = A\frac{dh}{dt} \qquad \text{...(7.97)}$$

Substituting equation (7.96) in equation (7.97), we get

$$q - Ch^{1/2} = A\frac{dh}{dt} \qquad \text{...(7.98)}$$

We can not define deviation variable of the equation (7.98) as usual because of the presence of the non-linear term $h^{1/2}$ in the equation for which no simple transform is known.

We know that q_o is a function of h. We can expand q_o around the steady state level 'h_s' by Tyler series as

$$q_o = q_o(h_s) + q_o'[h_s(h-h_s)] + q_o''[h_s(h-h_s)^2] + \ldots \qquad \text{...(7.99)}$$

where, $q_o(h_s) = q_{os}$ at $h = h_s$

$q_o'(h_s) = \dfrac{dq_o}{dh}$ at $h = h_s$

$q_o''(h_s) = \dfrac{d^2q_o}{dh^2}$ at $h = h_s$

If $(h - h_s)$ is very small, we can neglect the higher order terms from the equation (7.99).

$$\Rightarrow \quad q_o = q_{os} + q_o'(h_s)(h-h_s) \qquad \text{...(7.100)}$$

$$\Rightarrow \quad q_o = q_{os} + \frac{(h-h_s)}{R} \qquad \text{...(7.101)}$$

where, $\dfrac{1}{R} = \dfrac{dq_o}{dh}$ at $h = h_s$ ...(7.102)

$$\Rightarrow \quad \frac{1}{R} = \frac{d}{dh}(Ch^{1/2})$$

$$\Rightarrow \quad \frac{1}{R} = C.\frac{1}{2}\times\frac{1}{\sqrt{h}}$$

$$\Rightarrow \quad R = \frac{2\sqrt{h}}{C} \qquad \text{...(7.103)}$$

Substituting equation (7.101) in equation (7.97), we get

$$q - \left(q_{os} + \frac{h-h_s}{R}\right) = A\frac{dh}{dt} \qquad \text{...(7.104)}$$

At steady state $\qquad q_s = q_{os}$

$$\Rightarrow \quad (q - q_s) - \frac{(h-h_s)}{R} = A\frac{dh}{dt}$$

Defining deviation variable

$$Q = q - q_s$$

$$H = h - h_s$$

$$\Rightarrow \quad Q - \frac{H}{R} = A\frac{dH}{dt} \qquad \text{...(7.105)}$$

Taking Laplace transform of equation (7.105), we get

$$Q(s) - \frac{H(s)}{R} = As\,H(s)$$

$$Q(s) = As\,H(s) + \frac{H(s)}{R}$$

$$Q(s) = \left[As + \frac{1}{R}\right] H(s)$$

$$\frac{H(s)}{Q(s)} = \frac{R}{ARs + 1}$$

$$\frac{H(s)}{Q(s)} = \frac{R}{\tau s + 1} \qquad \text{...(7.106)}$$

where, $\tau = AR$

A = Cross-sectional area, m^2

$R = \frac{2\sqrt{h}}{C}$, resistance.

It is observed from the equation (7.106) that the transfer function obtained by linearizing a non-linear liquid level system is in the same form as that of the linear liquid level system. Only the difference is the resistance R which in this case depends on the steady state level around which it operates.

It can be concluded that the linearization is valid only when the system operates very near to the steady state value. If the variation is large, when the linear approximation will not give accurate result, in such case we have to go for the complicated non-linear method of analysis.

The linearization technique can be used for any system, which can be expressed by 'Tyler Series' or which can be expressed by the equation of the tangent line at the operating point.

■ **Example 7.16** A liquid level system as shown in Fig. 7.25 has a cross-sectional area of $3\ m^2$.

The valve characteristic is given by the equation

$$q_o = 8\sqrt{h}$$

where, q_o = Exit flow rate, m^3/min

h = Liquid level above the valve, m.

Fig. 7.26 Liquid level system.

Calculate the time constant for this system if the average operating level is :

(i) 3 m (ii) 9 m

Solution : From the given data, we have

$$q = q_o = 8\sqrt{h} \qquad \text{...(7.108)}$$

From the equation (7.102), we have

$$\frac{1}{R} = \frac{dq_o}{dh} \qquad \text{...(7.102)}$$

Substituting equation (7.108) in equation (7.102), we get

$$\frac{1}{R} = \frac{d(8\sqrt{h})}{dh}$$

$$\Rightarrow \qquad R = \frac{\sqrt{h}}{4}$$

(i) Average operating level is 3 m :

$$\tau = AR$$

$$\Rightarrow \qquad \tau = 3 \times \frac{\sqrt{1}}{4}$$

$$\Rightarrow \qquad \tau = 0.75 \text{ min}$$

(ii) Average operating level is 9 m :

$$\tau = AR$$

$$\Rightarrow \qquad \tau = 3 \times \frac{\sqrt{7}}{4}$$

$$\Rightarrow \qquad \tau = 1.98 \text{ min}$$

7.7.3 Mixing in a Tank

Consider a constant hold up stirred tank where a fluid is flowing in and out of the tank at a constant flow rate 'q'. Let $x(t)$ be the concentration of the fluid entering the tank and let us considei that it mixed well with the fluid in the tar'; and attains the concentration $y(t)$ and flows out. Let 'V' be the constant hold up of the tank. The following assumptions are made for the development of transfer function :

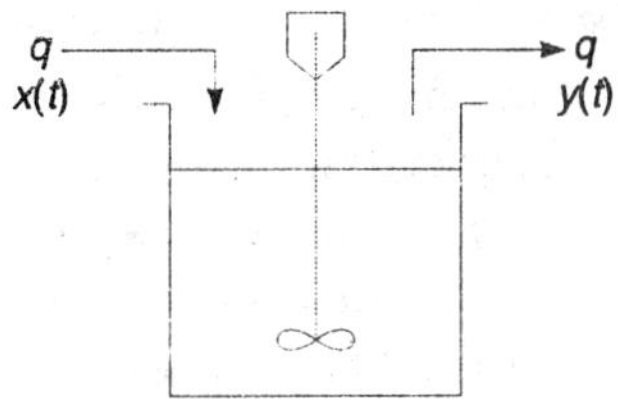

Fig. 7.27 Mixing process.

(i) The concentration of the fluid inside the tank is uniform everywhere and equal to out flow concentration.

(ii) The density of the solution is constant.

A typical mixing process is shown in Fig. 7.27.

We can write the transient state solute balance around the mixing tank as

$$qx - qy = \frac{d(Vy)}{dt} \qquad \ldots(7.109)$$

$$\Rightarrow \qquad qx - qy = V\frac{dy}{dt} \qquad \ldots(7.110)$$

We can write the steady state solute balance around the mixing tank as

$$qx_s - qy_s = 0 \qquad \ldots(7.111)$$

Subtracting equation (7.111) from equation (7.110), we get

$$q(x - x_s) - q(y - y_s) = V\frac{dy}{dt} \qquad \ldots(7.112)$$

Defining deviation variable

$$X = x - x_s$$

$$Y = y - y_s$$

$$\Rightarrow \qquad qX - qy = V\frac{dY}{dt}$$

$$\Rightarrow \qquad X - Y = \frac{V}{q}\frac{dY}{dt}$$

Let us assume, $$\tau = \frac{V}{q}$$

$$\Rightarrow \qquad X - Y = \tau \frac{dV}{dt} \qquad ...(7.113)$$

Taking Laplace transform of equation (7.113), we get

$$X(s) - Y(s) = \tau s\, Y(s)$$

$$X(s) = [\tau s + 1]\, Y(s)$$

$$\frac{Y(s)}{X(s)} = \frac{1}{\tau s + 1}$$

where, $\tau = \frac{V}{q}$

τ = Time constant, s

V = Volume, m^3

q = Volumetric flow rate, m^3/s

■ **Example 7.17** A gas tank is supplied air by a compressor at a rate of 0.12 m^3/min. The normal pressure in the tank is 3 kg/cm^2 absolute and the volume of the tank is 0.3 m^3. The gas is then throttled through a valve to a process down stream.

(i) The normal pressure of down stream process is 2.5 kg/cm^2 absolute. What is the time constant of the tank?

(ii) The normal pressure of the down stream process is 1.25 kg/cm^2 absolute. What is the time constant?

The temperature of the surrounding and tank may be taken as 30°C.

Solution : Data given : $P_1 = 3 \times 10^4$ kg/m^2

$$V_1 = 0.3\ m^3$$

$$P_2 = 2.5 \times 10^4\ kg/m^2$$

$$V_2 = ?$$

$$T_1 = T_2 = 30°C$$

$$q = 0.12\ m^3/min$$

(i) The normal pressure of down stream process is 2.5 kg/cm^2 :

From the ideal gas law, we have

$$\frac{P_1 V_1}{T_1} = \frac{P_2 V_2}{T_2}$$

$$\Rightarrow \qquad V_2 = V_1 \times \frac{P_1}{P_2}$$

$$\Rightarrow \qquad V_2 = 0.3 \times \frac{3 \times 10^4}{2.5 \times 10^4}$$

$$\Rightarrow \qquad V_2 = 0.36\ m^3$$

From the time constant of mixing process in a tank, we have

$$\tau = \frac{V}{q}$$

$$\Rightarrow \qquad \tau = \frac{0.36}{0.12}$$

$$\Rightarrow \qquad \tau = 3 \text{ min}$$

(ii) The normal pressure of down stream process is 1.25 kg/cm^2 :

From the ideal gas law, we have

$$V_2 = V_1 \times \frac{P_1}{P_2}$$

$$\Rightarrow \qquad V_2 = 0.3 \times \frac{3 \times 10^4}{1.25 \times 10^4}$$

$$\Rightarrow \qquad V_2 = 0.72 \text{ m}^3$$

From the time constant of mixing process in a tank, we have

$$\tau = \frac{V}{q}$$

$$\Rightarrow \qquad \tau = \frac{0.72}{0.12}$$

$$\Rightarrow \qquad \tau = 6 \text{ min}$$

7.7.4 Continuous Stirred Tank Reactor

Consider a CSTR as shown in Fig. 7.28, in which a first order reaction is carried out.

Let 'q' be the constant volumetric flow rate of the fluid. Let $x(t)$ be the concentration of A in the inlet stream and $y(t)$ be the concentration of A in the exit stream, where the first order reaction

$$A \rightarrow B$$

is carried out. Let the reaction rate be the

$$r = ky, \text{ moles of '}A\text{' reacting/(volume) (time)}$$

where k be the reaction rate constant and 'V' be the constant hold up of the tank. The following assumptions are made for the development of transfer function :

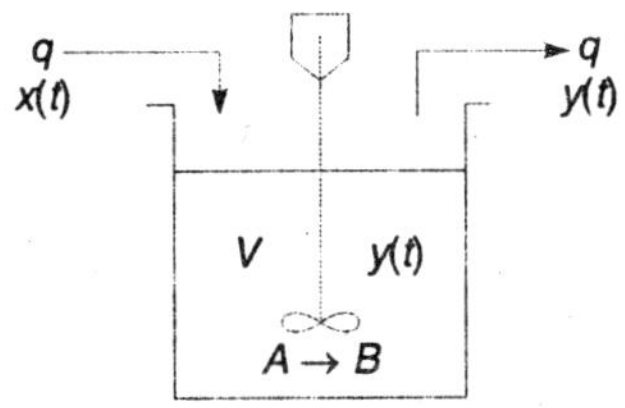

Fig. 7.28 Continuous stirred tank reactor.

(i) Density of the fluid is constant everywhere.
(ii) Tank is of constant hold up.
(iii) Tank is well stirred and concentration of the fluid inside the tank is uniform everywhere and equal to the exit concentration.

We can write the transient state solute balance around the CSTR as

$$\begin{pmatrix}\text{Rate of '}A\text{' moles}\\ \text{entering the tank}\end{pmatrix} - \begin{pmatrix}\text{Rate of '}A\text{' moles}\\ \text{leaving the tank}\end{pmatrix} - \begin{pmatrix}\text{Rate of '}A\text{' moles}\\ \text{reacted}\end{pmatrix} = \begin{pmatrix}\text{Rate of accumulation}\\ \text{of '}A\text{' in the tank}\end{pmatrix}$$

$$qx - qy - V.r = \frac{d(V.y)}{dt}$$

$$\Rightarrow \qquad qx - qy - Vky = V\frac{dy}{dt}$$

$$\Rightarrow \qquad qx - (q + Vk)\ y = V\frac{dy}{dt} \qquad \text{...(7.114)}$$

We can write the steady state solute balance around the CSTR as

$$qx_s - (q + Vk)\ y_s = 0 \qquad \text{...(7.115)}$$

Subtracting equation (7.115) from equation (7.114), we get

$$q(x - x_s) - (q + Vk)(y - y_s) = V\frac{dy}{dt} \qquad \text{...(7.116)}$$

Defining deviation variable

$$X(t) = (x - x_s)$$

$$Y(t) = (y - x_s)$$

$$\Rightarrow \qquad qX - (q + Vk)\ Y = V\frac{dY}{dt} \qquad \text{...(7.117)}$$

Taking Laplace transform of equation (7.117), we get

$$q \times (s) - (q + Vk)\ Y(s) = Vs\ Y(s) \qquad \text{...(7.118)}$$

Dividing equation (7.118) by $(q + Vk)$, we get

$$\left(\frac{q}{q + Vk}\right) \times (s) - Y(s) = \left(\frac{V}{q + Vk}\right) s\ Y(s)$$

Let us assume

$$k' = \frac{q}{q + Vk} \qquad \text{...(7.119)}$$

$$\tau' = \frac{V}{q + Vk} \qquad \text{...(7.120)}$$

$$\Rightarrow \qquad k'\ X(s) - Y(s) = \tau' s\ Y(s)$$

$$\Rightarrow \qquad k'\ X(s) = (\tau' s + 1)\ Y(s)$$

$$\Rightarrow \qquad \frac{Y(s)}{X(s)} = \frac{k'}{(\tau s + 1)} \qquad \text{...(7.121)}$$

where, k' = Steady stage gain
τ' = Time constant, s
q = Flow rate, m^3/s
V = Hold up volume, m^3
k = Reaction rate constant, s^{-1}
τ = Hold up time, hr.

■ **Example 7.18** A CSTR has the hold up of 1.5 hr and a reaction rate constant of hr^{-1}. Find the exit concentration as a function of time if there is a sudden change in the feed concentration from 0.4 to 0.5 moles/m^3. Assume inlet flow rate same as the exit flow rate and equal to 1 m^3/hr.

[Bangalore University, Oct. 1984]

Solution : Data given : $\tau = 1.5$ hr

$$q = 1 \text{m}^3/\text{hr}$$

$$k = 2 \text{ hr}^{-1}$$

$$x(t) = 0.1 \text{moles/m}^3$$

From the definition of hold up time of a reaction in CSTR, we have

$$\tau = \frac{V}{q}$$

$$1.5 = \frac{V}{1}$$

$$\Rightarrow \quad V = 1.5 \text{m}^3$$

From the equation (7.119), we have

$$k' = \frac{q}{q + Vk}$$

$$\Rightarrow \quad k' = \frac{1}{1 + (1.5)(2)}$$

$$\Rightarrow \quad k' = 0.25$$

From the equation (7.120), we have

$$\tau' = \frac{q}{q + Vk}$$

$$\Rightarrow \quad \tau' = \frac{1}{1 + (1.5)(2)}$$

$$\Rightarrow \quad \tau' = 0.375 \text{ hr}$$

From the transfer function of CSTR, we have

$$\frac{Y(s)}{X(s)} = \frac{k'}{(\tau' s + 1)} \qquad \text{...(7.121)}$$

$$X(t) = 0.1$$

$$X(s) = \frac{0.1}{s}$$

$$\Rightarrow \quad Y(s) = \frac{0.1}{s} \times \frac{0.25}{(0.375s + 1)}$$

$$\Rightarrow \quad Y(s) = \frac{0.1}{s} \times \frac{0.25}{0.375(s + 2.67)}$$

$$\Rightarrow \quad Y(s) = \frac{0.067}{(s)(s+2.67)}$$

$$\Rightarrow \quad Y(s) = \frac{0.067}{(s)(s+2.67)} = \frac{C_1}{s} + \frac{C_2}{(s+2.67)}$$

$$\Rightarrow \quad \frac{0.067}{(s)(s+2.67)} = \frac{C_1(s+2.67)+C_2(s)}{(s)(s+2.67)}$$

$$\Rightarrow \quad 0.067 = C_1(s+2.67)+C_2(s)$$

Put, $s = 0$; $C_1 = 0.025$

Put, $s = -2.67$; $C_2 = -0.025$

$$\Rightarrow \quad Y(s) = \frac{0.025}{s} - \frac{0.025}{(s+2.67)}$$

$$\Rightarrow \quad Y(t) = 0.025 - 0.025e^{-2.67t}$$

$$\Rightarrow \quad Y(t) = 0.025[1-e^{-2.67t}]$$

■ **Example 7.19** Consider the CSTR as shown in Fig. 7.29.

The reaction occurring is

$$A \rightarrow B$$

and it proceeds at the rate

$$r = kC_o$$

where, r = Moles of A reacting/(volume) (time)

k = Reaction rate constant, $(\text{time})^{-1}$

$C_o(t)$ = Concentration of A, moles/volume

V = Volume of reaction mixture in the reactor

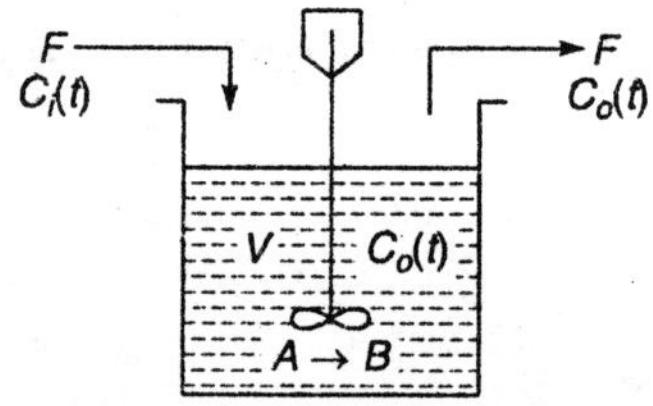

Fig. 7.29 Continuous stirred tank reactor.

Further, let

F = Constant feed rate, volume/time

$C_i(t)$ = Concentration of A in feed stream, moles/volume

Assuming constant density and constant volume, derive the transfer function relating the concentration in the reactor to the feed stream concentration. Sketch the response of the reactor to a unit step change in $C_i(t)$. **[Bangalore University, Nov. 1991]**

Solution : We can write the transient state solute balance around the CSTR as

$$FC_i - FC_o - V.r = \frac{d(VC_o)}{dt}$$

$$\Rightarrow \quad FC_i - FC_o - VkC_o = V\frac{dC_o}{dt}$$

$$\Rightarrow \quad FC_i - (F+Vk)C_o = V\frac{dC_o}{dt} \qquad \ldots(7.122)$$

We can write the steady state solute balance around the CSTR as

$$FC_{is} - (F+Vk)C_{os} = 0 \qquad \ldots(7.123)$$

Subtracting equation (7.123) from equation (7.122), we get

$$F(C_i - C_{is}) - (F + Vk)(C_o - C_{os}) = V \frac{dC_o}{dt} \quad ...(7.124)$$

Defining deviation variable

$$C_i'(t) = C_i - C_{is}$$

$$C_o'(t) = C_o - C_{os}$$

$$\Rightarrow \quad FC_i' - (F + Vk)C_o' = V \frac{dC_o'}{dt} \quad ...(7.125)$$

Taking Laplace transform of equation (7.125), we get

$$FC_i'(s) - (F + Vk)C_o' = VsC_o'(s) \quad ...(7.126)$$

Dividing equation (7.126) by $(F + Vk)$, we get

$$\left(\frac{F}{F + Vk}\right)C_i'(s) - C_o'(s) = \left(\frac{V}{F + Vk}\right)sC_o'$$

Let us assume

$$k' = \frac{F}{F + Vk}$$

$$\tau = \frac{V}{F + Vk}$$

$$\Rightarrow \quad k'C_i'(s) - C_o'(s) = \tau s C_o'(s)$$

$$\Rightarrow \quad k'C_i'(s) = (\tau s + 1)C_o'(s)$$

$$\Rightarrow \quad \frac{C_o'(s)}{C_i'(s)} = \frac{k'}{\tau s + 1} \quad ...(7.127)$$

$$C_i'(t) = 1$$

$$C_i'(s) = \frac{1}{s}$$

$$\Rightarrow \quad C_o'(s) = \frac{k'}{(s)(\tau s + 1)}$$

$$\Rightarrow \quad C_o'(s) = \frac{k'}{(s)\tau(s + 1/\tau)}$$

$$\Rightarrow \quad C_o'(s) = \frac{k'/\tau}{(s)(s + 1/\tau)} = \frac{C_1}{(s)} + \frac{C_2}{(s + 1/\tau)}$$

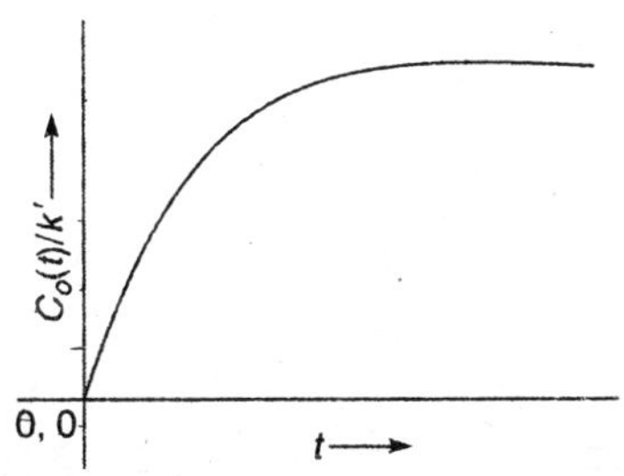

Fig. 7.30 Response of the reactor.

$$\Rightarrow \qquad \frac{k'}{\tau} = C_1(s+1/\tau) + C_2(s)$$

Put, $s=0$; $C_1 = k'$

Put, $s=-1/\tau$; $C_2 = -k'$

$$\Rightarrow \qquad C'_o(s) = \frac{k'}{s} - \frac{k'}{(s+1/\tau)}$$

$$\Rightarrow \qquad C'_o(t) = k'(1-e^{-t/\tau})$$

Example 7.20 In a CSTR, a reaction $A \rightarrow B$ is taking place with rate,

$$r = kC_o$$

where, r = Rate of reaction, mole/(volume) (time)

k = Reaction velocity constant, $(\text{time})^{-1}$

C_o = Exit concentration, moles/volume

The feed concentration is C_i. Assume constant density ρ, constant volume of reactor V and constant feed rate F, volume/time. Derive the transfer function relating concentration in the reactor to the feed stream concentration. Sketch the response of the reactor to a step change of magnitude 3 in C_i.

[Bangalore University, March 1989]

Solution : We can write the transient state solute balance around the CSTR as

$$FC_i - FC_o - V.r = \frac{d(VC_o)}{dt}$$

$$\Rightarrow \qquad FC_i - FC_o - VkC_o = V\frac{dC_o}{dt}$$

$$\Rightarrow \qquad FC_i - (F+Vk)C_o = V\frac{dC_o}{dt} \qquad \text{...(7.128)}$$

We can write the steady state solute balance around the CSTR as

$$FC_{is} - (F+Vk)\,C_{os} = 0 \qquad \text{...(7.129)}$$

Subtracting equation (7.129) from equation (7.128), we get

$$F(C_i - C_{is}) - (F+Vk)(C_o - C_{os}) = V\frac{dC_o}{dt} \qquad \text{...(7.130)}$$

Defining deviation variable

$$C'_i(t) = (C_i - C_{is})$$

$$C'_o(t) = (C_o - c_{os})$$

$$\Rightarrow \qquad FC'_i - (F+Vk)\,C'_o = V\frac{dC'_o}{dt} \qquad \text{...(7.131)}$$

Taking Laplace transform of equation (7.131), we get

$$FC'_i(s) - (F+Vk)\,C'_o = VsC'_o \qquad \text{...(7.132)}$$

Dividing equation (7.132) by $(F+Vk)$, we get

$$\left(\frac{F}{F+Vk}\right)C'_i(s) - C'_o(s) = \left(\frac{V}{F+Vk}\right)sC'_o(s)$$

Let us assume

$$k' = \frac{F}{F + Vk}$$

$$\tau = \frac{V}{F + Vk}$$

$$\Rightarrow \quad k' C_i'(s) - C_o'(s) = \tau s\, C_o'(s)$$

$$\Rightarrow \quad k' C_i'(s) = (\tau s + 1)\, C_o'(s)$$

$$\Rightarrow \quad \frac{C_o'(s)}{C_i'(s)} = \frac{k'}{\tau s + 1} \qquad \ldots(7.133)$$

$$C_i'(t) = 3$$

$$C_i'(s) = \frac{3}{s}$$

$$\Rightarrow \quad C_o'(s) = \frac{3k'}{(s)\,(\tau s + 1)}$$

$$\Rightarrow \quad C_o'(s) = \frac{3k'}{(s)\,\tau\,(s + 1/\tau)}$$

$$\Rightarrow \quad C_o'(s) = \frac{3k'/\tau}{(s)\,(s + 1/\tau)}$$

$$\Rightarrow \quad C_o'(s) = \frac{3k'/\tau}{(s)\,(s + 1/\tau)} = \frac{C_1}{s} + \frac{C_2}{(s + 1/\tau)}$$

$$\Rightarrow \quad \frac{3k'}{\tau} = C_1 (s + 1/\tau) + C_2 (s)$$

Put, $s = 0$; $C_1 = 3k'$

Put, $s = -1/\tau$; $C_2 = -3k'$

$$\Rightarrow \quad C_o'(s) = \frac{3k'}{s} - \frac{3k'}{(s + 1/\tau)}$$

$$\Rightarrow \quad C_o'(t) = 3k' - 3k'\, e^{-t/\tau}$$

$$\Rightarrow \quad C_o'(t) = 3k'\,(1 - e^{-t/\tau})$$

$$\Rightarrow \quad C_o'(t)/3k' = (1 - e^{-t/\tau})$$

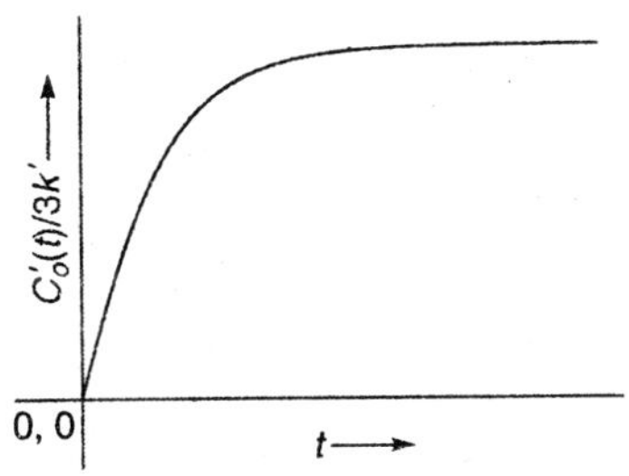

Fig. 7.31 Response of the reactor.

7.7.5 Resistance Capacitance Circuit

RC circuit contains a voltage source $V(t)$ applied to a series combination of resistance 'R' and capacitance 'C'. A typical RC circuit is shown in Fig. 7.32.

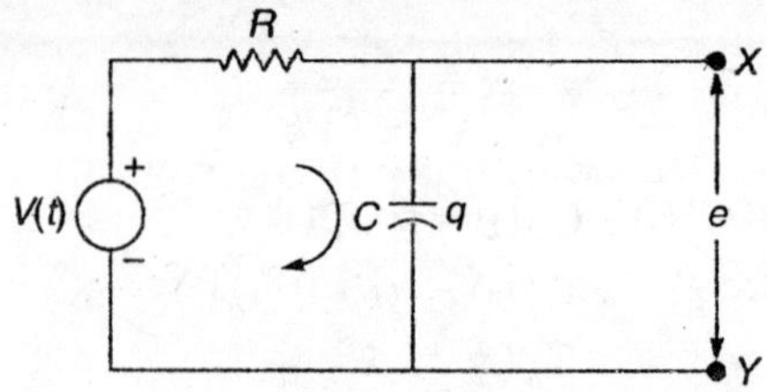

Fig. 7.32 Resistance capacitance circuit.

Because of the voltage applied across the circuit, current starts flowing through the resistance capacitor and the capacitance slowly get charged and a emf is developed across the capacitor. At steady state the capacitor is fully charged and the emf across the capacitor is equal to applied voltage.

Applying the Kirchoff law to the circuit which states that in any load the sum of voltage rises should be equal to sum of voltage drops.

$$\Rightarrow \qquad V = Ri + \frac{1}{C}\int_0^t i\,dt \qquad \ldots(7.134)$$

where, R = Resistance

$i = \frac{dq}{dt}$, rate of change of charge per time

q = Charge of capacitor

$$\Rightarrow \qquad V = R\frac{dq}{dt} + \frac{1}{C}\int_0^t \frac{dq}{dt}\,dt$$

$$\Rightarrow \qquad V = R\frac{dq}{dt} + \frac{q}{C} \qquad \ldots(7.135)$$

Therefore,

$$e = \frac{q}{C} \qquad \ldots(7.136)$$

We can write the steady state voltage balance as

$$V_s = \frac{q_s}{C} \qquad \ldots(7.137)$$

Subtracting equation (7.137) from equation (7.135), we get

$$V - V_s = R\frac{dq}{dt} + \frac{q - q_s}{C} \qquad \ldots(7.138)$$

Defining deviation variable

$$V(t) = V - V_s$$

$$Q(t) = q - q_s$$

$$\Rightarrow \qquad V = R\frac{dQ}{dt} + \frac{Q}{C} \qquad \ldots(7.139)$$

We can write equation (7.136) in steady state as

$$e_s = \frac{q_s}{C} \qquad \ldots(7.140)$$

Subtracting equation (7.140) from equation (7.136), we get

$$e - e_s = \frac{q - q_s}{C}$$

Defining deviation variable

$$E(t) = e - e_s$$

$$Q(t) = q - q_s$$

$$\Rightarrow \qquad E(t) = \frac{Q(t)}{C}$$

$$\Rightarrow \qquad Q = EC \qquad \ldots(7.141)$$

Substituting equation (7.141) in equation (7.139), we get

$$V = R\frac{d(EC)}{dt} + \frac{EC}{C}$$

$$\Rightarrow \qquad V(t) = RC\frac{dE}{dt} + E \qquad \ldots(7.142)$$

Taking Laplace transform of equation (7.142), we get

$$V(s) = RCs\,E(s) + E(s)$$

$$V(s) = (RCs + 1)\,E(s)$$

$$\frac{E(s)}{V(s)} = \frac{1}{\tau s + 1} \qquad \ldots(7.143)$$

where, $\tau = RC$

REVIEW QUESTIONS

7.1 Define the following :
(i) Controlled variable (ii) Process variable
(iii) Load variable (iv) Manipulated variable

7.2 Define the following :
(i) Controlled medium (ii) Controlled agent

7.3 Define open loop and closed loop control system. Give suitable examples.

7.4 Explain the action of feed forward control system. Give suitable examples.

7.5 Explain the action of feed back control system. Give suitable examples.

7.6 What are the basic components of a control system. Explain each of them in detail.

7.7 Using basic components draw the block diagram of a control system.

7.8 What is summing junction? Explain its action in process control.

7.9 Define transfer function. Give its applications.

7.10 Explain the procedure to obtain the transfer function.

7.11 What are the different type of input generally used in process control. Explain each of them in detail.

7.12 Define first order system. Give suitable examples.

7.13 Why the time constant of a thermometer in two different liquid media are different? Justify your answer.

7.14 Define time constant.

7.15 What do you mean by non-linear system? Explain with example.

7.16 What is the method by which non-linear system can be linearized? Explain.

7.17 What is the difference between pulse and impulse input? Explain.

EXERCISES

7.1 Starting from the basic derive the transfer function for thermometer. Discuss on the assumptions made for the derivation.

7.2 Derive the equation for response of a first order system for a step input of magnitude '*A*'.

7.3 Derive the equation for response of a first order system for a ramp input of magnitude '*A*'.

7.4 Derive the equation for response of a first order system for a impulse input of magnitude '*A*'.

7.5 Derive the equation for response of a first order system for a sinusoidal input of magnitude '*A*'.

7.6 A mercury thermometer of first order characteristics takes 6 sec to indicate 80% of the final temperature. Calculate time constant of the thermometer.

7.7 A vapor pressure thermometer of first order characteristics has time constant of 2 sec. Calculate the time required by the thermometer to indicate the 75% of the final temperature.

7.8 A thermometer having a time constant of 5 sec is at a steady state temperature of 28°C. At time $t = 0$, the thermometer is placed in a temperature bath maintained at 45°C. Determine the time needed for the thermometer to read 40°C.

7.9 A mercury thermometer bulb is 12 mm long and its diameter is 3.5 mm. If the glass envelope is very thin, calculate the time constant of thermometer placed in water flowing with a velocity of 3.2 m/s and its temperature is 45°C. Data : $h_w = 2500$ W/m^2°C; $C_{Pm} = 502$ J/kg°C.

7.10 A thermometer having a time constant of 0.2 sec is placed in a temperature bath. After the thermometer comes to equilibrium with bath, the temperature of the bath is increased linearly with time at a rate of 1°C per min. What is the difference between the indicated temperature and the bath temperature at $t = 0.5$ min after the change in temperature begins.

7.11 A thermometer is kept inside a constant temperature bath maintained at 70°C. This is suddenly transferred into another bath kept at 60°C at time $t = 0$. The following are the reading recorded.

Time (s)	Temperature (°C)
0	70
1	68
2	66
4	64
5	63
8	62
12	61

Calculate the time constant of the thermometer by theoretical method.

7.12 A thermometer is kept inside a constant temperature bath maintained at 75°C. At time $t = 0$, the thermometer is suddenly transferred into another bath kept at 65°C. The following readings are recorded.

Time (min)	Temperature (°C)
0	75
1	73
2	71
4	69
5	68
8	67
12	66

Calculate the time constant of the thermometer by graphical method.

7.13 A thermometer is at steady state temperature of 25°C. At time $t = 0$, it is placed inside a oil bath maintained at 45°C. The following readings are recorded.

Time (s)	Temperature (°C)
0	25
5	34
10	38
15	41
20	43
30	44
50	45

Calculate the time constant of the thermometer by tangent method.

7.14 A mercury thermometer, which has been on a table for some time is registering the room temperature 24°C. Suddenly, it is placed in a 204°C oil bath. The following data are obtained as response of the thermometer.

Time (s)	Temperature (°C)
0	24
1	42
2	56
5	96
8	118
10	139
15	164
30	196

Estimate the time constant of the thermometer by two independent method.

7.15 A water storage tank for a hostel incorporate a lever type of control system. A float is attached at one end of the lever while the other end is connected to the stem of a valve. When the tank is nearly full the water flow is automatically shut off. Draw a block diagram of the system. Identify the manipulated variable, control agent and controlled variable for this system.

7.16 The liquid level in a storage tank is controlled by controlling the flow rate of fluid to the tank. The fluid is pump out from the tank by constant displacement pump and the exit flow does not depend upon the level in the tank.

Derive expression for the transfer function relating liquid level with the inlet flow rate. Find the response of the system for a unit step input at the inlet flow rate.

[Bangalore University, Nov. 1990]

7.17 A mixing process in a tank having constant hold up of 5 m^3. If the constant volumetric flow rate of the fluid flowing is 0.5 m^3/s. What is the time constant of mixing process?

7.18 A liquid level system has the following properties. Tank dimensions 3 m high, 1.5 m dia steady state operating character :

Inflow (m^3/h)	Steady state level (m)
0	0
22.7	0.21
45.5	0.34
68.2	0.70
30.9	1.19
113.6	1.92
136.4	2.68

(i) Plot the level response of the tank under the following circumstances. The inlet flow rate is held at 13 m^3/min for 1 hr and then suddenly raised to 18 m^3/min.

(ii) How accurate is the steady state level calculated from the dynamic response in part (i) when compared with the value given in the table above?

7.19 A mixing process may be described as follows : A stream with solute concentration C_i (g/cm^3) is fed to a perfectly stirred tank at a constant flow rate of q (vol/time). The perfectly mixed product is withdrawn from the tank, also at the flow rate q at the same concentration as the material in the tank, C_o. The total volume of solution in the tank is constant at 'V'. Density may be considered to be independent of concentration. A trace of the tank concentration versus time appears as shown in Fig. 7.33.

(i) Plot on this same graph your best guess of the quantitative behaviour of the inlet concentration versus time. Be sure to level the graph with quantitative information regarding times and magnitude and any other data which will demonstrate your understanding of the situation.

(ii) Write an equation for C_i as a function of time. Data : Tank dimensions 2.5 m high by 1.5 m dia tank volume, V = 3.18 m^3. Flow rate, q = 0.45 m^3/min. Average density, ρ_{avg} = 1120 kg/m^3.

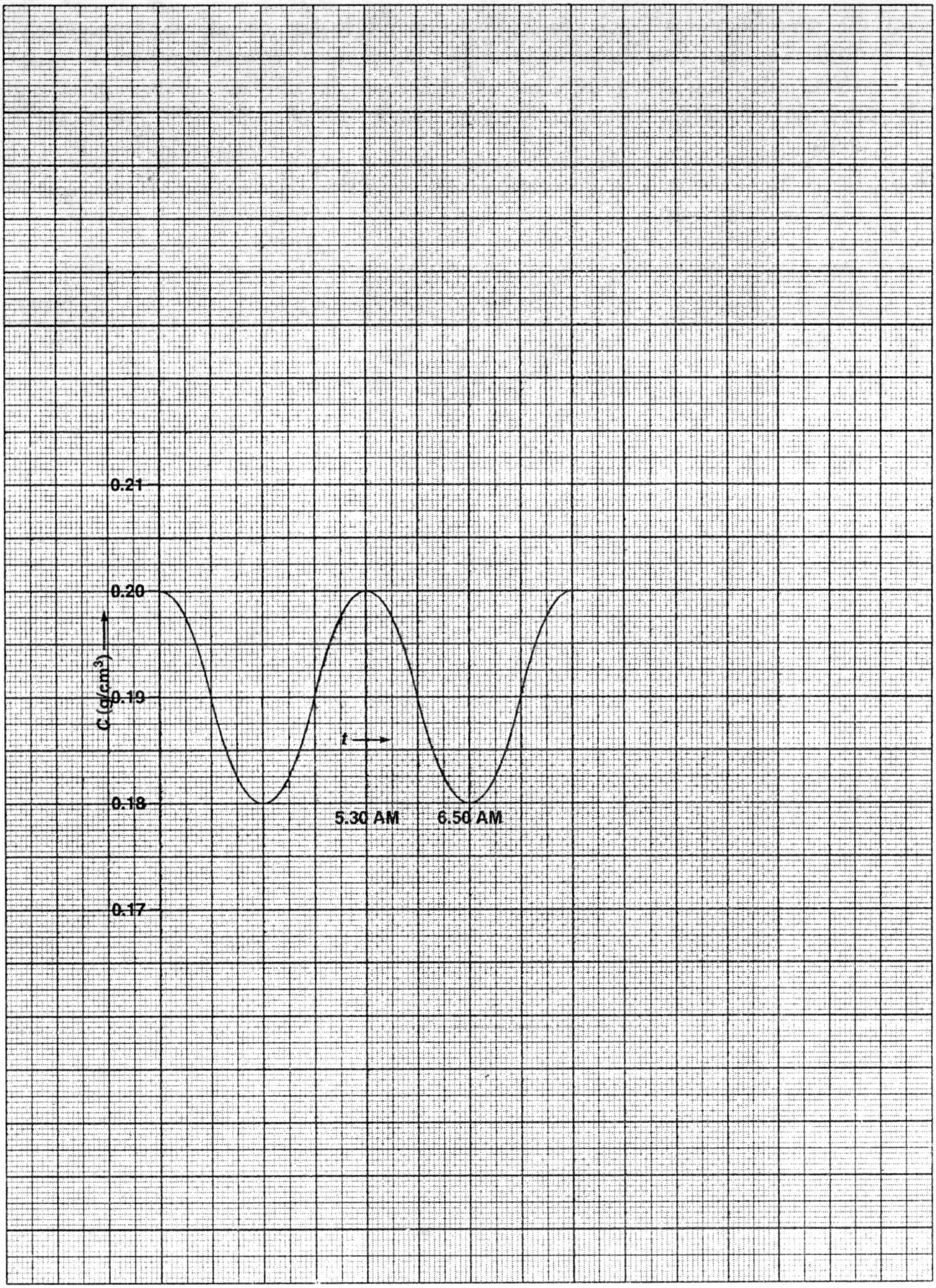

Fig. 7.33 Trace of the tank concentration vs time.

7.20 The temperature of a CSTR is measured by a mercury bulb thermometer with a time constant of 10 sec. The heat supply change such that the temperature is increased from a normal value of 100°C to 110°C at a rate of 20°C/min and maintained thereafter. Obtain the response of the thermometer as a function of time. Show the characteristics of the thermometer by means of a sketch. **[Bangalore University, July 1988]**

7.21 Derive the transfer function for jacketted vessel when the heat transfer occurs between kettle fluid and thermometer.

□□□

Chapter 8

First Order System in Series

8.1 INTRODUCTION

Most of the physical processes can be represented as a series of first order system. To illustrate this, consider two rectangular tank. The connection between the tank can be arranged in any one of the following way :

(i) Two tank non-interacting system in series (TTNIS).

(ii) Two tank interacting system in series (TTIS).

For example, in sugar industry evaporators are connected in this manner.

8.1.1 Two Tank Non-Interacting System

When the out flow from the tank 1 is discharged into the atmosphere before entering into tank 2 and the flow through R_1 depends only on level h_1 and is independent of the level in the tank 2 is called two tank non-interacting system. A typical TTNIS is shown in Fig. 8.1. The variation in level h_2 does not affect the transient response occurring in tank 1.

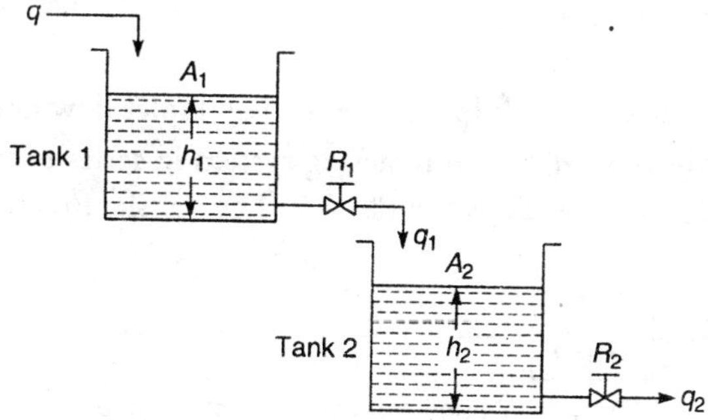

Fig. 8.1 Two tank non-interacting system.

We can write the transfer function for tank 1 relating inflow and outflow as

$$\frac{Q_1(s)}{Q(s)} = \frac{1}{\tau_1 s + 1} \quad \ldots(8.1)$$

We can also write the transfer function for tank 2 relating inflow and outflow as

$$\frac{Q_2(s)}{Q_1(s)} = \frac{1}{\tau_2 s + 1} \quad \ldots(8.2)$$

We can write the over all transfer function for TTNIS relating inflow and outflow as

$$\frac{Q_2(s)}{Q(s)} = \frac{Q_1(s)}{Q(s)} \times \frac{Q_2(s)}{Q_1(s)}$$

$$\Rightarrow \quad \frac{Q_2(s)}{Q(s)} = \frac{1}{(\tau_1 s+1)} \times \frac{1}{(\tau_2 s+1)}$$

$$\Rightarrow \quad \frac{Q_2(s)}{Q(s)} = \frac{1}{(\tau_1 s+1)(\tau_2 s+1)} \qquad \ldots(8.3)$$

It can be observed that overall transfer function is the product of individual transfer function of the system connected in series.

We can write the exit flow equation for tank 2 as

$$Q_2 = \frac{H_2}{R_2} \qquad \ldots(8.4)$$

Taking Laplace transform of equation (8.4), we get

$$Q_2(s) = \frac{H_2(s)}{R_2} \qquad \ldots(8.5)$$

Substituting equation (8.5) in equation (8.3), we get

$$\frac{H_2(s)}{Q(s)} = \frac{R_2}{(\tau_1 s+1)(\tau_2 s+1)} \qquad \ldots(8.6)$$

where,

$H_2(s)$ = Level in the tank 2, m

$Q(s)$ = Inlet flow rate, m^3/s

R_2 = Resistance of the tank 2, s/m^2

τ_1 = Time constant of tank 1, s

τ_2 = Time constant of tank 2, s

This is the transfer function of TTNIS relating level and inlet flow rate.

Overall transfer function for '*n*' non-interacting system in series : Consider '*n*' non-interacting first order system connected in series as shown in Fig. 8.2. The transfer function of the individual system are shown inside the blocks.

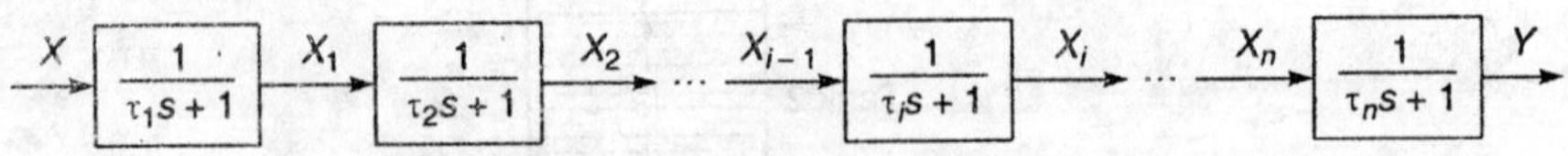

Fig. 8.2 '*n*' non-interacting system in series.

We can write the individual transfer function for each system as

$$\frac{X_1(s)}{X(s)} = \frac{1}{\tau_1 s+1} \qquad \ldots(8.7)$$

$$\frac{X_2(s)}{X_1(s)} = \frac{1}{\tau_2 s+1} \qquad \ldots(8.8)$$

$$\frac{X_i(s)}{X_{i-1}(s)} = \frac{1}{(\tau_i s+1)} \quad \text{...(8.9)}$$

$$\frac{Y(s)}{X_n(s)} = \frac{1}{\tau_n s+1} \quad \text{...(8.10)}$$

To obtain the overall transfer function, the equation (8.7), (8.8), (8.9) and (8.10) are multiplied.

$$\frac{Y(s)}{X(s)} = \frac{1}{\tau_1 s+1} \times \frac{1}{\tau_2 s+1} \times \ldots \times \frac{1}{(\tau_i s+1)} \times \ldots \times \frac{1}{\tau_n s+1}$$

$$\frac{Y(s)}{X(s)} = \prod_{i=1}^{n} \left[\frac{1}{\tau_i s+1}\right]$$

This is transfer function for '*n*' non-interacting system connected in series.

8.1.2 Two Tank Interacting System

When the outflow from the tank 1 without being discharge into atmosphere directly entered into the tank 2 and the flow through R_1 depends on the difference of liquid level between h_1 and h_2 is called two tank interacting system. A typical TTIS in series is shown in Fig. 8.3. The variation of level h_2 of tank 2 affect the transient response occurring in tank 1.

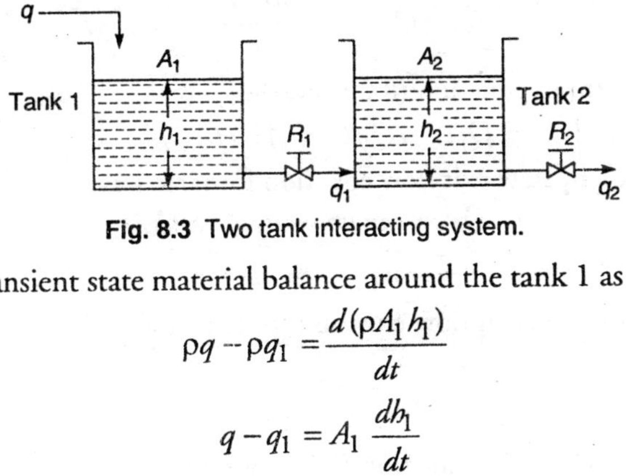

Fig. 8.3 Two tank interacting system.

We can write the transient state material balance around the tank 1 as

$$\rho q - \rho q_1 = \frac{d(\rho A_1 h_1)}{dt}$$

$$\Rightarrow \quad q - q_1 = A_1 \frac{dh_1}{dt} \quad \text{...(8.12)}$$

We can also write the steady state material balance around the tank 1 as

$$q_s - q_{1s} = 0 \quad \text{...(8.13)}$$

Subtracting equation (8.13) from equation (8.12), we get

$$(q - q_s) - (q_1 - q_{1s}) = A_1 \left(\frac{dh_1}{dt}\right) \quad \text{...(8.14)}$$

Defining deviation variables

$$Q = q - q_s$$

$$Q_1 = q_1 - q_{1s}$$

$$H_1 = h_1 - h_{1s}$$

$$\Rightarrow \quad Q - Q_1 = A_1 \left(\frac{dH_1}{dt}\right) \quad \text{...(8.15)}$$

Taking Laplace transform of equation (8.15), we get

$$Q(s) - Q_1(s) = A_1 s H_1(s) \quad \text{...(8.16)}$$

We can write the transient state material balance around the tank 2 as

$$\rho q_1 - \rho q_2 = \frac{d(\rho A_2 h_2)}{dt}$$

$$q_1 - q_2 = A_2 \left(\frac{dh_2}{dt} \right) \quad \text{...(8.17)}$$

We can write the steady state material balance around the tank 2 as

$$q_{1s} - q_{2s} = 0 \quad \text{...(8.18)}$$

Subtracting equation (8.18) from equation (8.17), we get

$$(q_1 - q_{1s}) - (q_2 - q_{2s}) = A_2 \frac{dh_2}{dt} \quad \text{...(8.19)}$$

Defining deviation variables

$$Q_1 = q_1 - q_{1s}$$
$$Q_2 = q_2 - q_{2s}$$
$$H_2 = h_2 - h_{2s}$$

$$\Rightarrow \quad Q_1 - Q_2 = A_2 \frac{dH_2}{dt} \quad \text{...(8.20)}$$

Taking Laplace transform of equation (8.20), we get

$$Q_1(s) - Q_2(s) = A_2 s H_2(s) \quad \text{...(8.21)}$$

We can write the transient state exit flow equation for tank 1 as

$$q_1 = \frac{h_1 - h_2}{R_1} \quad \text{...(8.22)}$$

We can write the steady state exit flow equation for tank 1 as

$$q_{1s} = \frac{h_{1s} - h_{2s}}{R_1} \quad \text{...(8.23)}$$

Subtracting equation (8.23) from equation (8.22), we get

$$q_1 - q_{1s} = \frac{(h_1 - h_{1s}) - (h_2 - h_{2s})}{R_1} \quad \text{...(8.24)}$$

Defining deviation variables

$$Q_1 = q_1 - q_{1s}$$
$$H_1 = h_1 - h_{1s}$$
$$H_2 = h_2 - h_{2s}$$

$$\Rightarrow \quad Q_1 = \frac{H_1 - H_2}{R_1} \quad \text{...(8.25)}$$

Taking Laplace transform of equation (8.25), we get

$$Q_1(s) = \frac{H_1(s) - H_2(s)}{R_1} \quad \text{...(8.26)}$$

We can write the transient state exit flow equation for tank 2 as

$$q_2 = \frac{h_2}{R_2} \quad \text{...(8.27)}$$

We can write the steady state exit flow equation for tank 2 as

$$q_{2s} = \frac{h_{2s}}{R_2} \quad \text{...(8.28)}$$

Subtracting equation (8.28) from equation (8.27), we get

$$(q_2 - q_{2s}) = \frac{h_2 - h_{2s}}{R_2} \quad \text{...(8.29)}$$

Defining deviation variables

$$Q_2 = q_2 - q_{2s}$$
$$H_2 = h_2 - h_{2s}$$

$$\Rightarrow \quad Q_2 = \frac{H_2}{R_2} \quad \text{...(8.30)}$$

Taking Laplace transform of equation (8.30), we get

$$Q_2(s) = \frac{H_2(s)}{R_2} \quad \text{...(8.31)}$$

Adding equation (8.16) and (8.21), we get

$$Q(s) - Q_2(s) = A_1 sH_1(s) + A_2 sH_2(s) \quad \text{...(8.32)}$$

Substituting equation (8.31) in equation (8.32), we get

$$Q(s) - \frac{H_2(s)}{R_2} = A_1 sH_1(s) + A_2 sH_2(s)$$

$$\Rightarrow \quad R_2\, Q(s) = A_1 R_2 sH_1(s) + A_2 R_2 sH_2(s) + H_2(s) \quad \text{...(8.33)}$$

From equation (8.26), we get

$$H_1(s) = R_1 Q_1(s) + H_2(s) \quad \text{...(8.34)}$$

Substituting equation (8.34) in equation (8.33), we get

$$R_2 Q(s) = A_1 R_2 s[R_1 Q_1(s) + H_2(s)] + A_2 R_2 sH_2(s) + H_2(s)$$
$$R_2 Q(s) = A_1 R_1 R_2 sQ_1(s) + A_1 R_2 sH_2(s) + A_2 R_2 sH_2(s) + H_2(s) \quad \text{...(8.35)}$$

From the equation (8.21), we get

$$Q_1(s) = Q_2(s) + A_2 sH_2(s) \quad \text{...(8.36)}$$

Substituting equation (8.36) in equation (8.35), we get

$$R_2 Q(s) = A_1 R_1 R_2 s[Q_2(s) + A_2 sH_2(s)] + A_1 R_2 sH_2(s) + A_2 R_2 sH_2(s) + H_2(s)$$

$$\Rightarrow \quad R_2 Q(s) = A_1 R_1 R_2 s\left[\frac{H_2(s)}{R_2} + A_2 sH_2(s)\right] + A_1 R_2 sH_2(s) + A_2 R_2 sH_2(s) + H_2(s)$$

$$\Rightarrow \quad R_2 Q(s) = A_1 R_1 sH_2(s) + A_1 R_1 A_2 R_2 s^2 H_2(s) + A_1 R_2 sH_2(s) + A_2 R_2 sH_2(s) + H_2(s)$$

$$\Rightarrow \quad R_2 Q(s) = [A_1 R_1 s + A_1 R_1 A_2 R_2 s^2 + A_1 R_2 s + A_2 R_2 s + 1]\, H_2(s)$$

$$\Rightarrow \quad \frac{H_2(s)}{Q(s)} = \frac{R_2}{A_1R_1A_2R_2s^2 + A_1R_1s + A_2R_2s + A_1R_2s + 1}$$

Let us assume

$$\tau_1 = A_1R_1$$

$$\tau_2 = A_2R_2$$

$$\Rightarrow \quad \frac{H_2(s)}{Q(s)} = \frac{R_2}{\tau_1\tau_2s^2 + (\tau_1 + \tau_2 + A_1R_2)s + 1} \qquad \text{...(8.37)}$$

This is the transfer function of two tank interacting system in series. It is observed from equation (8.37) that the overall transfer function is not just the multiplication of the individual transfer function as in the case of non-interacting systems. The difference between the transfer function of interacting and non-interacting systems are the coefficient A_1R_2 in the denominator of the transfer function of interacting system.

8.2 RESPONSE OF TWO TANK NON-INTERACTING SYSTEM IN SERIES FOR A UNIT STEP CHANGE IN INPUT

From the transfer function of two tank non-interacting system in series, we have

$$\frac{H_2(s)}{Q(s)} = \frac{R_2}{(\tau_1s+1)(\tau_2s+1)} \qquad \text{...(8.6)}$$

Let us consider that time constant of both the tanks are equal.

Therefore, $\tau_1 = \tau_2 = \tau$

$$\Rightarrow \quad \frac{H_2(s)}{Q(s)} = \frac{R_2}{(\tau s+1)(\tau s+1)} \qquad \text{...(8.38)}$$

For a unit step change in input, we have

$$Q(t) = 1$$

$$\Rightarrow \quad Q(s) = \frac{1}{s} \qquad \text{...(8.39)}$$

Substituting equation (8.39) in equation (8.38), we get

$$H_2(s) = \frac{R_2}{(s)(\tau s+1)(\tau s+1)}$$

$$\Rightarrow \quad H_2(s) = \frac{R_2/\tau^2}{(s)(s+1/\tau)(s+1/\tau)} = \frac{C_1}{s} + \frac{C_2}{(s+1/\tau)} + \frac{C_3}{(s+1/\tau)^2}$$

$$\Rightarrow \quad \frac{R_2}{\tau^2} = C_1(s+1/\tau)^2 + C_2(s)(s+1/\tau) + C_3(s) \qquad \text{...(8.40)}$$

Put, $s = 0$; $C_1 = R_2$

Put, $s = -1/\tau$; $C_3 = -R_2/\tau$

Comparing the coefficient of s^2, we get

$$0 = C_1 + C_2$$

$$0 = R_2 + C_2$$

$$C_2 = -R_2$$

$$\Rightarrow \qquad H_2(s) = \frac{R_2}{s} - \frac{R_2}{s+1/\tau} - \frac{R_2}{\tau(s+1/\tau)^2}$$

$$\Rightarrow \qquad H_2(t) = R_2\left[1 - e^{-t/\tau} - \frac{t}{\tau}e^{-t/\tau}\right] \qquad ...(8.41)$$

This is the response equation of two tank non-interacting system in series for a unit step change in input.

8.3 RESPONSE OF TWO TANK INTERACTING SYSTEM IN SERIES FOR A UNIT STEP CHANGE IN INPUT

From the transfer function of two tank interacting system in series, we have

$$\frac{H_2(s)}{Q(s)} = \frac{R_2}{\tau_1\tau_2 s^2 + (\tau_1 + \tau_2 + A_1 R_2)s + 1} \qquad ...(8.37)$$

Let us assume that time constant of both the tanks are equal. Let us also assume that both the tanks are of equal size and having equal resistance.

Therefore, $\tau_1 = \tau_2 = \tau$

$$\Rightarrow \qquad \frac{H_2(s)}{Q(s)} = \frac{R_2}{\tau^2 s^2 + 3\tau s + 1} \qquad ...(8.42)$$

For a unit step change in input, we have

$$Q(t) = 1$$

$$\Rightarrow \qquad Q(s) = \frac{1}{s} \qquad ...(8.39)$$

Substituting equation (8.39) in equation (8.42), we get

$$H_2(s) = \frac{R_2}{(s)(\tau^2 s^2 + 3\tau s + 1)} \qquad ...(8.43)$$

The denominator of equation (8.43) is a quadratic equation. It can be factorized as

$$\tau^2 s^2 + 3\tau s + 1 = 0$$

$$s = \frac{-3\tau \pm \sqrt{(3\tau)^2 - 4\tau^2}}{2\tau^2}$$

$$s = \frac{-3\tau \pm \sqrt{5\tau^2}}{2\tau^2}$$

$$s = \frac{-3\tau \pm \tau\sqrt{5}}{2\tau^2}$$

$$s = \frac{-3 \pm \sqrt{5}}{2\tau}$$

$$s = \frac{-3+\sqrt{5}}{2\tau} \quad \text{or} \quad \frac{-3-\sqrt{5}}{2\tau}$$

$$s = \frac{-0.382}{\tau} \quad \text{or} \quad \frac{-2.618}{\tau}$$

Therefore, the factors are

$$(0.382\,\tau s+1) \quad \text{and} \quad (2.618\,\tau s+1)$$

$$\Rightarrow \quad H_2(s) = \frac{R_2}{(s)(0.382\,\tau s+1)(2.618\,\tau s+1)} \qquad \ldots(8.44)$$

$$\Rightarrow \quad H_2(s) = \frac{R_2/\tau^2}{(s)(s+2.618/\tau)(s+0.382/\tau)} = \frac{C_1}{s} + \frac{C_2}{(s+2.618/\tau)} + \frac{C_3}{(s+0.382/\tau)}$$

$$\Rightarrow \quad \frac{R_2}{\tau^2} = C_1(s+2.618/\tau)(s+0.382/\tau) + C_2(s)(s+0.382/\tau) + C_3(s)(s+2.618/\tau) \qquad \ldots(8.45)$$

Put, $s = 0$; $C_1 = R_2$

Put, $s = -0.382/\tau$; $C_3 = -1.171R_2$

Put, $s = -2.618/\tau$; $C_2 = 0.171R_2$

$$\Rightarrow \quad H_2(s) = \frac{R_2}{s} + \frac{0.171R_2}{(s+2.618/\tau)} - \frac{1.171R_2}{(s+0.382/\tau)}$$

$$\Rightarrow \quad H_2(t) = R_2[1+0.171e^{-2.618t/\tau} - 1.171e^{-0.382t/\tau}] \qquad \ldots(8.46)$$

This is the response equation of two tank interacting system in series for a unit step change in input.

8.4 EFFECT OF INTERACTION ON STEP RESPONSE

The plot of $H_2(t)$ vs t/τ of two tank interacting and non-interacting system is shown in Fig. 8.4.

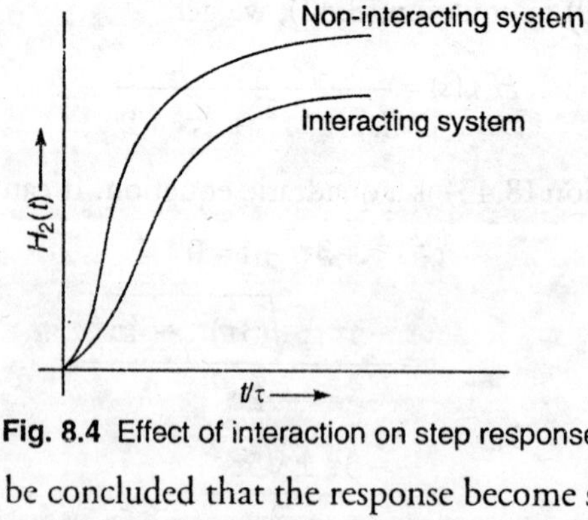

Fig. 8.4 Effect of interaction on step response.

From the Fig. 8.4 it can be concluded that the response become slower in the case of interacting system compared to response in the case of non-interacting system.

In general, the effect of interaction on a system containing two first order lags is to change the ratio of effective time constants in the interacting system. In terms of the transient response, this means that the interacting system is more sluggish than the non-interacting system.

■ **Example 8.1** Two non-interacting tanks are connected in series. The time constants are $\tau_1 = 1$ and $\tau_2 = 2$. Sketch the response of the tank level 2 if a unit step change is made in the inlet flow rate to tank 1.

Solution : From the transfer function of two tank non-interacting system in series, we have

$$\frac{H_2(s)}{Q(s)} = \frac{R_2}{(\tau_1 s+1)(\tau_2 s+1)} \quad \text{...(8.6)}$$

From the data given, we have

$$\tau_1 = 1 \text{ sec}$$

$$\tau_2 = 2 \text{ sec}$$

$$\Rightarrow \quad \frac{H_2(s)}{Q(s)} = \frac{R_2}{(s+1)(2s+1)} \quad \text{...(8.47)}$$

For a unit step change in input, we have

$$Q(t) = 1$$

$$Q(s) = \frac{1}{s} \quad \text{...(8.39)}$$

Substituting equation (8.39) in equation (8.47), we get

$$H_2(s) = \frac{R_2}{(s)(s+1)\,2(s+0.5)}$$

$$\Rightarrow \quad H_2(s) = \frac{R_2}{(s)(s+1)\,2(s+0.5)}$$

$$\Rightarrow \quad H_2(s) = \frac{0.5R_2}{(s)(s+1)(s+0.5)}$$

$$\Rightarrow \quad H_2(s) = \frac{0.5R_2}{(s)(s+1)(s+0.5)} = \frac{C_1}{s} + \frac{C_2}{(s+1)} + \frac{C_3}{(s+0.5)}$$

$$\Rightarrow \quad \frac{0.5R_2}{(s)(s+1)(s+0.5)} = \frac{C_1(s+1)(s+0.5) + C_2(s)(s+0.5) + C_3(s)(s+1)}{(s)(s+1)(s+0.5)}$$

$$\Rightarrow \quad 0.5R_2 = C_1(s+1)(s+0.5) + C_2(s)(s+0.5) + C_3(s)(s+1)$$

Put, $s = 0$; $C_1 = R_2$

Put, $s = -1$; $C_2 = R_2$

Put, $s = -0.5$; $C_3 = -2R_2$

$$\Rightarrow \quad H_2(s) = \frac{R_2}{s} + \frac{R_2}{(s+1)} - \frac{2R_2}{(s+0.5)}$$

$$\Rightarrow \quad H_2(t) = R_2 + R_2 e^{-t} - 2R_2 e^{-0.5t}$$

$$\Rightarrow \quad H_2(t) = R_2\,[1 + e^{-t} - 2e^{-0.5t}] \quad \text{...(8.48)}$$

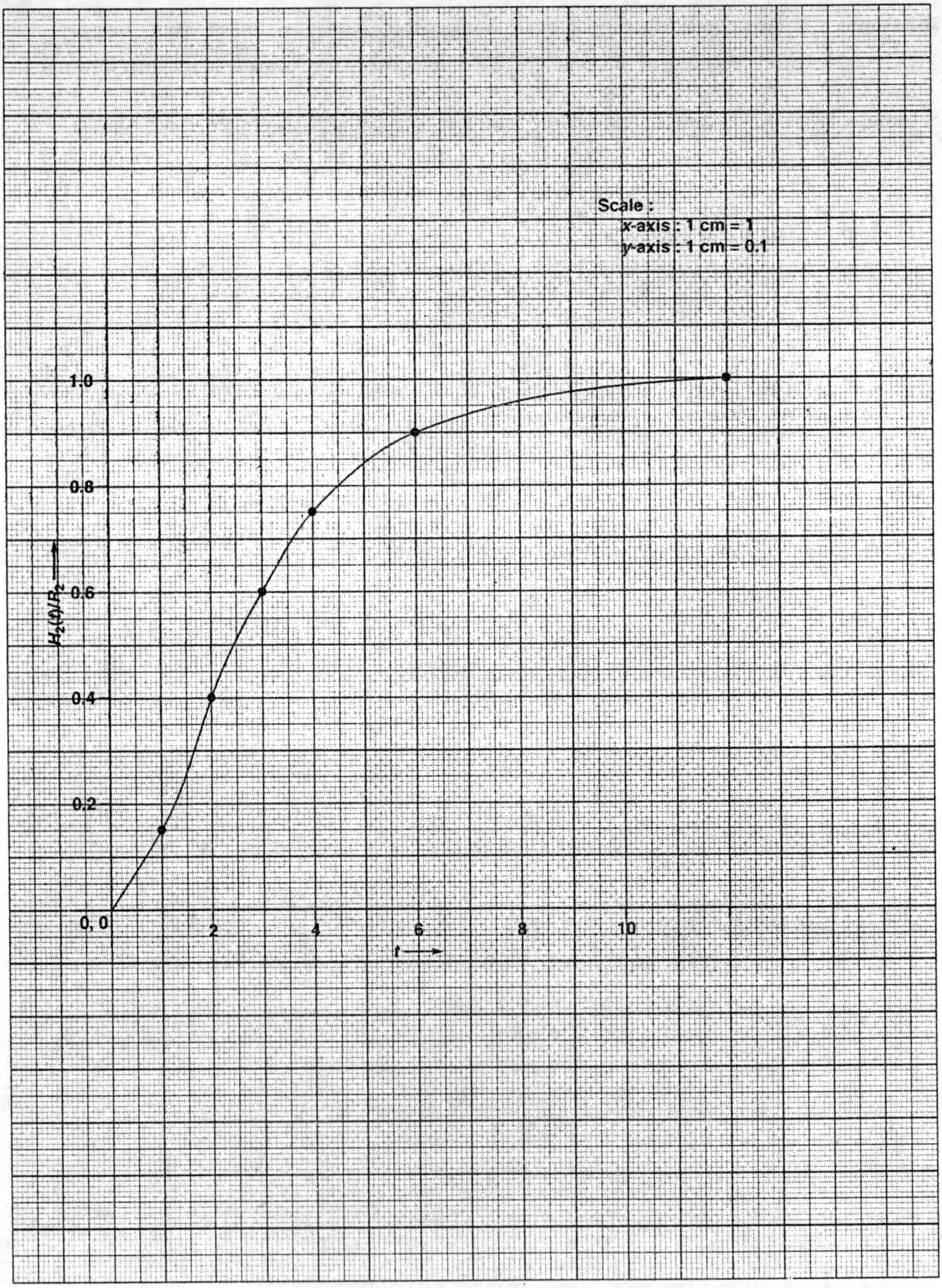

Fig. 8.5 Sketch of the response in tank 2.

Table 8.1 : Response of the level in tank 2.

t (sec)	$H_2(t)/R_2$
0	0
1	0.16
2	0.40
3	0.61
4	0.75
5	0.84
6	0.90
⋮	⋮
∞	1.00

■ **Example 8.2** The two tank liquid level system is operating at steady state. At time $t = 0$, $10\ \text{m}^3$ of water is quickly added to the first tank. Using appropriate figure and equation,. determine the maximum deviation in level of both tanks from the ultimate steady state values and the time at which each maximum occurs.

Data : $A_1 = 10\ \text{m}^2$

$A_2 = 10\ \text{m}^2$

$R_1 = 0.1\ \text{s/m}^2$

$R_2 = 0.35\ \text{s/m}^2$

Solution :

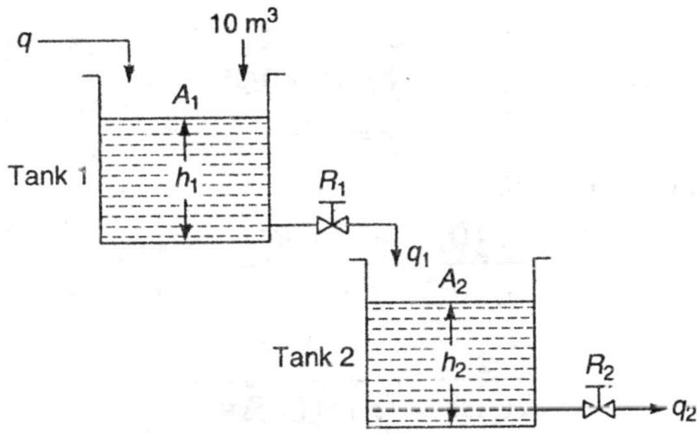

Fig. 8.6 Two tank liquid level system.

From the given data, we have

$$A_1 = 10\ \text{m}^2$$

$$R_1 = 0.1\ \text{s/m}^2$$

From the time constant of tank 1, we have

$$\tau_1 = A_1 R_1$$

$$\Rightarrow \quad \tau_1 = 10 \times 0.1$$

$$\Rightarrow \quad \tau_1 = 1\ \text{s}$$

From the transfer function of tank 1, we have

$$\frac{H_1(s)}{Q(s)} = \frac{R_1}{(\tau_1 s + 1)}$$

$$\Rightarrow \quad \frac{H_1(s)}{Q(s)} = \frac{0.1}{(s+1)} \qquad \text{...(8.49)}$$

From the given data, we have

$$Q(t) = 10\,\partial(t)$$

$$Q(s) = 10 \qquad \text{...(8.50)}$$

Substituting equation (8.50) in equation (8.49), we get

$$H_1(s) = \frac{0.1 \times 10}{(s+1)}$$

$$H_1(s) = \frac{1}{s+1}$$

$$\Rightarrow \quad H_1(t) = e^{-t} \qquad \text{...(8.51)}$$

The maximum deviation in level of tank 1 from the ultimate steady state value occurs at $t = 0$.

Therefore, $H_1(t) = 1\,\text{m}$

From the given data, we have

$$A_2 = 10\,\text{m}^2$$

$$R_2 = 0.35\,\text{s/m}^2$$

From the time constant of tank 2, we have

$$\tau_2 = A_2 R_2$$

$$\Rightarrow \quad \tau_2 = 10 \times 0.35$$

$$\Rightarrow \quad \tau_2 = 3.5\,\text{s}$$

From the transfer function of tank 2, we have

$$\frac{H_2(s)}{Q(s)} = \frac{R_2}{(\tau_1 s + 1)(\tau_2 s + 1)}$$

$$\Rightarrow \quad \frac{H_2(s)}{Q(s)} = \frac{0.35}{(s+1)(3.5s+1)}$$

$$\Rightarrow \quad \frac{H_2(s)}{Q(s)} = \frac{0.35}{(s+1)\,3.5(s+1/3.5)}$$

$$\Rightarrow \quad \frac{H_2(s)}{Q(s)} = \frac{0.1}{(s+1)(s+0.286)} \qquad \text{...(8.52)}$$

From the given data, we have

$$Q(t) = 10\,\partial(t)$$

$$\Rightarrow \quad Q(s) = 10 \qquad \text{...(8.50)}$$

Substituting equation (8.50) in equation (8.52), we get

$$H_2(s) = \frac{1}{(s+1)(s+0.286)} \qquad \ldots(8.53)$$

$$H_2(s) = \frac{1}{(s+1)(s+0.286)} = \frac{C_1}{(s+1)} + \frac{C_2}{(s+0.286)}$$

$$\frac{1}{(s+1)(s+0.286)} = \frac{C_1(s+0.286) + C_2(s+1)}{(s+1)(s+0.286)}$$

$$\Rightarrow \quad 1 = C_1(s+0.286) + C_2(s+1)$$

Put, $s = -0.286$; $C_2 = 1.4$

Put, $s = -1$; $C_1 = -1.4$

$$\Rightarrow \quad H_2(s) = \frac{-1.4}{(s+1)} + \frac{1.4}{(s+0.286)}$$

$$\Rightarrow \quad H_2(t) = -1.4e^{-t} + 1.4e^{-0.286t}$$

$$\Rightarrow \quad H_2(t) = 1.4e^{-0.286t} - 1.4e^{-t} \qquad \ldots(8.54)$$

Differentiating equation (8.54) w.r.t. 't' and equalizing to zero, we get

$$\frac{dH_2(t)}{dt} = 1.4\,(e^{-0.286t})(-0.286) - 1.4(e^{-t})(-1) = 0$$

$$e^{-t} = 0.286e^{-0.286t} \qquad \ldots(8.55)$$

Taking ln on both side of equation (8.55), we get

$$-t = \ln(0.286) - 0.286t$$

$$-0.714t = \ln(0.286)$$

$$\Rightarrow \quad t = 1.75 \text{ min}$$

Substituting the value of 't' in equation (8.54), we get

$$H_2(t) = 1.4\,[e^{-0.286 \times 1.75} - e^{-1.75}]$$

$$\Rightarrow \quad H_2(t) = 0.61 \text{ m}$$

■ **Example 8.3** In a process control laboratory, experiment is conducted on two tank non-interacting system connected in series and following readings are recorded.

Level 'h_1' (cm)	Level 'h_2' (cm)	Volumetric flow rate (m^3/min)
8.6	6.4	12×10^{-3}
9.6	7.6	15×10^{-3}
10.4	8.6	18×10^{-3}
11.4	9.8	21×10^{-3}
12.6	10.8	24×10^{-3}
13.6	12.0	27×10^{-3}

Data : $A_1 = 0.135\,\text{m}^2$

$A_2 = 0.135\,\text{m}^2$

Calculate the time constant of both tanks.

Solution :

Table 8.2 : Experimental data of TTNIS in series.

Level 'h_1' (m)	Level 'h_2' (m)	Volumetric flow rate (m^3/min)
8.6×10^{-2}	6.4×10^{-2}	12×10^{-3}
9.6×10^{-2}	7.6×10^{-2}	15×10^{-3}
10.4×10^{-2}	8.6×10^{-2}	18×10^{-3}
11.4×10^{-2}	9.8×10^{-2}	21×10^{-3}
12.6×10^{-2}	10.8×10^{-2}	24×10^{-3}
13.6×10^{-2}	12.0×10^{-2}	27×10^{-3}

From the given data, we have

$$A_1 = 0.135\,\text{m}^2$$

From the graph, we have

$$R_1 = 3.246\,\text{min/m}^2$$

From the time constant of tank 1, we have

$$\tau_1 = A_1 R_1$$

$$\Rightarrow \quad \tau_1 = 0.135 \times 3.246$$

$$\Rightarrow \quad \tau_1 = 0.438\,\text{min}$$

$$\Rightarrow \quad \tau_1 = 26\,\text{s}$$

From the given data, we have

$$A_2 = 0.135\,\text{m}^2$$

From the graph, we have

$$R_2 = 3.663\,\text{min/m}^2$$

From the time constant of tank 2, we have

$$\tau_2 = A_2 R_2$$

$$\Rightarrow \quad \tau_2 = 0.135 \times 3.663$$

$$\Rightarrow \quad \tau_2 = 0.425\,\text{min}$$

$$\Rightarrow \quad \tau_2 = 30\,\text{s}$$

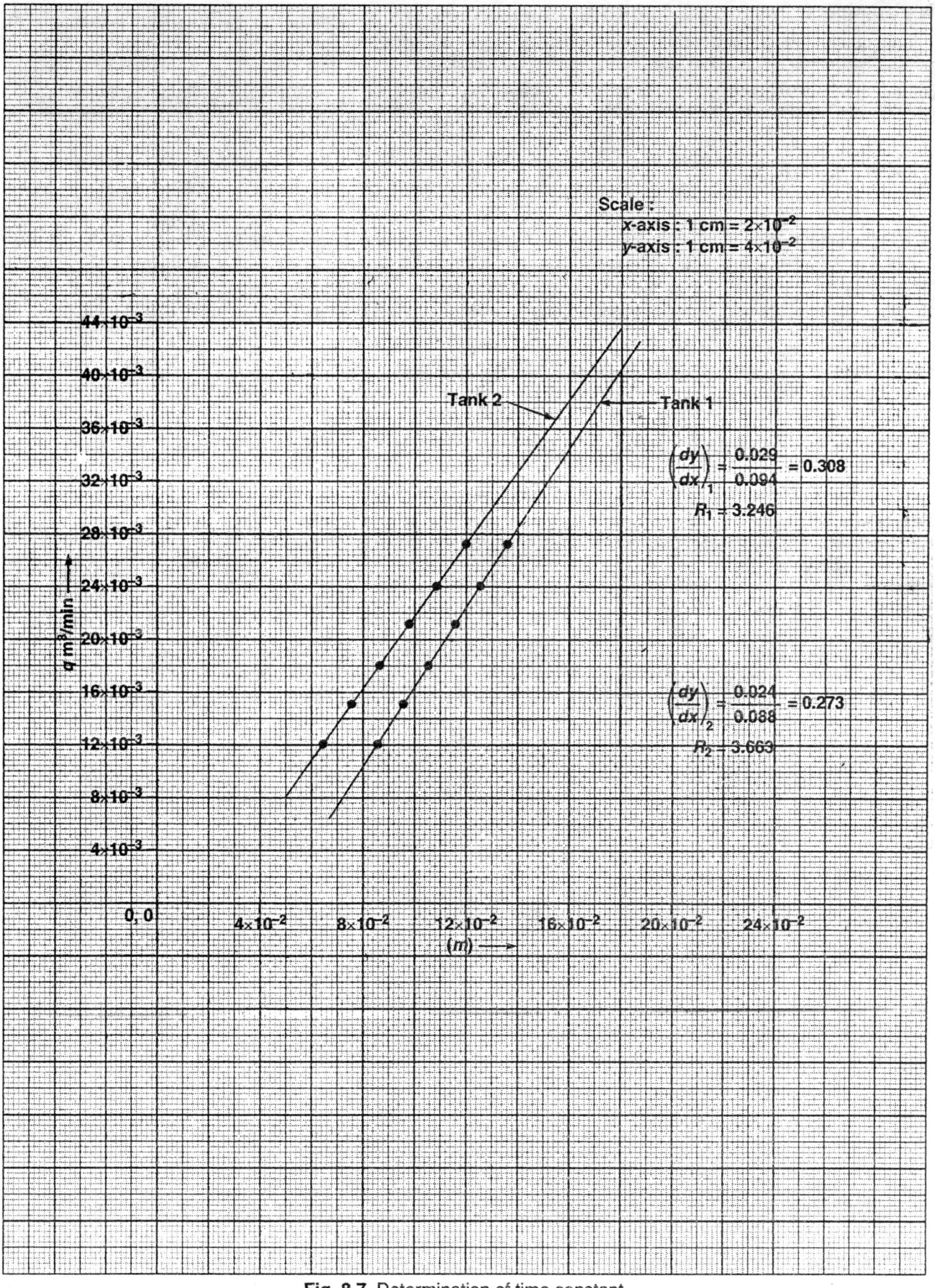

Fig. 8.7 Determination of time constant.

■ **Example 8.4** In a process control laboratory experiment is conducted on two tank interacting system connected in series and the following readings are recorded.

Level 'h_1' (cm)	Level 'h_2' (cm)	Volumetric flow rate (m^3/min)
9.2	8.6	12×10^{-3}
10.6	9.4	15×10^{-3}
12.2	10.5	18×10^{-3}
13.6	11.4	21×10^{-3}
15.0	12.6	24×10^{-3}
16.4	13.6	27×10^{-3}

Data : $A_1 = 0.125\,m^2$

$A_2 = 0.125\,m^2$

Calculate the time constant of both tanks.

Solution :

Level 'h_1' (m)	Level 'h_2' (m)	Volumetric flow rate (m^3/min)
9.2×10^{-2}	8.6×10^{-2}	12×10^{-3}
10.6×10^{-2}	9.4×10^{-2}	15×10^{-3}
12.2×10^{-2}	10.5×10^{-2}	18×10^{-3}
13.6×10^{-2}	11.4×10^{-2}	21×10^{-3}
15.0×10^{-2}	12.6×10^{-2}	24×10^{-3}
16.4×10^{-2}	13.6×10^{-2}	27×10^{-3}

From the given data, we have

$$A_1 = 0.126\,m^2$$

From the graph, we have

$$R_1 = 4.926\,\text{min}/m^2$$

From the time constant of tank 1, we have

$$\tau_1 = A_1 R_1$$

$$\Rightarrow \quad \tau_1 = 0.126 \times 4.926$$

$$\Rightarrow \quad \tau_1 = 0.62\,\text{min}$$

$$\Rightarrow \quad \tau_1 = 37\,\text{s}$$

From the given data, we have

$$A_2 = 0.126\,m^2$$

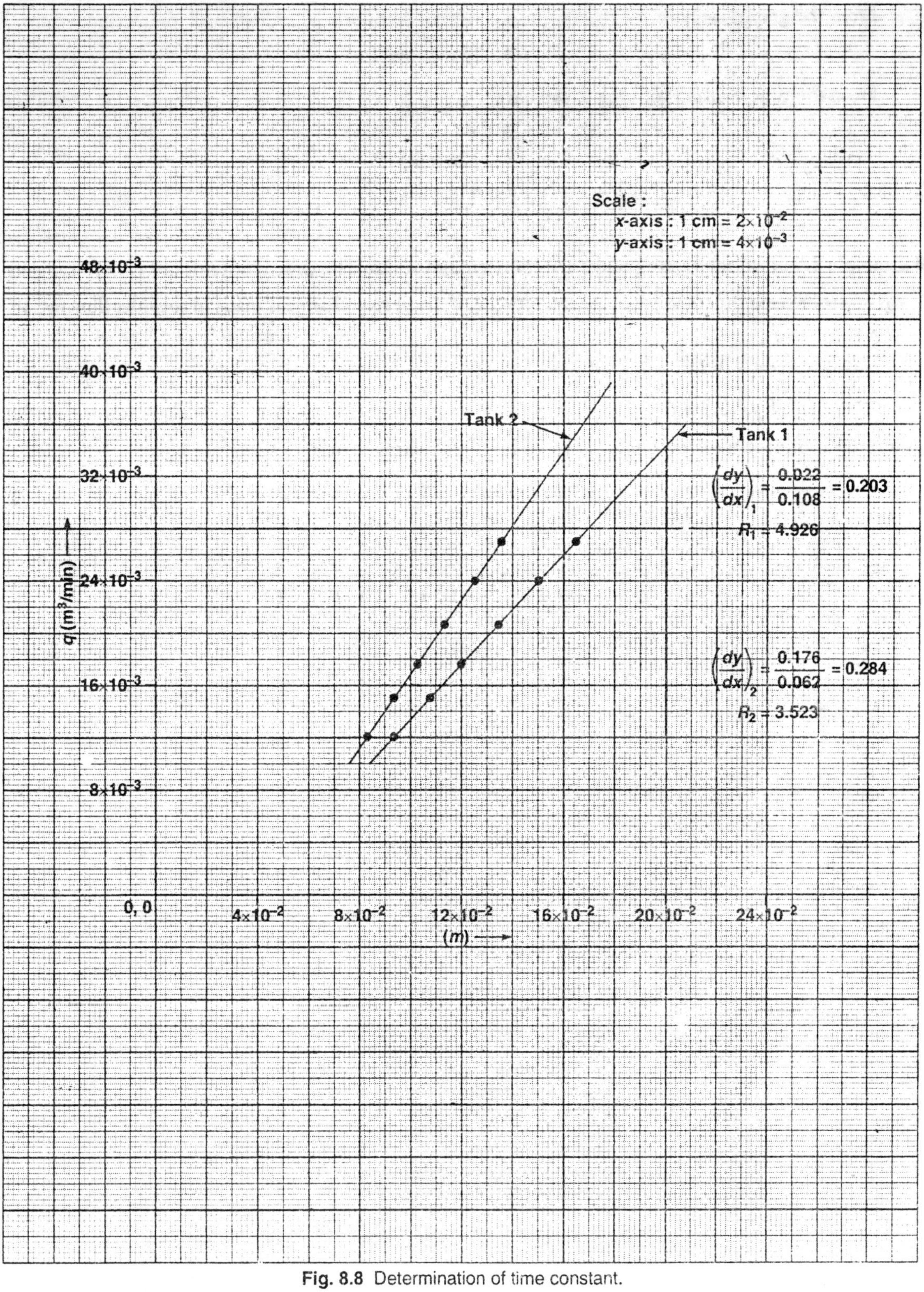

Fig. 8.8 Determination of time constant.

From the graph, we have

$$R_2 = 3.523 \text{ min/m}^2$$

From the time constant of tank 2, we have

$$\tau_2 = A_2 R_2$$

$$\Rightarrow \quad \tau_2 = 0.126 \times 3.523$$

$$\Rightarrow \quad \tau_2 = 0.444 \text{ min}$$

$$\Rightarrow \quad \tau_2 = 27 \text{ s}$$

■ **Example 8.5** Two interacting tanks are connected in series. The time constants are $\tau_1 = 10$ sec and $\tau_2 = 5$ sec. Sketch the response of the tank level 2, if a unit step change is made in the inlet flow rate to tank 1.

Data: $A_1 = 10 \text{ m}^2$

$R_2 = 0.5 \text{ s/m}^2$

Solution : From the transfer function of two tank interacting system in series, we have

$$\frac{H_2(s)}{Q(s)} = \frac{R_2}{\tau_1 \tau_2 s^2 + (\tau_1 + \tau_2 + A_1 R_2) s + 1} \qquad \ldots(8.37)$$

$$\Rightarrow \quad \frac{H_2(s)}{Q(s)} = \frac{0.5}{(10 \times 5) s^2 + (10 + 5 + 10 \times 0.5) s + 1}$$

$$\Rightarrow \quad \frac{H_2(s)}{Q(s)} = \frac{0.5}{50 s^2 + 20 s + 1} \qquad \ldots(8.56)$$

From the data given, we have

$$Q(t) = 1$$

$$Q(s) = \frac{1}{s} \qquad \ldots(8.39)$$

Substituting equation (8.39) in equation (8.56), we get

$$H_2(s) = \frac{0.5}{(s)(50 s^2 + 20 s + 1)} \qquad \ldots(8.57)$$

Factorizing the denominator of equation (8.57), we get

$$50 s^2 + 20 s + 1 = 0$$

$$s = \frac{-20 \pm \sqrt{(20)^2 - 4 \times 50 \times 1}}{2 \times 50}$$

$$s = \frac{-20 \pm \sqrt{400 - 200}}{100}$$

$$s = \frac{-20 \pm 10\sqrt{2}}{100}$$

$$s = \frac{-20 \pm \sqrt{2}}{10}$$

$$s = \frac{-2+\sqrt{2}}{10}, \quad \frac{-2-\sqrt{2}}{10}$$

$$s = -0.0585, \quad -0.3417$$

Therefore, the factors are

$$(17.092\, s + 1) = 0 \quad \text{and} \quad (2.926\, s + 1) = 0$$

$$\Rightarrow \quad H_2(s) = \frac{0.5}{(s)(s+0.0585)(s+0.3417)}$$

$$\Rightarrow \quad H_2(s) = \frac{0.5}{(s)(s+0.0585)(s+0.3417)} = \frac{C_1}{s} + \frac{C_2}{(s+0.0585)} + \frac{C_3}{(s+0.3417)} \quad \text{...(8.58)}$$

$$\frac{0.5}{(s)(s+0.0585)(s+0.3417)}$$

$$= \frac{C_1(s+0.0585)(s+0.3417) + C_2(s)(s+0.3417) + C_3(s)(s+0.0585)}{(s)(s+0.0585)(s+0.3417)}$$

$$0.5 = C_1(s+0.0585)(s+0.3417) + C_2(s)(s+0.3417) + C_3(s)(s+0.0585)$$

Put, $s = 0$; $C_1 = 25$

Put, $s = -0.0585$; $C_2 = -30.18$

Put, $s = -0.3417$; $C_3 = 5.16$

$$\Rightarrow \quad H_2(s) = \frac{25}{s} - \frac{30.18}{(s+0.0585)} + \frac{5.16}{(s+0.3417)}$$

$$\Rightarrow \quad H_2(t) = 25 - 30.18e^{-0.0585t} + 5.16e^{-0.3417t}$$

$$\Rightarrow \quad H_2(t) = 25[1 - 1.2e^{-0.0585t} + 0.206e^{-0.3417t}]$$

REVIEW QUESTIONS

8.1 With a neat block diagram explain the working of two tank non-interacting system connected in series.

8.2 Discuss on the application of two tank non-interacting system in industry.

8.3 With a neat block diagram explain the working of two tank interacting system connected in series.

8.4 Discuss on the application of two tank interacting system in industry.

8.5 What do you mean by non-interacting system? Explain.

8.6 What do you mean by interacting system? Explain.

8.7 In two tank non-interacting system $\tau_1 < \tau_2$ or $\tau_1 > \tau_2$? Explain.

8.8 In two tank interacting system $\tau_1 > \tau_2$ or $\tau_1 < \tau_2$? Explain.

8.9 Explain the effect of interaction on step response.

EXERCISES

8.1 Derive the transfer function for two tank non-interacting system.

8.2 Derive the transfer function for two tank interacting system.

8.3 Derive the over all transfer function for '*n*' non-interacting system connected in series.

8.4 Derive the response equation for two tank non-interacting system in series for a unit step change in input.

8.5 Derive the response equation for two tank non-interacting system in series for a unit impulse change in input.

8.6 Derive the response equation for two tank interacting system in series for a unit step change in input.

8.7 Derive the response equation for two tank interacting system in series for a unit impulse change in input.

8.8 Two non-interacting tanks are connected in series. The time constants are $\tau_1 = 2$ sec and $\tau_2 = 4$ sec. Sketch the response of the tank level 2 if a unit step change is made in the inlet flow rate to tank 1.

8.9 The two tank non-interacting liquid level system is operating at steady state. At time $t = 0, 5\ m^3$ of water is quickly added to the first tank. Using appropriate figure and equation, determine the maximum deviation in level of both tanks from the ultimate steady state values and the time at which each maximum occurs.

Data : $A_1 = 8\ m^2$

$A_2 = 8\ m^2$

$R_1 = 0.1\ s/m^2$

$R_2 = 0.25\ s/m^2$

8.10 In a process control laboratory, experiment is conducted on two tank non-interacting system connected in series and following readings are recorded.

Level 'h_1' (cm)	Level 'h_2' (cm)	Volumetric flow rate (m^3/min)
9	6	12×10^{-3}
10	8	15×10^{-3}
11	9	18×10^{-3}
12	10	21×10^{-3}
13	11	24×10^{-3}
14	12	27×10^{-3}

Data : $A_1 = 0.13\ m^2$

$A_2 = 0.13\ m^2$

Calculate the time constant of tank 1 and tank 2.

8.11 In a process control laboratory experiment is conducted on two tank interacting system connected in series and the following readings are recorded.

Level 'h_1' (cm)	Level 'h_2' (cm)	Volumetric flow rate (m^3/min)
9	8	12×10^{-3}
11	9	15×10^{-3}
12	10	18×10^{-3}
14	12	21×10^{-3}
15	13	24×10^{-3}
17	14	27×10^{-3}

Data : $A_1 = 0.12\ m^2$

$A_2 = 0.12\ m^2$

Calculate the time constant of both tanks.

8.12 Two interacting tanks are connected in series. The time constants are $\tau_1 = 8$ sec and $\tau_2 = 4$ sec. Sketch the response of the tank level 2 if a unit step change is made in the inlet flow rate to tank 1.

□□□

Chapter 9

Second Order System

9.1 INTRODUCTION

A system which can be represented by a linear second order differential equation is called second order system. The examples of second order system are :

(i) Damped vibrator

(ii) U-tube manometer

System with second or higher order dynamics can arise from several physical situations. For examples multi-capacity processes, inherently second order system and a processing system with its controller.

The majority of the second or higher order systems encountered in a chemical plant comes from multi-capacity processes. Very rarely we will find systems with appreciable inherent or second or higher order dynamics.

9.1.1 Damped Vibrator

It consists of a block or mass '*M*' resting on a horizontal friction less table attached to a linear spring and a viscous damper. A typical damped vibrator is shown in Fig. 9.1.

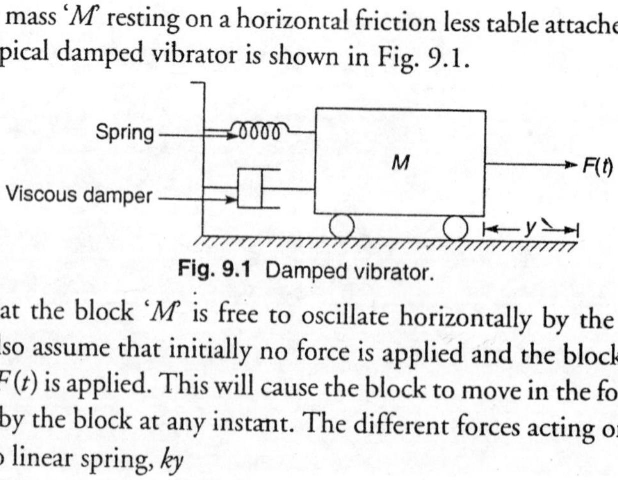

Fig. 9.1 Damped vibrator.

Let us assume that the block '*M*' is free to oscillate horizontally by the application of forcing function $F(t)$. Let us also assume that initially no force is applied and the block '*M*' is at rest. At time $t = 0$, an external force $F(t)$ is applied. This will cause the block to move in the forward direction. Let '*y*' be the distance moved by the block at any instant. The different forces acting on the block are :

(i) Force due to linear spring, ky

(ii) Viscous ffiction force of the damper, $C\,\dfrac{dy}{dt}$

(iii) The external force, $F(t)$.

From the Newton's second law of motion, we know that net forces acting on a mass is equal to the rate of change momentum.

$$\Rightarrow \quad M \times \frac{d^2 y}{dt^2} = F(t) - ky - C\frac{dy}{dt} \quad \ldots(9.1)$$

$$\Rightarrow \quad M \times \frac{d^2 y}{dt^2} + C\frac{dy}{dt} + ky = F(t)$$

$$\Rightarrow \quad \frac{M}{k}\frac{d^2 y}{dt^2} + \frac{C}{k}\frac{dy}{dt} + y = \frac{F(t)}{k} \quad \ldots(9.2)$$

Let,

$$\tau^2 = \frac{M}{k}$$

$$2\tau\xi = \frac{C}{k}$$

$$k' X(t) = \frac{F(t)}{k}$$

$$\Rightarrow \quad \tau^2 \frac{d^2 y}{dt^2} + 2\tau\xi\frac{dy}{dt} + y = k' X(t) \quad \ldots(9.3)$$

Taking Laplace transform of equation (9.3), we get

$$\tau^2 s^2 Y(s) + 2\tau\xi\, sY(s) + Y(s) = k' X(s)$$

$$Y(s)[\tau^2 s^2 + 2\tau\xi s + 1] = k' X(s)$$

$$\frac{Y(s)}{X(s)} = \frac{k'}{\tau^2 s^2 + 2\tau\xi s + 1} \quad \ldots(9.4)$$

where, $Y(s)$ = Output in deviation form
$X(s)$ = Input in deviation form
k' = Steady state gain
τ = Time constant of damped vibrator
ξ = Damping coefficient

This is the transfer function of damped vibrator.

9.1.2 U-Tube Manometer

It is a pressure measuring device. It consists of a U-shaped glass tube in which manometric fluid is present. The one limb of the manometer is connected to a pressure chamber at a pressure p_1 and the other limb is open to atmosphere at a pressure p_2. A typical U-tube manometer is shown in Fig. 9.2.

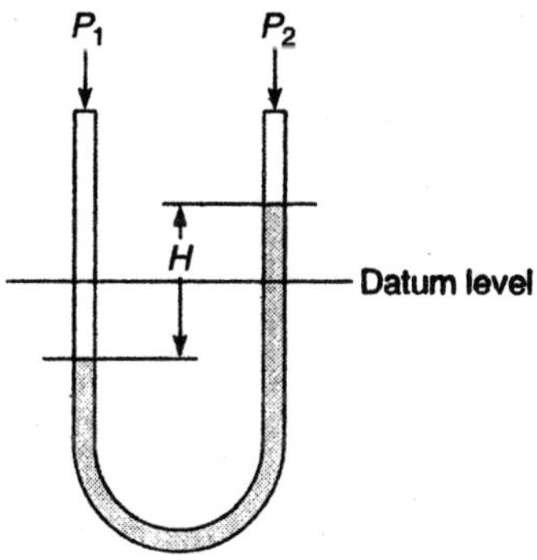

Fig. 9.2 U-tube manometer.

Let L = Total length of the liquid column
d = Inner diameter of the manometer tube
ρ = Density of the manometric fluid
μ = Viscosity of the manometric fluid
A = Cross-sectional area of the tube

The following assumptions are made for the development of transfer function :

(i) Flow in the manometer is laminar.

Therefore, $$R = \frac{\Delta P_f}{V} \quad \text{...(9.5)}$$

where, R = Frictional resistance
ΔP_f = Change in pressure
V = Velocity of fluid

From the Haugen Poiseuille's equation for laminar flow, we have

$$\Delta P_f = \frac{32\mu LV}{D^2} \quad \text{...(9.6)}$$

$$\Rightarrow \quad R = \frac{\Delta P_f}{V} = \frac{32\mu L}{D^2} \quad \text{...(9.7)}$$

(ii) Velocity profile is flat. Each molecule moved with same velocity.

(iii) The density of the fluid above the manometric fluid is negligible.

The different forces acting on the manometric fluid are :

(i) Force due to differential pressure, PA

(ii) Force due to gravity, $AH\rho g$

(iii) Frictional drag force, $RAV = \frac{32\mu L}{D^2} \times A \times \frac{d(H/2)}{dt}$

From the Newton's second law of motion, we know that net forces acting on a mass is equal to the rate of change of momentum.

$$\Rightarrow \quad (AL\rho) \times \frac{d^2(H/2)}{dt^2} = PA - \left(\frac{16\mu LA}{D^2}\right)\frac{dH}{dt} - AH\rho g \quad \text{...(9.8)}$$

$$\Rightarrow \quad \left(\frac{AL\rho}{2}\right)\frac{d^2H}{dt^2} + \left(\frac{16\mu LA}{D^2}\right)\frac{dH}{dt} + AH\rho g = PA$$

$$\Rightarrow \quad \left(\frac{L\rho}{2}\right)\frac{d^2H}{dt^2} + \left(\frac{16\mu L}{D^2}\right)\frac{dH}{dt} + H\rho g = P \quad \text{...(9.9)}$$

Dividing equation (9.9) by ρg, we get

$$\left(\frac{L}{2g}\right)\frac{d^2H}{dt^2} + \left(\frac{16\mu L}{\rho g D^2}\right)\frac{dH}{dt} + H = \frac{P}{\rho g}$$

Let, $$\tau^2 = \frac{L}{2g}$$

$$2\tau\xi = \frac{16\mu L}{\rho g D^2}$$

$$k' = \frac{1}{\rho g}$$

$$\Rightarrow \qquad \tau^2 \frac{d^2H}{dt^2} + 2\tau\xi \frac{dH}{dt} + H = k' P \qquad \text{...(9.10)}$$

Taking Laplace transform of equation (9.10), we get

$$\tau^2 s^2 \; H(s) + 2\tau\xi s H(s) + H(s) = k' P(s)$$

$$H(s)\,[\tau^2 s^2 + 2\tau\xi s + 1] = k' P(s)$$

$$\frac{H(s)}{P(s)} = \frac{k'}{(\tau^2 s^2 + 2\tau\xi s + 1)} \qquad \text{...(9.11)}$$

where, $H(s)$ = Output in deviation form
$P(s)$ = Input in deviation form
k' = Steady state gain
τ = Time constant of manometer
ξ = Damping coefficient

This is the transfer function of U-tube manometer.

9.1.2.1 Response of U-Tube Manometer for a Unit Step Change in Input

From the transfer function of U-tube manometer, we have

$$\frac{H(s)}{P(s)} = \frac{k'}{(\tau^2 s^2 + 2\tau\xi s + 1)} \qquad \text{...(9.11)}$$

Let $\qquad k' = 1$

$$\Rightarrow \qquad \frac{H(s)}{P(s)} = \frac{1}{(\tau^2 s^2 + 2\tau\xi s + 1)} \qquad \text{...(9.12)}$$

$$\Rightarrow \qquad \frac{H(s)}{P(s)} = \frac{1}{\tau^2\left(s^2 + \frac{2\xi s}{\tau} + \frac{1}{\tau^2}\right)} \qquad \text{...(9.13)}$$

$$\Rightarrow \qquad \frac{H(s)}{P(s)} = \frac{1/\tau^2}{\left(s^2 + \frac{2\xi s}{\tau} + \frac{1}{\tau^2}\right)} \qquad \text{...(9.14)}$$

Let us consider $(s - P_1)$ and $(s - P_2)$ are the factors of the denominator of equation (9.14).

$$\Rightarrow \qquad \frac{H(s)}{P(s)} = \frac{1/\tau^2}{(s - P_1)(s - P_2)} \qquad \text{...(9.15)}$$

From the given data, we have

$$P(t) = 1$$

$$P(s) = \frac{1}{s}$$

$$\Rightarrow \qquad H(s) = \frac{1/\tau^2}{(s)(s - P_1)(s - P_2)}$$

$$\Rightarrow \qquad H(s) = \frac{1/\tau^2}{(s)(s-P_1)(s-P_2)} = \frac{C_1}{s} + \frac{C_2}{s-P_1} + \frac{C_3}{s-P_2} \qquad \ldots(9.16)$$

$$\Rightarrow \qquad \frac{1/\tau^2}{(s)(s-P_1)(s-P_2)} = \frac{C_1(s-P_1)(s-P_2) + C_2(s)(s-P_2) + C_3(s)(s-P_1)}{(s)(s-P_1)(s-P_2)}$$

$$\Rightarrow \qquad \frac{1}{\tau^2} = C_1(s-P_1)(s-P_2) + C_2(s)(s-P_2) + C_3(s)(s-P_1)$$

Put, $s = 0 \quad ; \quad C_1 = \dfrac{1}{\tau^2 P_1 P_2}$

Put, $s = P_1 \quad ; \quad C_2 = \dfrac{1}{\tau^2 (P_1)(P_1 - P_2)}$

Put, $s = P_2 \quad ; \quad C_3 = \dfrac{1}{\tau^2 (P_2)(P_2 - P_1)}$

$$\Rightarrow \qquad H(s) = \frac{1}{\tau^2 P_1 P_2 s} + \frac{1}{\tau^2 P_1 (P_1 - P_2)(s - P_1)} + \frac{1}{\tau^2 P_2 (P_2 - P_1)(s - P_2)} \qquad \ldots(9.17)$$

Solving the denominator of equation (9.14), we get

$$s^2 + \frac{2\xi s}{\tau} + \frac{1}{\tau^2} = 0$$

$$a = 1$$

$$b = \frac{2\xi}{\tau}$$

$$c = \frac{1}{\tau^2}$$

$$s = \frac{-b \pm \sqrt{b^2 - 4ac}}{2a}$$

$$\Rightarrow \qquad s = \frac{-2\xi/\tau \pm \sqrt{(2\xi/\tau)^2 - 4 \times 1 \times (1/\tau^2)}}{2 \times 1}$$

$$\Rightarrow \qquad s = \frac{-2\xi/\tau \pm \sqrt{4\xi^2/\tau^2 - 4/\tau^2}}{2}$$

$$\Rightarrow \qquad s = \frac{-2\xi/\tau \pm \sqrt{4/\tau^2 (\xi^2 - 1)}}{2}$$

$$\Rightarrow \qquad s = \frac{-2\xi/\tau \pm 2/\tau \sqrt{(\xi^2 - 1)}}{2}$$

$$\Rightarrow \qquad s = \frac{-\xi + \sqrt{(\xi^2 - 1)}}{\tau}; \quad \frac{-\xi - \sqrt{(\xi^2 - 1)}}{\tau}$$

$$P_1 = \frac{-\xi + \sqrt{(\xi^2 - 1)}}{\tau} \qquad \ldots(9.18)$$

$$P_2 = \frac{-\xi - \sqrt{(\xi^2 - 1)}}{\tau} \qquad \ldots(9.19)$$

From equation (9.18) and (9.19), we get

$$P_1 P_2 = \frac{-\xi + \sqrt{(\xi^2 - 1)}}{\tau} \times \frac{-\xi - \sqrt{(\xi^2 - 1)}}{\tau}$$

$$\Rightarrow \quad P_1 P_2 = \frac{(-\xi)^2 - (\sqrt{(\xi^2 - 1)})^2}{\tau^2}$$

$$\Rightarrow \quad P_1 P_2 = \frac{\xi^2 - (\xi^2 - 1)}{\tau^2}$$

$$\Rightarrow \quad P_1 P_2 = \frac{1}{\tau^2} \qquad \ldots(9.20)$$

From the equation (9.18) and (9.19), we get

$$P_1 (P_1 - P_2) = \left[\frac{-\xi + \sqrt{(\xi^2 - 1)}}{\tau}\right]\left[\left(\frac{-\xi + \sqrt{(\xi^2 - 1)}}{\tau}\right) - \left(\frac{-\xi - \sqrt{(\xi^2 - 1)}}{\tau}\right)\right]$$

$$P_1 (P_1 - P_2) = \left[\frac{-\xi + \sqrt{(\xi^2 - 1)}}{\tau}\right]\left[\frac{-\xi + (\xi^2 - 1) + \xi + (\xi^2 - 1)}{\tau}\right]$$

$$P_1 (P_1 - P_2) = \frac{(2\sqrt{(\xi^2 - 1)})\,(-\xi + \sqrt{(\xi^2 - 1)})}{\tau^2} \qquad \ldots(9.21)$$

From the equation (9.18) and (9.19), we get

$$P_2 (P_2 - P_1) = \left[\frac{-\xi - \sqrt{(\xi^2 - 1)}}{\tau}\right]\left[\left(\frac{-\xi - \sqrt{(\xi^2 - 1)}}{\tau}\right) - \left(\frac{-\xi + \sqrt{(\xi^2 - 1)}}{\tau}\right)\right]$$

$$P_2 (P_2 - P_1) = \left[\frac{-\xi - \sqrt{(\xi^2 - 1)}}{\tau}\right]\left[\frac{-\xi - \sqrt{(\xi^2 - 1)} + \xi + \sqrt{(\xi^2 - 1)}}{\tau}\right]$$

$$P_2 (P_2 - P_1) = \frac{(2\sqrt{(\xi^2 - 1)})\,(\xi + \sqrt{(\xi^2 - 1)})}{\tau^2} \qquad \ldots(9.22)$$

Substituting equations (9.20), (9.21) and (9.22) in equation (9.17), we get

$$H(s) = \frac{1}{s} + \frac{1}{2\sqrt{(\xi^2 - 1)}\,(-\xi + \sqrt{(\xi^2 - 1)})\,(s - P_1)} + \frac{1}{2\sqrt{(\xi^2 - 1)}\,(\xi + \sqrt{(\xi^2 - 1)})\,(s - P_2)}$$

$$\ldots(9.23)$$

Taking Laplace inverse of equation (9.23), we get

$$H(t) = 1 + \frac{e^{P_1 t}}{2\sqrt{(\xi^2-1)}\,(-\xi+\sqrt{(\xi^2-1)})} + \frac{e^{P_2 t}}{2\sqrt{(\xi^2-1)}\,(\xi+\sqrt{(\xi^2-1)})} \quad \ldots(9.24)$$

Substituting equations (9.18) and (9.19) in equation (9.24), we get

$$H(t) = 1 + \frac{e^{(-\xi+\sqrt{(\xi^2-1)})t/\tau}}{2\sqrt{(\xi^2-1)}\,(-\xi+\sqrt{(\xi^2-1)})} + \frac{e^{(-\xi-\sqrt{(\xi^2-1)})t/\tau}}{2\sqrt{(\xi^2-1)}\,(\xi+\sqrt{(\xi^2-1)})} \quad \ldots(9.25)$$

$$H(t) = 1 + \frac{e^{-\xi t/\tau}.e^{(\sqrt{(\xi^2-1)})t/\tau}}{2\sqrt{(\xi^2-1)}\,(-\xi+\sqrt{(\xi^2-1)})} + \frac{e^{-\xi t/\tau}.e^{-(\sqrt{(\xi^2-1)})t/\tau}}{2\sqrt{(\xi^2-1)}\,(\xi+\sqrt{(\xi^2-1)})}$$

$$H(t) = 1 + \frac{e^{-\xi t/\tau}}{2\sqrt{(\xi^2-1)}}\left[\frac{e^{(\sqrt{(\xi^2-1)})t/\tau}}{(-\xi+\sqrt{(\xi^2-1)})} + \frac{e^{-(\sqrt{(\xi^2-1)})t/\tau}}{(\xi+\sqrt{(\xi^2-1)})}\right]$$

$$H(t) = 1 + \frac{e^{-\xi t/\tau}}{2\sqrt{(\xi^2-1)}}\left[\frac{(\xi+\sqrt{(\xi^2-1)})\,e^{(\sqrt{(\xi^2-1)})t/\tau} + (-\xi+\sqrt{(\xi^2-1)})\,e^{-(\sqrt{(\xi^2-1)})t/\tau}}{(-\xi+\sqrt{(\xi^2-1)})\,(\xi+\sqrt{(\xi^1-1)})}\right]$$

$$H(t) = 1 + \frac{e^{-\xi t/\tau}}{2\sqrt{(\xi^2-1)}}$$

$$\left[\frac{\xi\{e^{(\sqrt{(\xi^2-1)})t/\tau} - e^{-(\sqrt{(\xi^2-1)})t/\tau}\} + \sqrt{(\xi^2-1)}\,\{e^{(\sqrt{(\xi^2-1)})t/\tau} - e^{-(\sqrt{(\xi^2-1)})t/\tau}\}}{(\sqrt{(\xi^2-1)})^2 - (-\xi)^2}\right]$$

$$H(t) = 1 - \frac{e^{-\xi t/\tau}}{\sqrt{(\xi^2-1)}}$$

$$\left[\xi \times \left\{\frac{e^{(\sqrt{(\xi^2-1)})t/\tau} - e^{-(\sqrt{(\xi^2-1)})\,t/\tau}}{2}\right\} + \sqrt{(\xi^2-1)} \times \left\{\frac{e^{(\sqrt{(\xi^2-1)})t/\tau} + e^{-(\sqrt{(\xi^2-1)})t/\tau}}{2}\right\}\right]$$

$$H(t) = 1 - \frac{e^{-\xi t/\tau}}{\sqrt{(\xi^2-1)}}[\xi \times \sinh(\sqrt{(\xi^2-1)})t/\tau + \sqrt{(\xi^2-1)} \times \cosh(\sqrt{(\xi^2-1)})t/\tau]$$

$$H(t) = 1 - e^{-\xi t/\tau}\left[\cosh(\sqrt{(\xi^2-1)})t/\tau + \frac{\xi}{\sqrt{(\xi^2-1)}}\sinh(\sqrt{(\xi^2-1)})t/\tau\right] \quad \ldots(9.26)$$

This is the response of U-tube manometer for a unit step change in input. The response depends on the value of damping coefficient 'ξ'. The unit step response of second order system for different values of 'ξ' are discussed here.

Case 1 : $\xi > 1$

For this case, the response is given by the expression

$$H(t) = 1 - e^{-\xi t/\tau}\left[\cosh\left(\sqrt{(\xi^2 - 1)}\right)t/\tau + \frac{\xi}{\sqrt{(\xi^2 - 1)}}\sinh\left(\sqrt{(\xi^2 - 1)}\right)t/\tau\right] \quad \ldots(9.26)$$

The response is non-oscillatory and becomes more sluggish as 'ξ' increases. This type of non-oscillatory response of the second order system for $\xi > 1$ is called over damped system (ODS). A typical response of over damped system is shown in Fig. 9.3.

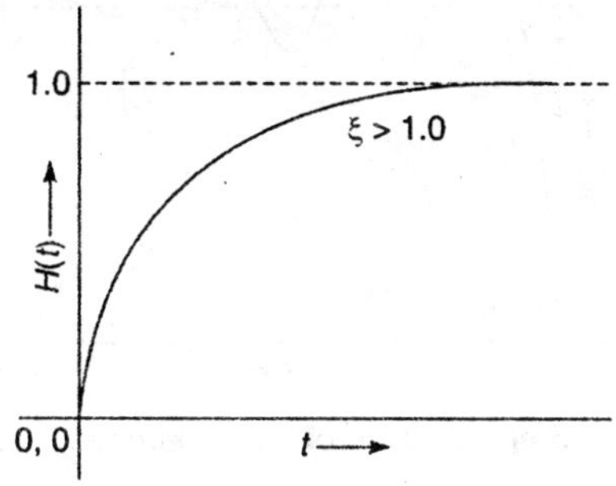

Fig. 9.3 Step response of over damped system.

Case 2 : $\xi < 1$

From the response of second order system, we have

$$H(t) = 1 - e^{-\xi t/\tau}\left[\cosh\left(\sqrt{(\xi^2 - 1)}\right)t/\tau + \frac{\xi}{\sqrt{(\xi^2 - 1)}}\sinh\left(\sqrt{(\xi^2 - 1)}\right)t/\tau\right] \quad \ldots(9.26)$$

$$\Rightarrow \quad H(t) = 1 - e^{-\xi t/\tau}\left[\cosh\left(\sqrt{-(1 - \xi^2)}\right)t/\tau + \frac{\xi}{\sqrt{-(1 - \xi^2)}}\sinh\left(\sqrt{-(1 - \xi^2)}\right)t/\tau\right]$$

$$\Rightarrow \quad H(t) = 1 - e^{-\xi t/\tau}\left[\cosh j\left(\sqrt{(1 - \xi^2)}\right)t/\tau + \frac{\xi}{j\sqrt{(1 - \xi^2)}}\sinh j\left(\sqrt{(1 - \xi^2)}\right)t/\tau\right]$$

From the mathematics, we have

$$\cosh jx = \frac{e^{jx} + e^{-jx}}{2}$$

$$\cosh jx = \frac{(\cos x + j\sin x) + (\cos x - j\sin x)}{2}$$

$$\cosh jx = \cos x$$

Similarly, we have

$$\sinh jx = j\sin x$$

$$\Rightarrow \quad H(t) = 1 - e^{-\xi t/\tau}\left[\cos\left(\sqrt{(1 - \xi^2)}\right)t/\tau + \frac{\xi}{\sqrt{(1 - \xi^2)}}\sin\left(\sqrt{(1 - \xi^2)}\right)t/\tau\right] \quad \ldots(9.27)$$

$$\Rightarrow \quad H(t) = 1 - e^{-\xi t/\tau}\left[\sin\left\{\frac{\sqrt{(1-\xi^2)}}{\tau} + \tan^{-1}\left(\frac{\sqrt{(1-\xi^2)}}{\xi^2}\right)\right\}\right] \quad \text{...(9.28)}$$

The response is oscillatory and the oscillation decreases with increase in time. This type of oscillatory response of the second order system for $\xi < 1$ is called under damped system. A typical response of under damped system is shown in Fig. 9.4.

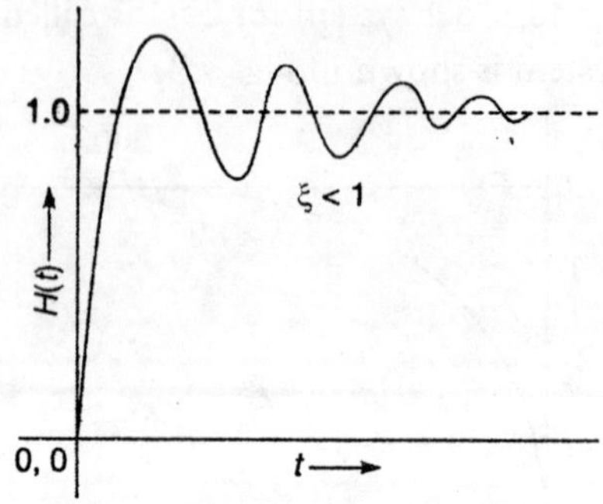

Fig. 9.4 Step response of under damped system.

Case 3 : $\xi = 1$

From the response of second order under damped system, we have

$$H(t) = 1 - e^{-\xi t/\tau}\left[\cos(\sqrt{(1-\xi^2)})t/\tau + \frac{\xi}{\sqrt{(1-\xi^2)}}\sin(\sqrt{(1-\xi^2)})t/\tau\right] \quad \text{...(9.27)}$$

Substituting $\xi = 1$, we get

$$H(t) = 1 - e^{-t/\tau}\,[1 + \text{Infinite}] \quad \text{...(9.29)}$$

Therefore, L. Hospital rule is applicable. Applying L. Hospital rule to equation (9.27), we get

$$H(t) = 1 - e^{-t/\tau}\left[1 + \frac{\underset{\xi\to 1}{Lt}\dfrac{d}{d\xi}\{\xi \times \sin(\sqrt{(1-\xi^2)})t/\tau\}}{\underset{\xi\to 1}{Lt}\dfrac{d}{d\xi}(\sqrt{(1-\xi^2)})}\right]$$

$$\Rightarrow H(t) = 1 - e^{-t/\tau}$$

$$\left[1 + \underset{\xi\to 1}{Lt}\frac{\xi.\cos(\sqrt{(1-\xi^2)})t/\tau \times \dfrac{1}{2\sqrt{(1-\xi^2)}} \times \dfrac{t}{\tau} \times (-2\xi) + \sin(\sqrt{(1-\xi^2)})t/\tau}{\dfrac{1}{2\sqrt{(1-\xi^2)}} \times (-2\xi)}\right]$$

$$\Rightarrow H(t) = 1 - e^{-t/\tau}\left[1 + \underset{\xi\to 1}{Lt}\left\{\xi.\cos(\sqrt{(1-\xi^2)})t/\tau \times \frac{t}{\tau} - \sin(\sqrt{(1-\xi^2)})t/\tau \times \frac{2\sqrt{(1-\xi^2)}}{(2\xi)}\right\}\right]$$

$$\Rightarrow \quad H(t) = 1 - e^{-t/\tau}\left[1 + \frac{t}{\tau}\right] \quad \ldots(9.30)$$

The response is non-oscillatory and approaches to 1 without any oscillation. This type of response at $\xi = 1$ is known as critically damped system. A typical response of critically damped system is shown in Fig. 9.5.

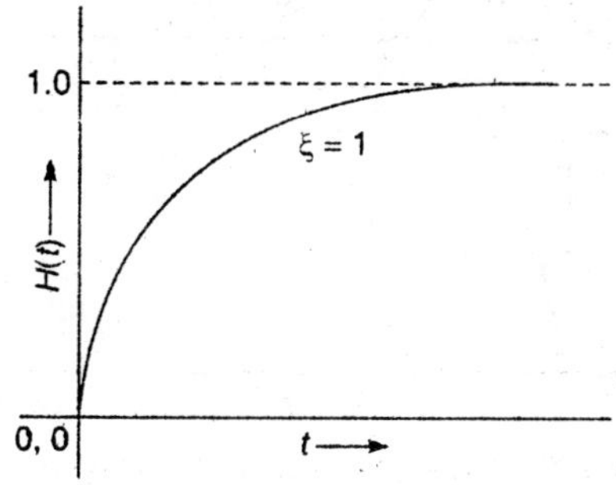

Fig. 9.5 Step response of critically damped system.

Case 4 : $\xi = 0$

From the response of second order under damped system, we have

$$H(t) = 1 - e^{-\xi t/\tau}\left[\cos\left(\sqrt{(1-\xi^2)}\right)t/\tau + \frac{\xi}{\sqrt{(1-\xi^2)}}\sin\left(\sqrt{(1-\xi^2)}\right)t/\tau\right] \quad \ldots(9.27)$$

Substituting $\xi = 0$ in equation (9.27), we get

$$H(t) = 1 - [\cos(t/\tau) + 0]$$

$$H(t) = 1 - \cos t/\tau \quad \ldots(9.31)$$

The response is oscillatory and the oscillation is continuous. This type of response of second order system for $\xi = 0$ is called oscillatory system. A typical response of oscillatory system is shown in Fig. 9.6.

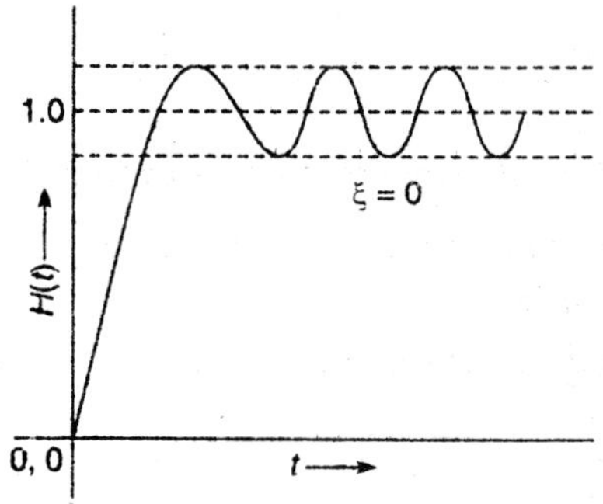

Fig. 9.6 Step response of oscillatory system.

Case 5 : $\xi < 0$

From the response of second order under-damped system, we have

$$H(t) = 1 - e^{-\xi t/\tau}\left[\cos\left(\sqrt{(1-\xi^2)}\right)t/\tau + \frac{\xi}{\sqrt{(1-\xi^2)}}\sin\left(\sqrt{(1-\xi^2)}\right)t/\tau\right] \quad \ldots(9.27)$$

Substituting $\xi < 0$ in equation (9.27), we find no change occurs in the existing equation. The response is oscillatory and the oscillation is continuous. Also, the oscillation become larger with increase of time. This type of response of second order system for $\xi < 0$ is called unstable system. A typical response of unstable system is shown in Fig. 9.7.

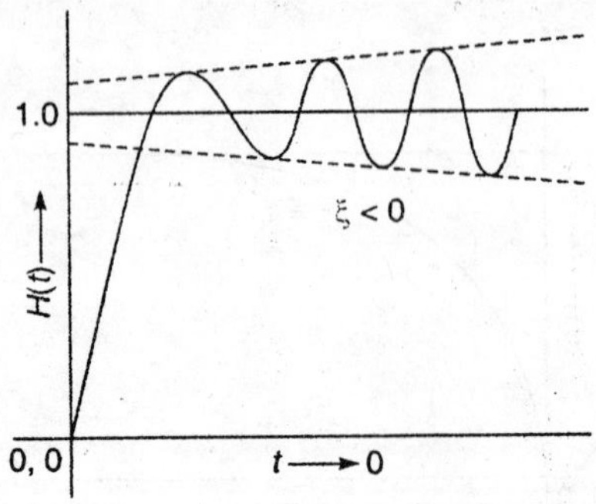

Fig. 9.7 Step response of unstable system.

Basic Concept of Under-damped Response of Second Order System

In the previous topic of this chapter, we have discussed several cases of response of second order system. Of these cases, the under-damped response and critically damped response are having more importance in chemical process industries. Therefore, a number of terms are used to describe the basic concept of under-damped oscillatory step response of second order system as shown Fig. 9.8.

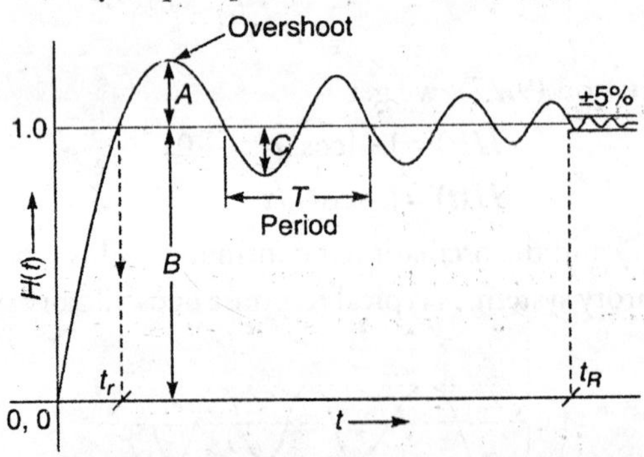

Fig. 9.8 Under damped response of second order system.

(i) **Overshoot :** It is a measure of how much the response exceed the ultimate value following a step change. It is expressed as the ratio of A/B.

From the step response of under-damped system, we have

$$H(t) = 1 - e^{-\xi t/\tau}\left[\cos\left(\sqrt{(1-\xi^2)}\right)t/\tau + \frac{\xi}{\sqrt{(1-\xi^2)}}\sin\left(\sqrt{(1-\xi^2)}\right)t/\tau\right] \quad \ldots(9.27)$$

Differentiating equation (9.27) w.r.t. 't' and equalizing to zero, we get

$$\frac{dH(t)}{dt} = \frac{d}{dt}\left[1 - e^{-\xi t/\tau}\left\{\cos\left(\sqrt{(1-\xi^2)}\right)t/\tau + \frac{\xi}{\sqrt{(1-\xi^2)}}\sin\left(\sqrt{(1-\xi^2)}\right)t/\tau\right\}\right]$$

$$0 = (-e^{-\xi t/\tau})$$

$$\left[-\sin(\sqrt{(1-\xi^2)})t/\tau \times \sqrt{(1-\xi^2)}/\tau + \frac{\xi}{\sqrt{(1-\xi^2)}} \times \cos(\sqrt{(1-\xi^2)})t/\tau \times \sqrt{(1-\xi^2)}/\tau\right]$$

$$+\left[\cos(\sqrt{(1-\xi^2)})t/\tau + \frac{\xi}{\sqrt{(1-\xi^2)}} \sin(\sqrt{(1-\xi^2)})t/\tau\right](-e^{-\xi t/\tau})(-\xi/\tau)$$

$$0 = e^{-\xi t/\tau}.\sin(\sqrt{(1-\xi^2)})t/\tau.\sqrt{(1-\xi^2)}/\tau - \frac{\xi}{\tau}.e^{-\xi t/\tau}.\cos(\sqrt{(1-\xi^2)})t/\tau$$

$$+\frac{\xi}{\tau}.e^{-\xi t/\tau}.\cos(\sqrt{(1-\xi^2)})t/\tau + \frac{\xi^2}{\tau(\sqrt{(1-\xi^2)})}.\sin(\sqrt{(1-\xi^2)})t/\tau.e^{-\xi t/\tau}$$

$$0 = e^{-\xi t/\tau}.\sin(\sqrt{(1-\xi^2)}).\sqrt{(1-\xi^2)}/\tau + \frac{\xi^2}{\tau(\sqrt{(1-\xi^2)})}.\sin(\sqrt{(1-\xi^2)})t/\tau.e^{-\xi t/\tau}$$

$$0 = e^{-\xi t/\tau}.\sin(\sqrt{(1-\xi^2)})t/\tau\left[\sqrt{(1-\xi^2)}/\tau + \frac{\xi^2}{\tau(\sqrt{(1-\xi^2)})}\right]$$

Applying algebraical mathematics rule, we get

$$\sin(\sqrt{(1-\xi^2)})t/\tau = 0 \quad \text{...(9.32)}$$

Substituting equation (9.32) in equation (9.27), we get

$$H(t) = 1 - e^{-\xi t/\tau}[\cos(\sqrt{(1-\xi^2)})t/\tau] \quad \text{...(9.33)}$$

Let, $n\pi = (\sqrt{(1-\xi^2)})t/\tau$

At, $n = 1$

$$\Rightarrow \quad t = \frac{\pi\tau}{\sqrt{(1-\xi^2)}}$$

$$\Rightarrow \quad H(t) = 1 - e^{-\xi t/\tau}.\cos\pi$$

$$\Rightarrow \quad H(t) = 1 + e^{-\xi t/\tau}$$

$$\Rightarrow \quad H(t) - 1 = e^{-\xi t/\tau}$$

$$\Rightarrow \quad \text{Overshoot} = e^{-\pi\xi/\sqrt{(1-\xi^2)}} \quad \text{...(9.34)}$$

As 'ξ' increases overshoot decreases. At $\xi = 1$, overshoot is equal to zero, *i.e.*, there is no oscillation. It is desirable to have systems with very less overshoot, *i.e.*, 'ξ' very nearly equal to one.

(ii) Decay Ratio : It is defined as the ratio of the successive peaks. Therefore, it may be expressed as C/A.

By the definition, we have

$$\text{Decay ratio} = \frac{C}{A}$$

$$\text{D.R.} = \frac{e^{-3\pi\xi/\sqrt{(1-\xi^2)}}}{e^{-\pi\xi/\sqrt{(1-\xi^2)}}}$$

$$\text{D.R.} = e^{-2\pi\xi/\sqrt{(1-\xi^2)}}$$

$$\text{D.R.} = [e^{-\pi\xi/\sqrt{(1-\xi^2)}}]^2 \qquad \text{...(9.35)}$$

$$\text{D.R.} = (\text{Overshoot})^2$$

As 'ξ' increases D.R. decreases. It is desirable to have decay ratio as small as possible.

(iii) Rise Time : It is the time required for the response to first reach its ultimate value. It is denoted by t_r.

From the step response of under-damped system, we have

$$H(t) = 1 - e^{-\xi t/\tau}\left[\cos(\sqrt{(1-\xi^2)})t/\tau + \frac{\xi}{\sqrt{(1-\xi^2)}}\sin(\sqrt{(1-\xi^2)})t/\tau\right] \qquad \text{...(9.27)}$$

At $H(t) = 1$

$$\Rightarrow \qquad 1 = 1 - e^{-\xi t/\tau}\left[\cos(\sqrt{(1-\xi^2)})t/\tau + \frac{\xi}{\sqrt{(1-\xi^2)}}\sin(\sqrt{(1-\xi^2)})t/\tau\right]$$

$$\Rightarrow \qquad 0 = -e^{-\xi t/\tau}\left[\cos(\sqrt{(1-\xi^2)})t/\tau + \frac{\xi}{\sqrt{(1-\xi^2)}}\sin(\sqrt{(1-\xi^2)})t/\tau\right]$$

$$\Rightarrow \qquad 0 = \cos(\sqrt{(1-\xi^2)})t/\tau + \frac{\xi}{\sqrt{(1-\xi^2)}}\sin(\sqrt{(1-\xi^2)})t/\tau$$

$$\Rightarrow \qquad \cos(\sqrt{(1-\xi^2)})t/\tau = -\frac{\xi}{\sqrt{(1-\xi^2)}}\sin(\sqrt{(1-\xi^2)})t/\tau$$

$$\Rightarrow \qquad \tan(\sqrt{(1-\xi^2)})t/\tau = -\frac{\sqrt{(1-\xi^2)}}{\xi}$$

$$\Rightarrow \qquad (\sqrt{(1-\xi^2)})t/\tau = -\tan^{-1}(\sqrt{(1-\xi^2)}/\xi)$$

$$\Rightarrow \qquad t_r = -\frac{\tau}{\sqrt{(1-\xi^2)}}\tan^{-1}(\sqrt{(1-\xi^2)}/\xi) \qquad \text{...(9.36)}$$

As 'ξ' increases rise time increases.

(iv) Response Time : It is the time required for the response to reach within $\pm 5\%$ of its ultimate value and thereafter remains within the range. It is denoted by t_R. The $\pm 5\%$ limit is only arbitrary and it is fixed according to requirement.

(v) Period of Oscillation : It is the time required for the response to complete one cycle. It is denoted by 'T'. The radian frequency 'ω' is the coefficient of 't' of sine term.

$$\Rightarrow \quad \omega = \frac{\sqrt{(1-\xi^2)}}{\tau} \quad \text{...(9.37)}$$

where, ω = Radian frequency.

Radian frequency may be represented in terms of cyclical frequency 'f' by the equation

$$\omega = 2\pi f \quad \text{...(9.38)}$$

Comparing equations (9.37) and (9.38), we get

$$2\pi f = \frac{\sqrt{(1-\xi^2)}}{\tau}$$

$$f = \frac{\sqrt{(1-\xi^2)}}{2\pi\tau}$$

We know that

$$f = \frac{1}{T}$$

$$\Rightarrow \quad \frac{1}{T} = \frac{\sqrt{(1-\xi^2)}}{2\pi\tau}$$

$$\Rightarrow \quad T = \frac{2\pi\tau}{\sqrt{(1-\xi^2)}} \quad \text{...(9.39)}$$

(vi) Natural Period of Oscillation : If the damping is eliminated from the system, *i.e.,* $\xi = 0$, then the system will oscillate continuously without any change in the amplitude of oscillation. This type of oscillation is knows as natural period of oscillation.

From equation (9.39), we have

$$T = \frac{2\pi\tau}{\sqrt{(1-\xi^2)}} \quad \text{...(9.39)}$$

At $\xi = 0$

$$\Rightarrow \quad T_n = 2\pi\tau \quad \text{...(9.40)}$$

■ **Example 9.1** The two tank liquid level system is operating at steady state. When a unit step change is made in the flow rate to tank 1, the transient response is critically damped and it takes 1 min for the change in liquid level of the second tank to reach 60% of the total change.

If the ratio of cross-sectional area of the tank is $A_1/A_2 = 2$. Calculate the ratio R_1/R_2. Also, calculate the time constant of each tank. How long does it take for the change in level of the first tank to reach 95% of the total change?

Solution :

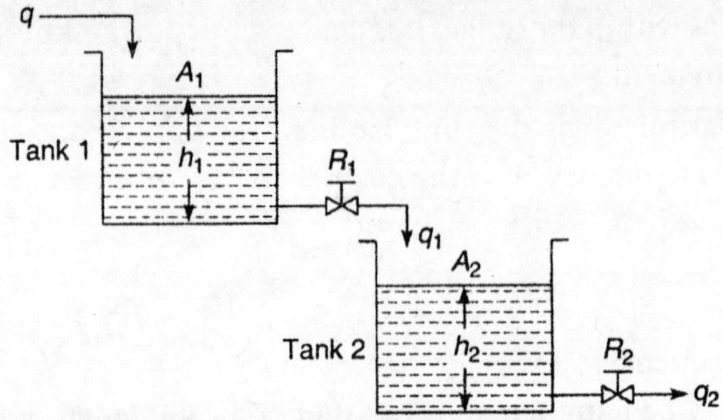

Fig. 9.9 Two tank non-interacting system.

From the transfer function of two tank non-interacting system, we have

$$\frac{H_2(s)}{Q(s)} = \frac{R_2}{(\tau_1 s + 1)(\tau_2 s + 1)} \quad \ldots(8.6)$$

$$\Rightarrow \quad \frac{H_2(s)}{Q(s)} = \frac{R_2}{\tau_1\tau_2 s^2 + (\tau_1 + \tau_2)s + 1} \quad \ldots(9.41)$$

From the standard transfer function of second order system, we have

$$\frac{Y(s)}{X(s)} = \frac{k'}{\tau^2 s^2 + 2\tau\xi s + 1} \quad \ldots(9.4)$$

Comparing the equations (9.41) and (9.4), we get

$$\tau^2 = \tau_1\tau_2$$

$$\Rightarrow \quad \tau = \sqrt{\tau_1\tau_2}$$

$$\Rightarrow \quad 2\tau\xi = \tau_1 + \tau_2$$

For critically damped system, $\xi = 1$

$$\Rightarrow \quad 2\tau = \tau_1 + \tau_2$$

$$\Rightarrow \quad 2\sqrt{\tau_1\tau_2} = \tau_1 + \tau_2$$

$$\Rightarrow \quad 2\sqrt{\tau_1\tau_2} = (\sqrt{\tau_1})^2 + (\sqrt{\tau_2})^2$$

$$\Rightarrow \quad 0 = (\sqrt{\tau_1})^2 - 2\sqrt{\tau_1}\sqrt{\tau_2} + (\sqrt{\tau_2})^2$$

$$\Rightarrow \quad (\sqrt{\tau_1} - \sqrt{\tau_2})^2 = 0$$

$$\Rightarrow \quad \sqrt{\tau_1} - \sqrt{\tau_2} = 0$$

$$\Rightarrow \quad \sqrt{\tau_1} = \sqrt{\tau_2}$$

$$\Rightarrow \quad \tau_1 = \tau_2$$

$$\Rightarrow \quad A_1 R_1 = A_2 R_2$$

$$\Rightarrow \quad \frac{A_1}{A_2} = \frac{R_2}{R_1} = 2$$

$$\Rightarrow \quad \frac{R_1}{R_2} = 0.5$$

Time constants τ_1 and τ_2 are equal.

Therefore,
$$\frac{H_2(s)}{Q(s)} = \frac{R_2}{(\tau s+1)^2} \qquad \text{...(9.42)}$$

From the given data, we have

$$Q(t) = 1$$
$$Q(s) = \frac{1}{s}$$

$$\Rightarrow \quad H_2(s) = \frac{R_2}{(s)(\tau s+1)^2}$$

$$\Rightarrow \quad H_2(s) = \frac{R_2/\tau^2}{(s)(s+1/\tau)^2}$$

$$\Rightarrow \quad H_2(s) = \frac{R_2/\tau^2}{(s)(s+1/\tau)^2} = \frac{C_1}{s} + \frac{C_2}{(s+1/\tau)} + \frac{C_3}{(s+1/\tau)^2}$$

$$\Rightarrow \quad \frac{R_2}{\tau^2} = C_1(s+1/\tau)^2 + C_2(s)(s+1/\tau) + C_3(s)$$

Put, $s = 0$; $C_1 = R_2$

Put, $s = -1/\tau$; $C_3 = -R_2/\tau$

Comparing the coefficient of s^2, we get

$$C_1 + C_2 = 0 \; ; \; C_2 = -R_2$$

$$\Rightarrow \quad H_2(s) = \frac{R_2}{s} - \frac{R_2}{(s+1/\tau)} - \frac{R_2}{\tau(s+1/\tau)^2}$$

$$\Rightarrow \quad H_2(t) = R_2 - R_2 e^{-t/\tau} - R_2 \cdot \frac{t}{\tau} e^{-t/\tau}$$

$$\Rightarrow \quad H_2(t) = R_2 \left[1 - e^{-t/\tau} - t/\tau \cdot e^{-t/\tau}\right]$$

Applying limit $t \to \infty$, we get ultimate change in the level of second tank.

$$\underset{t\to\infty}{Lt} \; H_2(t) = R_2$$

Therefore, 60% of the total change

$$H_2(t) = 0.6R_2$$

$$\Rightarrow \quad 0.6R_2 = R_2\left[1 - e^{-t/\tau} - t/\tau \, e^{-t/\tau}\right]$$

$$\Rightarrow \quad 0.4 = e^{-t/\tau} + t/\tau \, e^{-t/\tau}$$

From the given data, we have

$$t = 1 \text{ min}$$

$$\Rightarrow \quad 0.4 = e^{-1/\tau} + \frac{1}{\tau} e^{-1/\tau} \qquad \text{...(9.43)}$$

Solving equation (9.48) by trial and error method, we get

τ (min)	LHS	RHS
0.4	0.4	0.287
0.5	0.4	0.405

Therefore, time constant of tank 1 and tank 2 is 0.5 min.

From the transfer function of tank 1, we have

$$\frac{H_1(s)}{Q(s)} = \frac{R_1}{(\tau_1 s + 1)} \quad \ldots(7.70)$$

From the given data, we have

$$Q(t) = 1$$

$$Q(s) = \frac{1}{s}$$

$$\Rightarrow \quad H_1(s) = \frac{R_1}{(s)(\tau s + 1)}$$

$$\Rightarrow \quad H_1(s) = \frac{R_1/\tau}{(s)(s + 1/\tau)}$$

$$\Rightarrow \quad H_1(s) = \frac{R_1/\tau}{(s)(s+1/\tau)} = \frac{C_1}{(s)} + \frac{C_2}{(s+1/\tau)}$$

$$\Rightarrow \quad \frac{R_1}{\tau} = C_1(s + 1/\tau) + C_2(s)$$

Put, $s = 0$; $C_1 = R_1$

Put, $s = -1/\tau$; $C_2 = -R_1$

$$\Rightarrow \quad H_1(s) = \frac{R_1}{s} - \frac{R_1}{(s+1/\tau)}$$

$$\Rightarrow \quad H_1(t) = R_1 - R_1\, e^{-t/\tau}$$

$$\Rightarrow \quad H_1(t) = R_1\,[1 - e^{-t/\tau}]$$

Applying limit $t \to \infty$, we get ultimate change in the level of first tank.

$$\underset{t \to \infty}{Lt}\; H_1(t) = R_1$$

Therefore, 95% of the total change

$$H_1(t) = 0.95R_1$$

$$\Rightarrow \quad 0.95R_1 = R_1\,[1 - e^{-t/\tau}]$$

$$\Rightarrow \quad 0.95 = 1 - e^{-t/\tau}$$

$$e^{-t/\tau} = 0.05$$

From the calculated data, we have

$$\tau = 0.5$$

$$\Rightarrow \quad e^{-t/0.5} = 0.05$$

$$\Rightarrow \quad -t/0.5 = \ln(0.05)$$

$$\Rightarrow \quad t = 1.49 \text{ min}$$

It takes 1.49 min for the change in level of the first tank to reach 95% of the total change.

■ **Example 9.2** A step change of magnitude 4 is introduced into a system having transfer function

$$\frac{Y(s)}{X(s)} = \frac{10}{s^2 + 1.6s + 4}$$

Determine :

(i) % overshoot

(ii) Rise time

(iii) Ultimate value of $Y(t)$

(iv) Maximum value of $Y(t)$

(v) Period of oscillation.

Solution : From the given data of transfer function, we have

$$\frac{Y(s)}{X(s)} = \frac{10}{s^2 + 1.6s + 4}$$

$$\Rightarrow \quad \frac{Y(s)}{X(s)} = \frac{10}{4(0.25s^2 + 0.4s + 1)}$$

$$\Rightarrow \quad \frac{Y(s)}{X(s)} = \frac{2.5}{(0.25s^2 + 0.4s + 1)} \qquad \text{...(9.44)}$$

From the second order standard transfer function, we have

$$\frac{Y(s)}{X(s)} = \frac{k'}{\tau^2 s^2 + 2\tau\xi s + 1} \qquad \text{...(9.4)}$$

Comparing equations (9.44) and (9.4), we get

$$\tau^2 = 0.25$$

$$\tau = 0.5$$

$$2\tau\xi = 0.4$$

$$2 \times 0.5 \times \xi = 0.4$$

$$\Rightarrow \quad \xi = 0.4$$

(i) % overshoot : From the overshoot equation, we have

$$\% \text{ overshoot} = e^{-\pi\xi/\sqrt{(1-\xi^2)}} \qquad \text{...(9.34)}$$

$$= e^{-\pi \times 0.4/\sqrt{(1-0.4^2)}} = 25.38$$

(ii) Rise time : From the rise time equation, we have

$$t_r = \left| -\frac{\tau}{\sqrt{(1-\xi^2)}} \tan^{-1}\left(\sqrt{(1-\xi^2)}/\xi\right) \right|$$

$$= \left| -\frac{0.5}{\sqrt{(1-0.4^2)}} \tan^{-1}\left(\sqrt{(1-0.4^2)}/0.4\right) \right|$$

$$= 36.72 \text{ units}$$

(iii) Ultimate value of $Y(t)$: For the step change of magnitude 4, we have

$$H(t) = 10\left[1 - e^{-\xi t/\tau}\left\{\cos\left(\sqrt{(1-\xi^2)}\right)t/\tau + \frac{\xi}{\sqrt{(1-\xi^2)}}\sin\left(\sqrt{(1-\xi^2)}\right)t/\tau\right\}\right]$$

$$\underset{t\to\infty}{Lt}\ H(t) = \underset{t\to\infty}{Lt}\ 10\left[1 - e^{-\xi t/\tau}\left\{\cos\left(\sqrt{(1-\xi^2)}\right)t/\tau + \frac{\xi}{\sqrt{(1-\xi^2)}}\sin\left(\sqrt{(1-\xi^2)}\right)t/\tau\right\}\right]$$

$$H(t)_{\text{ultimate}} = 0$$

$$B = 10 \text{ units}$$

(iv) Maximum value of $Y(t)$:

$$\text{Overshoot} = \frac{A}{B}$$

$$0.2538 = \frac{A}{10}$$

$$A = 2.538$$

Maximum value of $Y(t) = A + B$

$$= 2.538 + 10 = 12.538 \text{ units}$$

(v) Period of oscillation : From the equation of period of oscillation, we have

$$T = \frac{2\pi\tau}{\sqrt{(1-\xi^2)}}$$

$$\Rightarrow \quad T = \frac{2\times\pi\times 0.5}{\sqrt{(1-0.4^2)}}$$

$$\Rightarrow \quad T = 3.43 \text{ units}$$

9.1.2.2 Response of U-Tube Manometer for a Impulse Change in Input

From the transfer function of second order system, we have

$$\frac{H(s)}{P(s)} = \frac{k'}{(\tau^2 s^2 + 2\tau\xi s + 1)} \qquad \ldots(9.11)$$

At
$$k' = 1$$

$$\Rightarrow \qquad \frac{H(s)}{P(s)} = \frac{1}{(\tau^2 s^2 + 2\tau\xi s + 1)} \qquad \ldots(9.45)$$

From the given data, we have

$$P(t) = 1\partial(t)$$
$$P(s) = 1$$

$$\Rightarrow \qquad H(s)_{\text{impulse}} = \frac{1}{(\tau^2 s^2 + 2\tau\xi s + 1)} \qquad \ldots(9.46)$$

For a unit step change in input, we have

$$P(t) = 1$$
$$P(s) = \frac{1}{s}$$

$$\Rightarrow \qquad H(s)_{\text{step}} = \frac{1}{(s)(\tau^2 s^2 + 2\tau\xi s + 1)} \qquad \ldots(9.47)$$

Comparing equation (9.45) and (9.47), we get

$$sH(s)_{\text{step}} = H(s)_{\text{impulse}}$$

$$\Rightarrow \qquad \frac{d}{dt}[H(t)_{\text{step}}] = H(t)_{\text{impulse}} \qquad \ldots(9.48)$$

From the equation (9.48), we can conclude that the impulse response can be obtained simply by differentiating the corresponding step response.

The unit impulse response of second order system for different values of 'ξ' are discussed here.

Case 1 : $\xi > 1$

For this case, the response is given by the expression

$$H(t) = 1 - e^{-\xi t/\tau}\left[\cosh\left(\sqrt{(\xi^2 - 1)}\right) t/\tau + \frac{\xi}{\sqrt{(\xi^2 - 1)}} \sinh\left(\sqrt{(\xi^2 - 1)}\right) t/\tau\right]$$

$$H(t)_{\text{impulse}} = \frac{d}{dt}\left[1 - e^{-\xi t/\tau}\left\{\cosh\left(\sqrt{(\xi^2 - 1)}\right) t/\tau + \frac{\xi}{\sqrt{(\xi^2 - 1)}} \sinh\left(\sqrt{(\xi^2 - 1)}\right) t/\tau\right\}\right]$$

$$H(t)_{\text{impulse}} = 0 + (-e^{-\xi t/\tau})$$

$$\left\{\sinh\left(\sqrt{(\xi^2 - 1)}\right) t/\tau \,.\left(\sqrt{(\xi^2 - 1)}\right)/\tau + \frac{\xi}{\sqrt{(\xi^2 - 1)}} \cosh\left(\sqrt{(\xi^2 - 1)}\right) t/\tau \,.\left(\sqrt{(\xi^2 - 1)}\right)/\tau\right\}$$

$$+\left\{\cosh\left(\sqrt{(\xi^2 - 1)}\right) t/\tau + \frac{\xi}{\sqrt{(\xi^2 - 1)}} \sinh\left(\sqrt{(\xi^2 - 1)}\right) t/\tau\right\}(-e^{-\xi t/\tau})(-\xi/\tau)$$

$$H(t)_{\text{impulse}} = -e^{-\xi t/\tau}.\sinh(\sqrt{(\xi^2-1)})\,t/\tau.(\sqrt{(\xi^2-1)})/\tau - e^{-\xi t/\tau}.\xi/\tau.\cosh(\sqrt{(\xi^2-1)})\,t/\tau$$

$$+e^{-\xi t/\tau}.\xi/\tau.\cosh(\sqrt{(\xi^2-1)})\,t/\tau + \frac{\xi^2}{\tau(\sqrt{(\xi^2-1)})}.e^{-\xi t/\tau}.\sinh(\sqrt{(\xi^2-1)})\,t/\tau$$

$$H(t)_{\text{impulse}} = (-e^{-\xi t/\tau}).\sinh(\sqrt{(\xi^2-1)})\,t/\tau\left[\frac{\sqrt{(\xi^2-1)}}{\tau} - \frac{\xi^2}{\tau(\sqrt{(\xi^2-1)})}\right]$$

$$H(t)_{\text{impulse}} = \frac{e^{-\xi t/\tau}}{\sqrt{(\xi^2-1)}}\sinh(\sqrt{(\xi^2-1)})\,t/\tau \qquad \ldots(9.49)$$

This type of response is oscillatory and approaches to zero as $t \to \infty$.

Case 2 : $\xi < 1$

From the response of second order under-damped system, we have

$$H(t)_{\text{step}} = 1 - e^{-\xi t/\tau}\left[\cos(\sqrt{(1-\xi^2)})\,t/\tau + \frac{\xi}{\sqrt{(1-\xi^2)}}\sin(\sqrt{(1-\xi^2)})\,t/\tau\right] \qquad \ldots(9.27)$$

From the impulse response of second order system, we have

$$H(t)_{\text{impulse}} = \frac{d}{dt}[H(t)_{\text{step}}] \qquad \ldots(9.48)$$

$$H(t)_{\text{impulse}} = \frac{d}{dt}\left[1 - e^{-\xi t/\tau}\left\{\cos(\sqrt{(1-\xi^2)})\,t/\tau + \frac{\xi}{\sqrt{(1-\xi^2)}}\sin(\sqrt{(1-\xi^2)})\,t/\tau\right\}\right]$$

$$H(t)_{\text{impulse}} = 0 + (-e^{-\xi t/\tau})$$

$$\left\{-\sin(\sqrt{(1-\xi^2)})\,t/\tau.(\sqrt{(1-\xi^2)})/\tau + \frac{\xi}{\sqrt{(1-\xi^2)}}\cos(\sqrt{(1-\xi^2)})\,t/\tau.(\sqrt{(1-\xi^2)})/\tau\right\}$$

$$+\left\{\cos(\sqrt{(1-\xi^2)})\,t/\tau + \frac{\xi}{\sqrt{(1-\xi^2)}}\sin(\sqrt{(1-\xi^2)})\,t/\tau\right\}(-e^{-\xi t/\tau})(-\xi/\tau)$$

$$H(t)_{\text{impulse}} = e^{-\xi t/\tau}.\sin(\sqrt{(1-\xi^2)})\,t/\tau.(\sqrt{(1-\xi^2)})/\tau - (\xi/\tau).e^{-\xi t/\tau}.\cos(\sqrt{(1-\xi^2)})\,t/\tau$$

$$+(\xi/\tau).e^{-\xi t/\tau}.\cos(\sqrt{(1-\xi^2)})\,t/\tau + \frac{\xi^2}{\tau(\sqrt{(1-\xi^2)})}.e^{-\xi t/\tau}.\sin(\sqrt{(1-\xi^2)})\,t/\tau$$

$$H(t)_{\text{impulse}} = e^{-\xi t/\tau}.\sin(\sqrt{(1-\xi^2)})\,t/\tau.(\sqrt{(1-\xi^2)})/\tau$$

$$+\frac{\xi^2}{\tau(\sqrt{(1-\xi^2)})}.e^{-\xi t/\tau}.\sin(\sqrt{(1-\xi^2)})\,t/\tau$$

$$H(t)_{\text{impulse}} = (e^{-\xi t/\tau}).\sin(\sqrt{(1-\xi^2)})\, t/\tau \left[\frac{\sqrt{(1-\xi^2)}}{\tau} + \frac{\xi^2}{\tau(\sqrt{(1-\xi^2)})}\right]$$

$$H(t)_{\text{impulse}} = e^{-\xi t/\tau}.\sin(\sqrt{(1-\xi^2)})\, t/\tau \left[\frac{(\sqrt{(1-\xi^2)})^2 + \xi^2}{\tau(\sqrt{(1-\xi^2)})}\right]$$

$$H(t)_{\text{impulse}} = \frac{e^{-\xi t/\tau}}{\tau\sqrt{(1-\xi^2)}}.\sin(\sqrt{(1-\xi^2)})\, t/\tau \qquad \ldots(9.50)$$

This type of response is oscillatory and approaches to zero as $t \rightarrow \infty$.

Case 3 : $\xi = 1$

From the response of critically damped system, we have

$$H(t) = 1 - e^{-t/\tau}[1 + t/\tau] \qquad \ldots(9.30)$$

From the impulse response of second order system, we have

$$H(t)_{\text{impulse}} = \frac{d}{dt}[H(t)_{\text{step}}] \qquad \ldots(9.48)$$

$$\Rightarrow \quad H(t)_{\text{impulse}} = \frac{d}{dt}\left[1 - e^{-t/\tau}\left\{1 + \frac{t}{\tau}\right\}\right]$$

$$\Rightarrow \quad H(t)_{\text{impulse}} = 0 + (-e^{-t/\tau})\left\{0 + \frac{1}{\tau}\right\} + \left\{1 + \frac{t}{\tau}\right\}(-e^{-t/\tau})\left(-\frac{1}{\tau}\right)$$

$$\Rightarrow \quad H(t)_{\text{impulse}} = -\frac{e^{-t/\tau}}{\tau} + \frac{e^{-t/\tau}}{\tau} + \frac{t.e^{-t/\tau}}{\tau^2}$$

$$\Rightarrow \quad H(t)_{\text{impulse}} = \frac{t.e^{-t/\tau}}{\tau^2} \qquad \ldots(9.51)$$

This type of response is non-oscillatory and approaches to zero as $t \rightarrow \infty$.

Example 9.3 Derive the transfer function for control valve associated with controller.

Solution : The function of the control valve is to change the flow according to change in the controller output pressure. Most of the pneumatic control valve has some dynamic lag, which means that the change in flow is not instantaneous with the change in applied pressure. From the experiments conducted on the pneumatic control valve, it has been found that the relation between the flow and the valve top pressure is represented by a first order transfer function

$$\frac{Q(s)}{P(s)} = \frac{k_v}{(\tau_v s + 1)} \qquad \ldots(9.52)$$

where, k_v = Steady state gain of pneumatic control valve

τ_v = Time constant of pneumatic control valve.

In many practical systems the time constant of the valve is very small when compare with the time constant of the other components of the control system. In such a case, we can approximate the transfer function of the valve as

$$\frac{Q(s)}{P(s)} \approx k_v \qquad \ldots(9.53)$$

i.e., the valve is set to have negligible dynamic lag. This approximation can be justified as follows : Consider a first order process and a control valve connected in series as shown in Fig. 9.10.

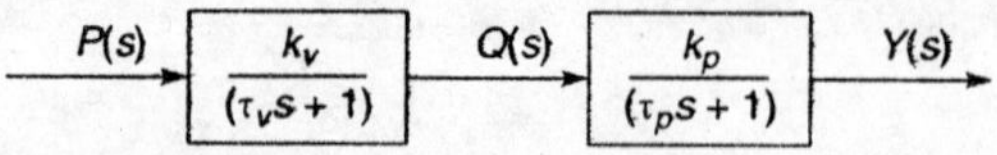

Fig. 9.10 Control valve and a first order process connected in series.

We can write the overall transfer function for this system as

$$\frac{Y(s)}{P(s)} = \frac{k_v . k_p}{(\tau_v s+1)(\tau_p s+1)} \qquad \ldots(9.54)$$

This is the transfer function of control valve associated with controller.

■ **Example 9.4** Derive the response equation of pneumatic control valve associated with controller for a unit step change in input.

Solution : From the transfer function of control valve associated with controller, we have

$$\frac{Y(s)}{P(s)} = \frac{k_v . k_p}{(\tau_v s+1)(\tau_p s+1)} \qquad \ldots(9.54)$$

From the data given, we have

$$P(t) = 1$$

$$P(s) = \frac{1}{s}$$

$$\Rightarrow \quad Y(s) = \frac{k_v . k_p}{(s)(\tau_v s+1)(\tau_p s+1)} \qquad \ldots(9.55)$$

$$\Rightarrow \quad Y(s) = \frac{k_v . k_p}{(s)(\tau_v s+1)(\tau_p s+1)} = \frac{C_1}{s} + \frac{C_2}{(\tau_v s+1)} + \frac{C_3}{(\tau_p s+1)}$$

$$\Rightarrow \quad k_v . k_p = C_1(\tau_v s+1)(\tau_p s+1) + C_2\,(s)(\tau_p s+1) + C_3\,(s)(\tau_v s+1)$$

Put, $s = 0$; $C_1 = k_v . k_p$

Put, $s = -\dfrac{1}{\tau_v}$; $C_2 = \dfrac{k_v . k_p . \tau_v^2}{(\tau_p - \tau_v)}$

Put, $s = -\dfrac{1}{\tau_p}$; $C_3 = \dfrac{k_v . k_p . \tau_p^2}{(\tau_v - \tau_p)}$

$$\Rightarrow \quad Y(s) = \frac{k_v . k_p}{(s)} + \frac{k_v . k_p . \tau_v^2}{(\tau_p - \tau_v)(\tau_v s+1)} + \frac{k_v . k_p . \tau_p^2}{(\tau_v - \tau_p)(\tau_p s+1)}$$

$$\Rightarrow \quad Y(s) = \frac{k_v . k_p}{(s)} + \frac{k_v . k_p . \tau_v}{(\tau_p - \tau_v)(s+1/\tau_v)} + \frac{k_v . k_p . \tau_p}{(\tau_v - \tau_p)(s+1/\tau_p)}$$

$$\Rightarrow \quad Y(s) = k_v . k_p \left[\frac{1}{s} + \frac{\tau_v}{(\tau_p - \tau_v)} \times \frac{1}{(s + 1/\tau_v)} + \frac{\tau_p}{(\tau_v - \tau_p)} \times \frac{1}{(s + 1/\tau_p)} \right]$$

$$\Rightarrow \quad Y(t) = k_v . k_p \left[1 + \frac{\tau_v . e^{-t/\tau_v}}{(\tau_p - \tau_v)} + \frac{\tau_p . e^{-t/\tau_p}}{(\tau_v - \tau_p)} \right]$$

$$\Rightarrow \quad Y(t) = k_v . k_p \left[1 + \frac{\tau_v . e^{-t/\tau_v}}{(\tau_p - \tau_v)} - \frac{\tau_p . e^{-t/\tau_p}}{(\tau_p - \tau_v)} \right]$$

$$\Rightarrow \quad Y(t) = k_v . k_p \left[1 - \frac{\tau_v . \tau_p}{(\tau_p - \tau_v)} \left\{ \frac{e^{-t/\tau_p}}{\tau_v} - \frac{e^{-t/\tau_v}}{\tau_p} \right\} \right] \qquad ...(9.56)$$

If the time constant of the control valve is much less than time constant of the process, *i.e.*, $\tau_v <<< ... < \tau_p$.

$$\Rightarrow \quad Y(t) = k_v . k_p \, [1 - e^{-t/\tau_p}] \qquad ...(9.57)$$

This is the response equation of control valve associated with process for a unit step change in input.

■ **Example 9.5** Starting from the basic derive the transfer function for a jacketed kettle.

Case 1 : When the heat transfer is in between kettle fluid and thermometer.

Case 2 : When the heat transfer is in between jacket and kettle fluid.

Solution : Consider a jacketed kettle as shown in Fig. 9.11. A thermometer is used to measure the temperature of jacketed kettle fluid..

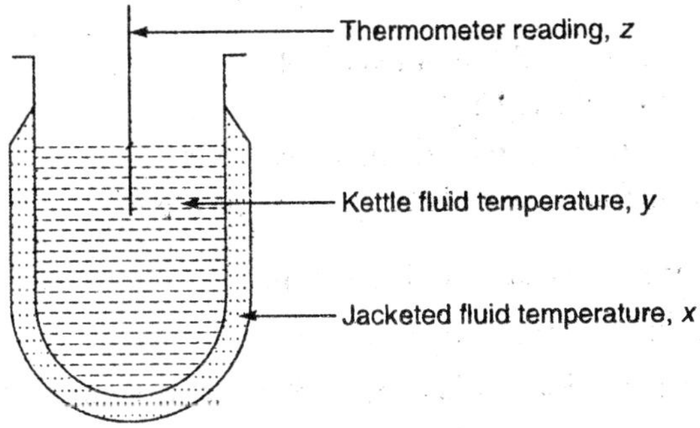

Fig. 9.11 Jacketed kettle.

Let x be the temperature of jacketed fluid

Let y be the temperature of kettle fluid

Let z be the temperature of thermometer reading

Case 1 : Heat transfer is in between kettle fluid and thermometer

At transient state heat energy balance, we have

Rate of heat in – Rate of heat out = Rate of heat accumulation

$$h_{kf} A_{tb}(y - z) - 0 = m_m c_{pm} \frac{dz}{dt} \quad ...(9.58)$$

At steady state, we have

$$h_{kf} A_{tb} (y_s - z_s) - 0 = 0 \quad ...(9.59)$$

Subtracting equation (9.59) from equation (9.58), we get

$$h_{kf} A_{tb} [(y - y_s) - (z - z_s)] = m_m C_{pm} \frac{dz}{dt} \quad ...(9.60)$$

Defining deviation variable, we have

$$Y = y - y_s$$
$$Z = z - z_s$$

$$\Rightarrow \quad h_{kf} A_{tb} [Y - Z] = m_m C_{pm} \frac{dZ}{dt} \quad ...(9.61)$$

$$\Rightarrow \quad Y - Z = \frac{m_m C_{pm}}{h_{kf} A_{tb}} \frac{dZ}{dt}$$

Let,

$$\tau_1 = \frac{m_m C_{pm}}{h_{kf} A_{tb}}$$

$$\Rightarrow \quad Y - Z = \tau_1 \frac{dZ}{dt} \quad ...(9.62)$$

$$\Rightarrow \quad Y = Z + \tau_1 \frac{dZ}{dt} \quad ...(9.63)$$

$$\Rightarrow \quad \frac{Z(s)}{Y(s)} = \frac{1}{\tau_1 s + 1} \quad ...(9.64)$$

where, m_m = Mass of mercury in the thermometer bulb, kg
C_{pm} = Specific heat of mercury, J/kg°C
h_{kf} = Heat transfer coefficient of kettle fluid, W/m^2°C
A_{tb} = Heat transfer surface area of thermometer bulb, m^2

Case 2 : Heat transfer is in between jacket and kettle fluid

At transient state heat energy balance, we have

Rate of heat in – Rate of heat out = Rate of heat accumulation

$$h_{jf} A_k (x - y) - h_{kf} A_{tb} (y - z) = m_{kf} C_{pkf} \frac{dy}{dt} \quad ...(9.65)$$

At steady state, we have

$$h_{jf} A_k (x_s - y_s) - h_{kf} A_{tb} (y_s - z_s) = 0 \quad ...(9.66)$$

Subtracting equation (9.66) from equation (9.65), we get

$$h_{jf} A_k [(x - x_s) - (y - y_s)] - h_{kf} A_{tb} [(y - y_s) - (z - z_s)] = m_{kf} C_{pkf} \frac{dy}{dt} \quad ...(9.67)$$

Defining deviation variable, we have

$$X = x - x_s$$
$$Y = y - y_s$$
$$Z = z - z_s$$

$$\Rightarrow \quad h_{jf} A_k [X - Y] - h_{kf} A_{tb} [Y - Z] = m_{kf} C_{pkf} \frac{dY}{dt}$$

$$\Rightarrow \quad X - Y - \frac{h_{kf} A_{tb}}{h_{jf} A_k} [Y - Z] = \frac{m_{kf} C_{pkf}}{h_{jf} A_k} \frac{dY}{dt} \quad \ldots(9.68)$$

Let,

$$\alpha = \frac{h_{jf} A_{tb}}{h_{jf} A_k}$$

$$\tau_2 = \frac{m_{kf} C_{pkf}}{h_{jf} A_k}$$

$$\Rightarrow \quad X - Y - \alpha Y + \alpha Z = \tau_2 \frac{dY}{dt}$$

$$\Rightarrow \quad X = Y + \alpha Y - \alpha Z + \tau_2 \frac{dY}{dt} \quad \ldots(9.69)$$

Substituting equation (9.63) in equation (9.69), we get

$$\Rightarrow \quad X = Z + \tau_1 \frac{dz}{dt} + \alpha \left[Z + \tau_1 \frac{dz}{dt} \right] - \alpha Z + \tau_2 \left[Z + \tau_1 \frac{dz}{dt} \right]$$

$$\Rightarrow \quad X = Z + \tau_1 \frac{dz}{dt} + \alpha Z + \alpha \tau_1 \frac{dz}{dt} - \alpha Z + \tau_2 z + \tau_1 \tau_2 \frac{d^2 z}{dt^2}$$

$$\Rightarrow \quad X(s) = Z(s) + \tau_1 s Z(s) + \alpha \tau_1 s Z(s) + \tau_2 Z(s) + \tau_1 \tau_2 s^2 Z(s)$$

$$\Rightarrow \quad X(s) = [\tau_1 \tau_2 s^2 + (\tau_1 + \tau_2 + \alpha \tau_1) s + 1] \, Z(s)$$

$$\Rightarrow \quad \frac{Z(s)}{X(s)} = \frac{1}{[\tau_1 \tau_2 s^2 + (\tau_1 + \tau_2 + \alpha \tau_1) s + 1]} \quad \ldots(9.70)$$

where, m_{kf} = Mass of kettle fluid, kg
C_{pkf} = Specific heat of kettle fluid, J/kg°C
h_{jf} = Heat transfer coefficient of jacketed fluid, W/m^2°C
A_k = Heat transfer surface area of kettle, m^2

This is the transfer function of jacketed kettle.

9.2 TRANSPORTATION LAG

Transportation lag also known as dead time or time delay or distance velocity lag. It is present in most of the flow system. It is present in the flow system because of the reason that the fluids are transported through the pipes to distance places. It is the time delay incorporated the information about the change of variable from one place to another place during the course of transportation.

Let us consider an insulated pipe of length '*L*' and uniform cross-sectional area '*A*' as shown in Fig. 9.12.

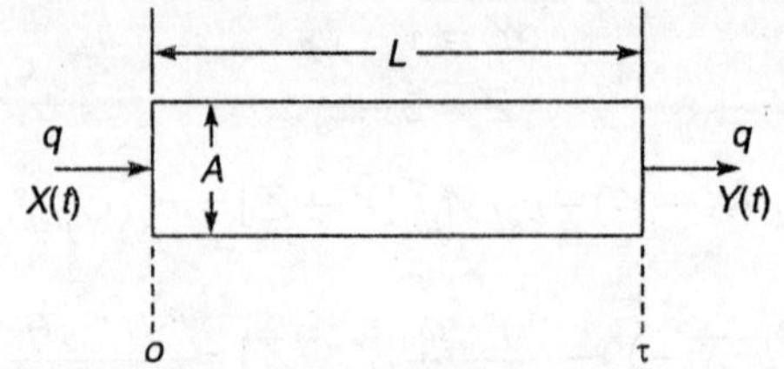

Fig. 9.12 Fluid flowing through insulated pipe.

Let us also consider a fluid is flowing through the insulated pipe with a constant volumetric flow rate '*q*' has density 'ρ' and specific heat C_p.

Let, $x(t)$ and $y(t)$ be the temperature of the fluid at inlet and outlet of the pipe at any instant.

We can write relation between the inlet and outlet temperature as

$$y = x(t - \tau) \qquad ...(9.71)$$

At steady state, we have

$$y_s = x_s (t - \tau) \qquad ...(9.72)$$

Subtracting equation (9.72) from equation (9.71), we get

$$y - y_s = x(t - \tau) - x_s (t - \tau) \qquad ...(9.73)$$

Defining deviation variable, we get

$$Y(t) = X(t - \tau) \qquad ...(9.74)$$

Taking Laplace transform of equation (9.74), we get

$$Y(s) = e^{-\tau s} \times (s)$$

$$\Rightarrow \qquad \frac{Y(s)}{X(s)} = e^{-\tau s} \qquad ...(9.75)$$

where, τ = Time delay, s.

This is the transfer function of the system having transportation lag.

REVIEW QUESTIONS

9.1 Define second order system.

9.2 Give the examples of second order system.

9.3 What is damping coefficient? Discuss.

9.4 Does damping coefficient exists in U-tube manometer? Why?

9.5 Flow of manometric fluid in U-tube manometer is laminar or turbulent? Discuss.

9.6 What is the effect of 'ξ' on the response of U-tube manometer.

9.7 When the system becomes critically damped? Explain.

9.8 What do you mean by overshoot? Explain.

9.9 What is decay ratio? Explain.

9.10 With a neat diagram explain the basic concept of under-damped response of second order system.

9.11 Define settling time.

9.12 Explain the term natural period of oscillation.

EXERCISES

9.1 Starting from the basic derive the transfer function for damped vibrator. State all the assumptions made.

9.2 Starting from the basic derive the transfer function for U-tube manometer. State all the assumption made.

9.3 Derive the response equation of U-tube manometer for a unit step change in input when the system is over-damped.

9.4 Derive the response equation of U-tube manometer for a unit step change in input when the system is under-damped.

9.5 Derive the response equation of U-tube manometer for a unit step change in input when the system is critically damped.

9.6 Derive the response equation of U-tube manometer for a unit impulse change in input when the system is over-damped.

9.7 Derive the response equation of U-tube manometer for a unit impulse change in input when the system is under-damped.

9.8 Derive the response equation of U-tube manometer for a unit impulse change in input when the system is critically damped.

9.9 Starting from the basic derive an equation for overshoot.

9.10 The two tank liquid level system is operating at steady state. When a unit step change is made in the flow rate to tank 1, the transient response is critically damped and it takes 0.1 min for the change in liquid level of the second tank to reach 50% of the total change.
If the ratio of cross-sectional area of the tank is $A_1/A_2 = 2$. Calculate the ratio R_1/R_2. Also, calculate the time constant of each tank. How long does it take for the change in level of the first tank to reach 90% of the total change?

9.11 A step change of magnitude 8 is introduced into a system having transfer function

$$\frac{Y(s)}{X(s)} = \frac{10}{2s^2 + 4s + 8}$$

Determine :

(i) % overshoot

(ii) Ultimate value of $Y(t)$.

□□□

Chapter 10

Industrial Bioprocesses

10.1 INTRODUCTION

Industrial bioprocess engineering is a special branch of chemical engineering. It is the conglomerate of mathematics, biology and industrial process. It deals with design and development of processes for the manufacturing of products from biological materials. For example - pharmaceutical products. It also, deals with studying of various biotechnological process used in industries for large scale production.

10.2 BASIC CONCEPT OF BIOCHEMICAL PROCESS

The basic concept of biochemical process involves following important terms :

10.2.1 Cell

It is the tiny unit of life that can replicate independently. Therefore, cell is also called building block of life. The study of cell is known as cell biology. It consists of a protoplasm enclosed within a membrane which, contain nucleic acid and proteins.

10.2.2 Bacteria

It is a living thing that has only one cell. Under the microscope it look like a ball, rod or spiral. It is present in most of the soil, water, acedic hot spring and radio active waste material. The study of the bacteria is known as bacterialogy. All bacteria are not harmful. Some bacteria helps to digest food and give the body needful vitamins. It is also used to make healthy food such as yogurt and cheese. But, some bacteria like streptococcus, staphytococcus and E-coli are harmful that can causes infectious deseases.

10.2.3 Algae

It is a very diverse group of photosynthetic organism that ranges from microscopic size to large seaweed. The large seaweed algae grows upto 20 m long is also known as giant kelp. Algae supply oxygen and consume nutrients in several different processes for biochemical waste treatment. Outdoor ponds containing brackish, sea and waste water are good for growing algae, because it needed vast surface area.

10.2.4 Fungi

Fungi are the member of a large group of eukaryotic organism that include microorganism such as yeasts, molds and mushrooms. These organisms are separate from plants, animals and bacteria. Fungi are heterotrophs. They can not manufacture their own food themselves. They decompose dead matter and some species are parasites.

10.2.5 Yeast

Yeast are unicellular microorganism surrounded by a cell wall, although some species are multicellular. It posses a distinct nucleus. With a very few exceptions, yeast reproduces by a process known as budding.

10.2.6 Viruses

Viruses are not cellular in structure. It is mainly composed of nucleic acid and surrounded by a protein sheath. The size of the virus particles are below the resolution of light microscope. Viruses exist only as intracellular, highly host specific parasite. Certain mould and many bacteria are subjected to invasion by virus particles.

10.3 BIOREACTOR

A bioreactor is a device that support biologically active environment to carried out chemical process which involves organisms. Generally, fermenters are called bioreactor. Some example of bioreactors are batch fermenter, horizontal fermenter, tower fermenter, air lift fermenter. A typical bioreactor is shown in Fig. 10.1.

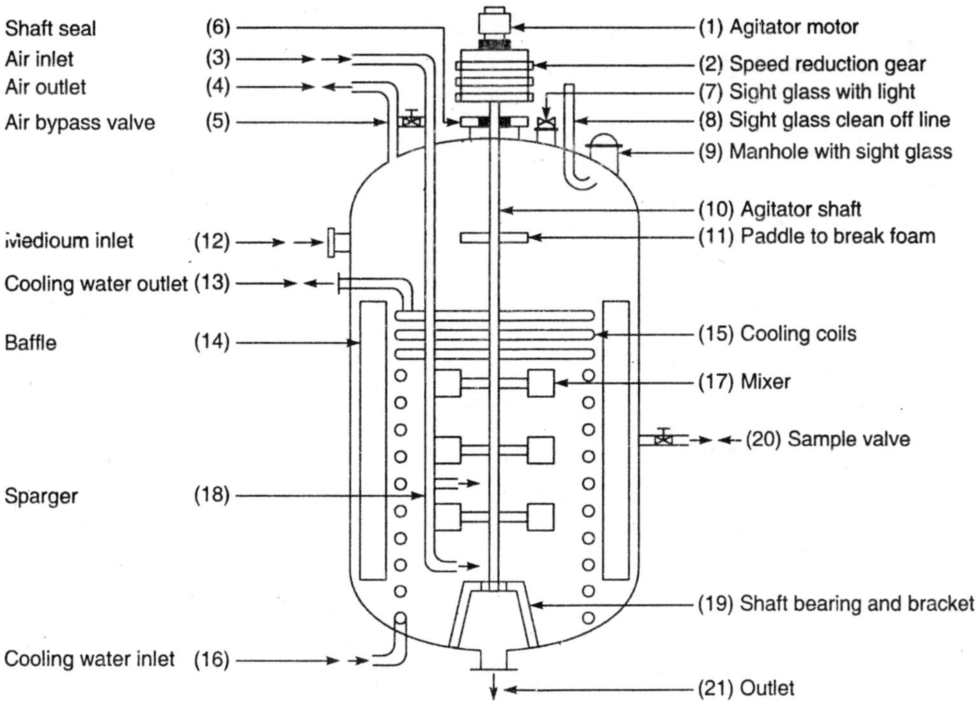

Fig. 10.1 Bioreactor.

These bioreactors are commonly cylindrical vessel consisting agitator, baffle, cooling coil, sparger and are made of stainless steel.

The functions desired in bioreactors are gas-liquid contacting, feed of nutrients, online sensing of concentration, foam control, mixing and heat transfer. The foam control is achieved either by mechanical destruction or by chemical anti foam agents. Several other factors such as temperature, pressure, shaft power and viscosity are monitored.

When a particular organism is introduced into a selected growth medium, it get inoculated with the particular organism. The growth of the inoculum does not take place immediately. It is a slow process. The adaptation take place during this period and called lag phase. Following the lag phase, the rate of growth of the organism increases exponentially for a certain period. This period is called exponential phase.

After certain time of exponential phase, the rate of growth of the organism slows down that is because of the reason that the concentration of the nutrients decreases where as the concentration of the toxic substances increases continuously. During this period, the rate of growth of organism is checked. This period is called deceleration phase. After the deceleration phase, the growth of the organism ceases and the culture enters a stationary phase.

The microbes used for biochemical process grow in a selected medium which supply the nutrients required by the organism. There are variety of media exists, but a media contains a carbon source, a nitrogen source, water, salt and micro nutrients are most preferable. For example, in the production of wine, the grape is used as medium for the growth of organism.

10.4 STIRRED TANK HEATER

A stirred tank heater is a device that used to heat the liquid in a tank by an electric heater. It is commonly a cylindrical vessel consisting agitator, heater, sensor and are made of stainless steel. It also has a temperature controller and a temperature converter. A typical industrial stirred tank heater is shown in Fig. 10.2.

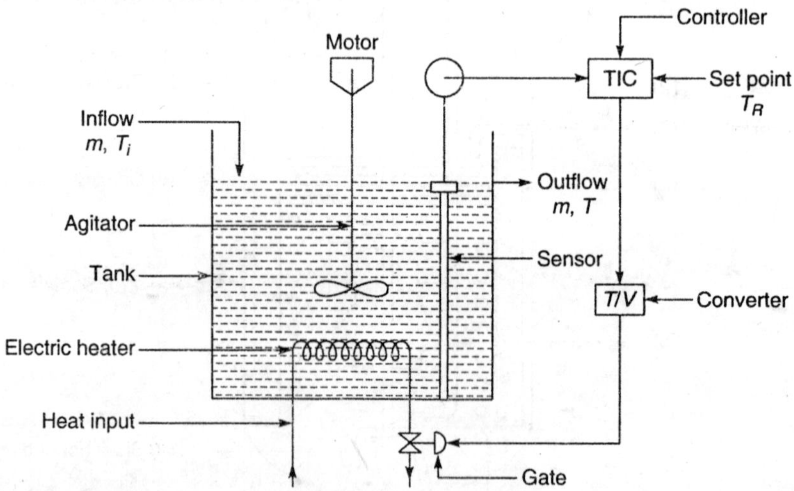

Fig. 10.2 Stirred tank heater.

The process fluid stream flows into a well agitated tank equipped with a electric heater at temperature T_i and constant flow rate m kg/s. It flows out of the tank by overflow. Thus, the volume of the tank is maintained constant.

The objectives of the stirred tank heater is to heat the fluid stream to a desired temperature T_i when the disturbance occurs in the system. The disturbances may occurs due to the change in inlet fluid flow rate and temperature. It is assumed that the agitation is sufficient to ensure fluid at any point in the tank is at same temperature T.

The automatic control for stirred tank heater can be achieved by measuring the outlet fluid temperature using a temperature sensor and comparing the same with the desired value and thereafter, adjusting the heat input to the heater accordingly.

10.5 ON-LINE DATA ANALYSIS

On-line data analysis for state and parameter estimation for a biochemical process is a technique which can provide improved control performance with enhanced productivity. For example, consider a bioreactor. The deficiency of online information of bioreactor concerning the process state impose limitation for effective control action. Hence, for better monitoring and control of a bioreactor on-line estimation of the process variable is essential.

The different methods by which state and parameters of a bioreactor can be estimated are as follows :

(i) Kalman filter
(ii) Observer
(iii) Balanced equation
(iv) Neural network
(v) Fuzzy reasoning
(vi) Other methods

In this section, Kalman filter method is discussed.

10.5.1 Kalman Filter

The Kalman filter is used to provide most favourable estimate of measured and unmeasured state for non-linear system in the presence of noise by combining information using a mathematical model of the process. A block diagram of state estimation scheme by Kalman filter is shown in Fig. 10.3.

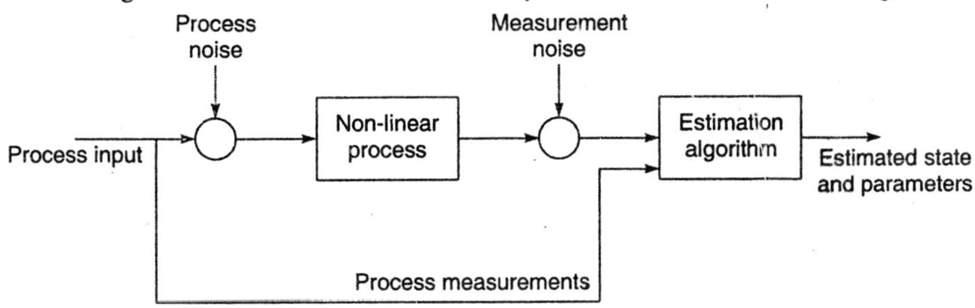

Fig. 10.3 Block diagram of Kalman filter.

The process estimate obtained by Kalman filter at any time is the optimum estimate condition among all observations up to that time.

The algorithm used in the Kalman filter consists of two recursive steps. In the first step the process model propagate the initial state estimate and in the second step the propagated model estimate is combined with the process measurement to provide an updated estimate.

Advantages : It can provide accurate estimate of the process. It can also deal with stochastic disturbances.

Disadvantages : It require much time and effort.

REVIEW QUESTIONS

10.1 Discuss on industrial bioprocess.

10.2 Explain the basic concept of biochemical process.

10.3 With a neat diagram explain the working of a bioreactor.

10.4 With a neat diagram explain the working of a stirred tank heater.

10.5 With a block diagram explain the on-line data analysis of a bioreactor.

□□□

Chapter 11

Controllers and Final Control Elements

11.1 INTRODUCTION

The controller is the active element that receives the information from the measurements and takes appropriate control action to adjust the value of the manipulated variables.

The controller mechanism reacts to the error signal and produces the output signal proportional to some function of the error. The output signal of the controller is the activating force, positioning the final control element.

11.2 BASIC MODES OF CONTROL ACTION

The basic modes of control action are as follows :

(i) Two position control
(ii) Proportional control
(iii) Derivative control
(iv) Integral control

These modes describe the action of the controller, *i.e.*, relation of the output signal to the input or error signal. Actual industrial controllers may be employed either separately or in combination.

11.2.1 Two Position Control

The two position control signifies that the output signal is either zero or 100%. This is often referred to as on-off or open-closed control.

This controller acts drastically because it attempts to correct any error presented in the system by shifting the output from one extreme to the other. That is because, when the measured variable value is above the desired variable value the output signal is zero and when the measured variable value is below the desired variable value the output signal is at its maximum. A typical two position control action is shown in Fig. 11.1.

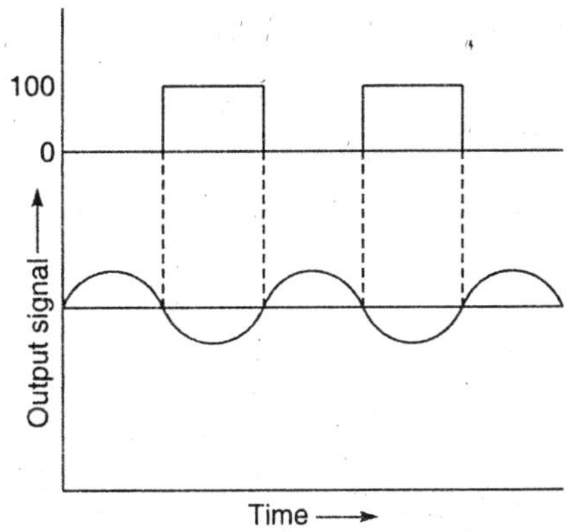

Fig. 11.1 Two position control action.

In actual controller, there is a small interval or differential gap between the points at which the output signal changes from zero to maximum or vice-versa. In this the output signal does not go to its maximum until the measured variable decreases to the lower line and does not go to the zero until the measured variable passes the upper limit.

Advantages : The advantages of on-off control are as follows :

(i) Simplest and cheapest.

(ii) Easy to maintain and operate.

(iii) Acts as a safety device in control loop.

Disadvantages : The disadvantages of on-off control are as follows :

(i) Off set is more.

(ii) Output has no relation whatever may be the input.

(iii) More wear and tear of the controlling equipment.

11.2.2 Proportional Control

The proportional control action is the action in which there is a continuous linear relationship between the controlled variable value and the position of the final control element, *i.e.*, the proportional controller produces an output signal which is proportional to the error 'ε'. This action may be expressed as

$$p = p_s + k_c \varepsilon \qquad \ldots(11.1)$$

where, p = Output signal from the controller

p_s = Steady state controller signal

k_c = Controller gain

ε = Error

$$\Rightarrow \quad p - p_s = k_c \varepsilon \qquad \ldots(11.2)$$

Let $P = p - p_s$

$$\Rightarrow \quad P = k_c \varepsilon$$

$$\Rightarrow \quad P(s) = k_c \varepsilon(s)$$

$$\Rightarrow \quad \frac{P(s)}{\varepsilon(s)} = k_c \qquad \ldots(11.3)$$

This is the transfer function of proportional controller.

Proportional Band (PB) : It is defined as the error required to move the valve position from fully open to fully closed. The term proportional band is commonly used among process control engineers in place of the term gain. This is also called as band width. It is expressed as the percentage of the range of measured variable.

$$\text{PB} = \frac{\Delta\varepsilon \text{ required to move the control valve to move from the fully open to fully closed position}}{\text{Range of measured variable}} \times 100$$

The proportional band 'b' is also related to gain k_c and may be expressed as

$$b = \frac{1}{k_c} \qquad \ldots(11.4)$$

11.2.3 Derivative Control

The derivative controller produces a continuous linear relation between rate of change of controlled variable and position of the final control element. This is also called preact, booster, rate or anticipating control action.

The derivative controller produces an output which is proportional to the derivative of the error. This action may be mathematically expressed as

$$p = p_s + k_c \tau_D \frac{d\varepsilon}{dt} \qquad \text{...(11.5)}$$

$$\Rightarrow \quad p - p_s = k_c \tau_D \frac{d\varepsilon}{dt}$$

Let, $$P = p - p_s$$

$$\Rightarrow \quad P = k_c \tau_D \frac{d\varepsilon}{dt}$$

$$\Rightarrow \quad P(s) = k_c \tau_D s\, \varepsilon(s)$$

$$\Rightarrow \quad \frac{P(s)}{\varepsilon(s)} = k_c \tau_D s \qquad \text{...(11.6)}$$

where, τ_D = Derivative time constant.

This is the transfer function of derivative controller. It is not satisfactory to use this response alone because of its inability to recognise a steady state deviation. So, the rate of response is used in conjunction with proportional control action.

11.2.4 Integral Control

The integral controller produces a continuous linear relation between the value of controlled variable and rate of motion of final control element.

The integral controller produces an output which is proportional to integral of the error. This action may be mathematically expressed as

$$p = p_s + \frac{k_c}{\tau_I} \int_0^t \varepsilon \, dt \qquad \text{...(11.7)}$$

$$\Rightarrow \quad p - p_s = \frac{k_c}{\tau_I} \int_0^t \varepsilon \, dt$$

Let, $$P = p - p_s$$

$$\Rightarrow \quad P = \frac{k_c}{\tau_I} \int_0^t \varepsilon \, dt$$

$$\Rightarrow \quad P(s) = \frac{k_c}{\tau_I} \frac{\varepsilon(s)}{s}$$

$$\Rightarrow \quad \frac{P(s)}{\varepsilon(s)} = \frac{k_c}{\tau_I s} \qquad \text{...(11.8)}$$

where, τ_I = Integral time constant.

11.3 COMBINATION OF CONTROL MODES

Most of the process control, where quality control is desired under conditions of significant load changes or large lags, is done by the use of combination of basic modes of control system. The combination mode of control systems are as follows :

(i) Proportional Integral (PI) Controller
(ii) Proportional Derivative (PD) Controller
(iii) Proportional Integral Derivative (PID) Controller

11.3.1 Proportional Integral Controller

This is also called proportional reset controller. In this case, in addition to the proportional controller there is an additional controller which is proportional to the integral of the error. This combination mode of control action may be expressed as

$$p = p_s + k_c \varepsilon + \frac{k_c}{\tau_I} \int_0^t \varepsilon \, dt \qquad \text{...(11.9)}$$

$$\Rightarrow \quad p - p_s = k_c \varepsilon + \frac{k_c}{\tau_I} \int_0^t \varepsilon \, dt$$

Let, $P = p - p_s$

$$\Rightarrow \quad P = k_c \varepsilon + \frac{k_c}{\tau_I} \int_0^t \varepsilon \, dt$$

$$\Rightarrow \quad P(s) = k_c \varepsilon(s) + \frac{k_c}{\tau_I} \frac{\varepsilon(s)}{s}$$

$$\Rightarrow \quad P(s) = k_c \left[1 + \frac{1}{\tau_I s}\right] \varepsilon(s)$$

$$\Rightarrow \quad \frac{P(s)}{\varepsilon(s)} = k_c \left[1 + \frac{1}{\tau_I s}\right]$$

This is the transfer function of the PI controller.

For a step change in input, we have

$$\varepsilon(t) = Au(t)$$

$$\Rightarrow \quad P(t) = k_c \, Au(t) + \frac{k_c}{\tau_I} \int_0^t Au(t) \, dt$$

$$\Rightarrow \quad P(t) = k_c A + \frac{k_c}{\tau_I} At \qquad \text{...(11.11)}$$

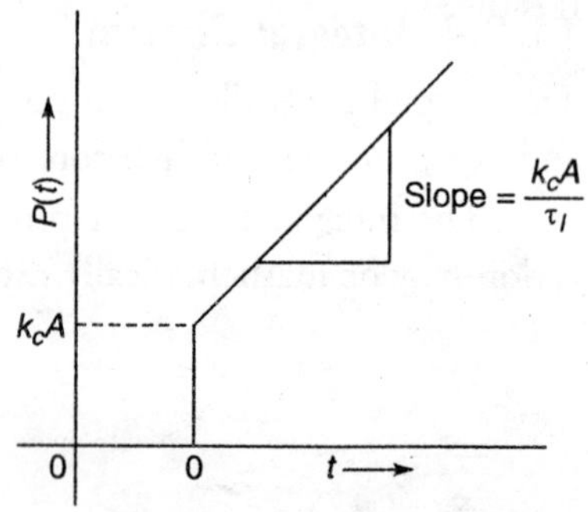

Fig. 11.2 Step response of PI controller.

The response of PI controller to a step change in error is shown in Fig. 11.2.

11.3.2 Proportional Derivative Controller

This type of controller action is also called proportional rate control. In this case, in addition to the proportional controller there is an additional controller which is proportional to the derivative of the error. This combination mode of control action may be expressed as

$$p = p_s + k_c\varepsilon + k_c\tau_D \frac{d\varepsilon}{dt} \quad ...(11.2)$$

$$\Rightarrow \quad p - p_s = k_c\varepsilon + k_c\tau_D \frac{d\varepsilon}{dt}$$

Let, $P = p - p_s$

$$\Rightarrow \quad P = k_c\varepsilon + k_c\tau_D \frac{d\varepsilon}{dt}$$

$$\Rightarrow \quad P(s) = k_c\varepsilon(s) + k_c\tau_D s\,\varepsilon(s)$$

$$\Rightarrow \quad P(s) = k_c[1 + \tau_D s]\varepsilon(s)$$

$$\Rightarrow \quad \frac{P(s)}{\varepsilon(s)} = k_c[1 + \tau_D s] \quad ...(11.13)$$

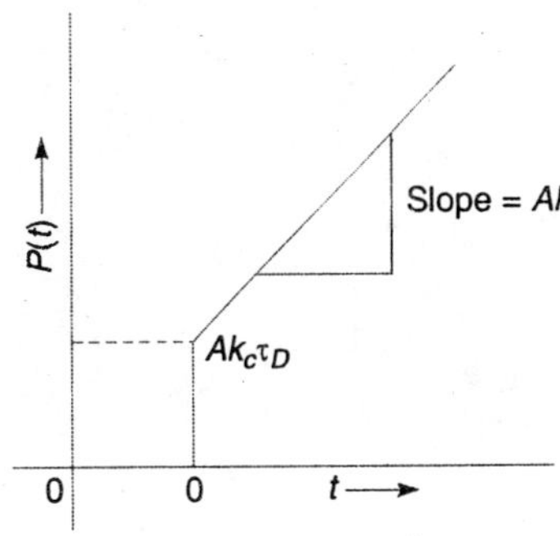

Fig. 11.3 Linear response of PD controller.

For a linear change in input, we have

$$\varepsilon(t) = At$$

$$\Rightarrow \quad P(t) = k_cAt + k_c\tau_D \frac{d(At)}{dt}$$

$$\Rightarrow \quad P(t) = k_cAt + k_c\tau_D A \quad ...(11.14)$$

Here, $P(t)$ changes suddenly by an amount $Ak_c\tau_D$ as a result of derivative action and then changes linearly at a rate Ak_c. The response of PD controller to a linear change in input is shown in Fig. 11.3.

11.3.3 Proportional Integral Derivative Controller

It is the combination of all the 3 modes of control action. This combination mode of control action may be expressed as

$$p = p_s + k_c\varepsilon + \frac{k_c}{\tau_I}\int_0^t \varepsilon\, dt + k_c\tau_D \frac{d\varepsilon}{dt} \quad ...(11.15)$$

$$\Rightarrow \quad p - p_s = k_c\varepsilon + \frac{k_c}{\tau_I}\int_0^t \varepsilon\, dt + k_c\tau_D \frac{d\varepsilon}{dt}$$

Let, $P = p - p_s$

$$\Rightarrow \quad P = k_c\varepsilon + \frac{k_c}{\tau_I}\int_0^t \varepsilon\, dt + k_c\tau_D \frac{d\varepsilon}{dt}$$

$$\Rightarrow \quad P(s) = k_c\varepsilon(s) + \frac{k_c}{\tau_I}\frac{\varepsilon(s)}{s} + k_c\tau_D s\,\varepsilon(s)$$

$$\Rightarrow \quad P(s) = k_c\left[1 + \frac{1}{\tau_I s} + \tau_D s\right]\varepsilon(s)$$

$$\Rightarrow \quad \frac{P(s)}{\varepsilon(s)} = k_c\left[1 + \frac{1}{\tau_I s} + \tau_D s\right] \quad ...(11.16)$$

For a linear change in input, we have

$$\varepsilon(t) = At$$

$$\Rightarrow \quad P(t) = k_cAt + \frac{k_cAt^2}{2\tau_I} + k_c\tau_D A \quad ...(11.17)$$

With the presence of the derivative term, the PID controller anticipates what the error will be in the immediate future and applies a control action which is proportional to the current rate of change in the error. The response of PID controller to a linear change in input is shown in Fig. 11.4.

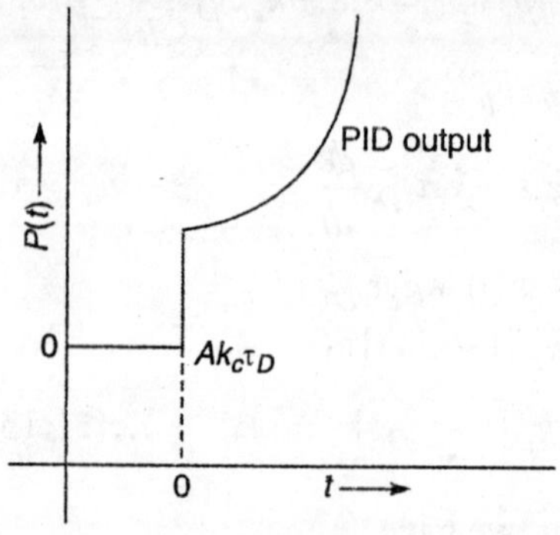

Fig. 11.4 Linear response of PID controller.

■ **Example 11.1** A proportional controller is used to control temperature within the range of 60 to 100°C. The controller is adjusted so that the output pressure goes from 3 Psi to 15 Psi. As the measured temperature goes from 71 to 75°C with the set point held constant. Find the proportional band and gain.

Solution :

$$PB = \frac{\Delta\varepsilon \text{ required to move the control valve from fully open to fully closed position}}{\text{Range of measured variable}} \times 100$$

$$PB = \frac{\Delta\varepsilon}{\Delta P}$$

$$PB = \frac{75-71}{100-60} \times 100$$

$$PB = 10\%$$

$$k_c = \frac{\Delta P}{\Delta\varepsilon}$$

$$k_c = \frac{15-3}{75-71}$$

$$k_c = \frac{12}{4}$$

$$k_c = 3 \text{ Psi/°C}$$

■ **Example 11.2** A proportional controller has an output changing linearly from 3 to 15 Psi when the deviation 'ε' changes from –100 to 100°C. Calculate the controller sensitivity.

Solution :

$$PB = \frac{\Delta\varepsilon \text{ required to move the control valve from fully open to fully closed position}}{\text{Range of measured variable}} \times 100$$

$$PB = \frac{15-3}{100-(-100)} \times 100$$

$$PB = \frac{12}{200} \times 100$$

$$PB = 6\%$$

■ **Example 11.3** An electric temperature control has an output range of 4 mA to 20 mA and a chart calibration of 10°C to 170°C. If the output is 5 mA at 80°C when the proportional gain is 0.5 mA/°C.

Determine :

(i) The proportional band

(ii) The output at 90°C.

Solution :

$$PB = \frac{\Delta\varepsilon \text{ required to move the control valve from fully open to fully closed position}}{\text{Range of measured variable}} \times 100$$

$$PB = \frac{\Delta\varepsilon}{\Delta P} \times 100$$

$$PB = \frac{32}{(170-10)} \times 100$$

$$PB = 20\%$$

From the equation of process gain, we have

$$k_c = \frac{\Delta mA}{\Delta\varepsilon}$$

$$0.5 = \frac{20-4}{\Delta\varepsilon}$$

$$\Delta\varepsilon = 32$$

(ii) Calculate the output at 90°C using the equation

$$mA = k_c\varepsilon + (mA)_s \qquad \text{...(11.18)}$$

where, ε = Set point – Measured value of controlled variable

$$\Rightarrow \qquad mA = 0.5\,(x - T) + (mA)_s$$

$$\Rightarrow \qquad 5 = 0.5\,(x - 80) + (mA)_s \qquad \text{...(11.19)}$$

$$\Rightarrow \qquad mA = 0.5\,(x - 90) + (mA)_s \qquad \text{...(11.20)}$$

Comparing equations (11.19) and (11.20), we get

$$mA = 0$$

■ **Example 11.4** A PI controller has an output pressure of 10 Psi when the set point and pen point are together. The set point and pen point are suddenly displaced by 0.5 inch and the following data are obtained.

Time (sec)	0	0	20	60	85
P (Psi)	10	8	7	5	3

Determine the actual gain and the integral time.

Solution : Using the given data plot P vs t.

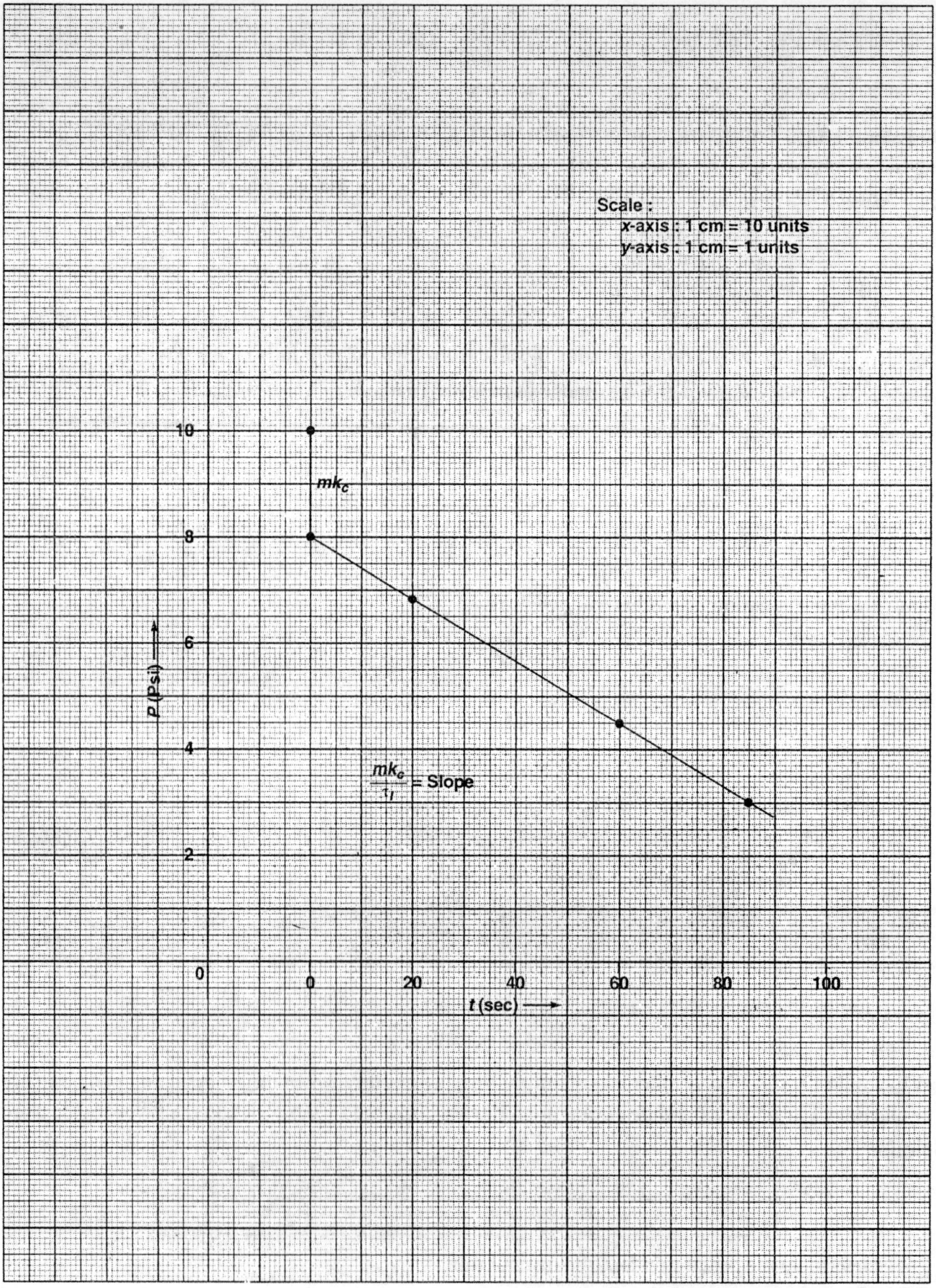

Fig. 11.5 Step response of PI controller.

From the Fig. 11.5, we get

$$mk_c = 8 - 10$$
$$mk_c = -2$$
$$k_c = \frac{-2}{m}$$

From the given data, we have

$$m = -0.5$$

$\Rightarrow$ $$k_c = \frac{-2}{-0.5}$$

$\Rightarrow$ $$k_c = 4 \text{ Psi/inch}$$

From the Fig. 11.5, we get

$$\frac{mk_c}{\tau_I} = -0.05$$
$$\tau_I = \frac{(-0.5)\times 4}{-0.05}$$
$$\tau_I = 40 \text{ sec}$$

■ **Example 11.5** A PD controller is given a linear input of 10% per min. The corresponding output is observed to undergo an initial step change of 5% followed by a gradual change at a steady rate of 2% per min. Find k_c and τ_D.

Solution : From the given data, we have

$$m = 10\%$$
$$mk_c = 2\%$$

$\Rightarrow$ $$k_c = \frac{2}{10}$$

$\Rightarrow$ $$k_c = 0.2$$

From the given data, we have

$$mk_c\tau_D = 5\%$$

$\Rightarrow$ $$\tau_D = \frac{5}{10\times 0.2}$$

$\Rightarrow$ $$\tau_D = 2.5 \text{ min}$$

■ **Example 11.6** A unit step change in error is introduced into a PID controller. If $k_c = 10, \tau_I = 1$ and $\tau_D = 0.5$. Plot the response of the controller against time. [Bangalore University, June/July 1989]

Solution : From the transfer function of PID controller, we have

$$\frac{P(s)}{\varepsilon(s)} = k_c\left[1 + \frac{1}{\tau_2 s} + \tau_D s\right] \quad \text{...(11.16)}$$

From the given data, we have

$$k_c = 10$$

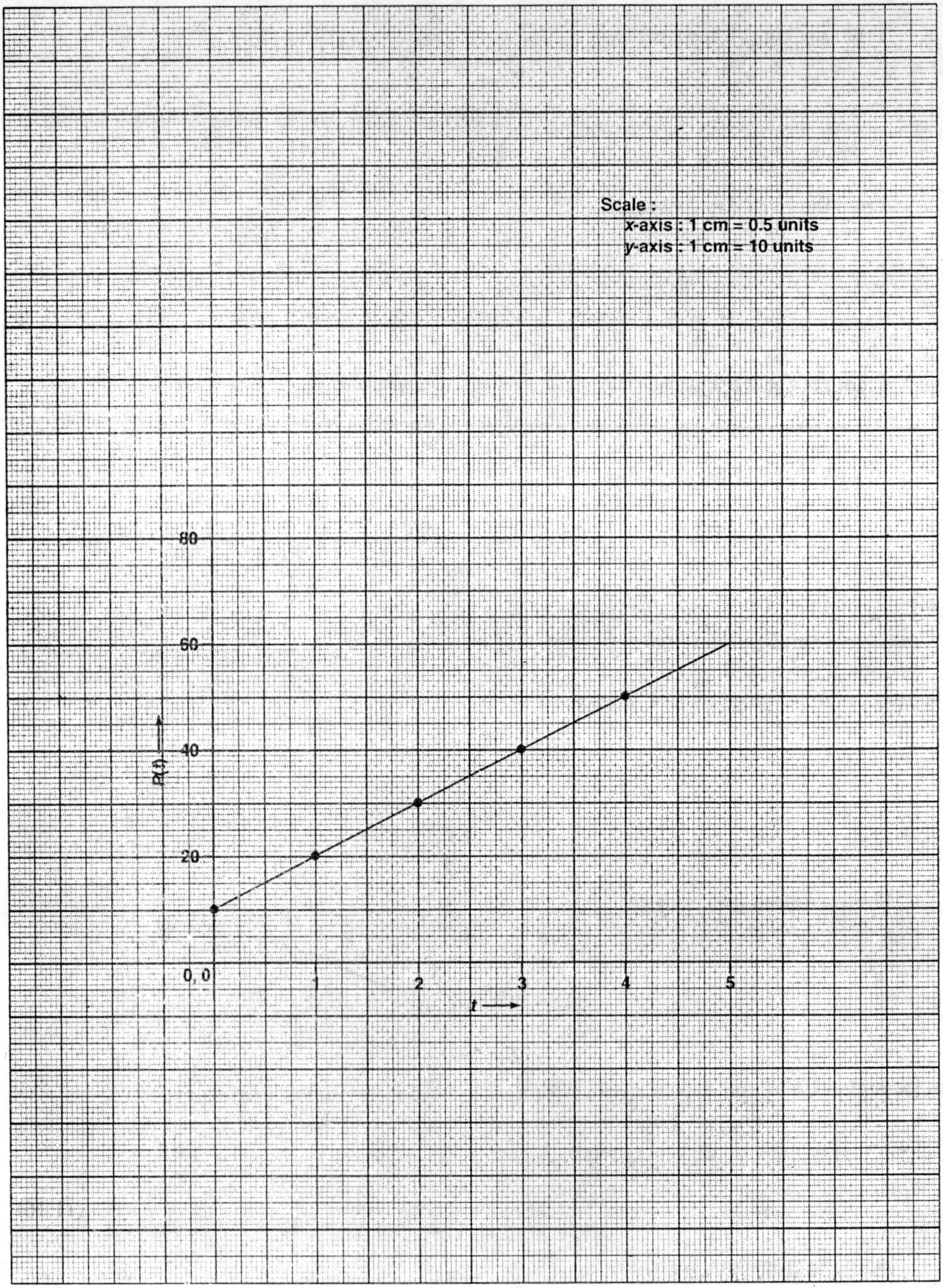

Fig. 11.6 Step response of PID controller.

$$\tau_I = 1$$
$$\tau_D = 0.5$$
$$\varepsilon(t) = 1$$
$$\varepsilon(s) = \frac{1}{s}$$

$$\Rightarrow \quad P(s) = 10\left[1 + \frac{1}{s} + 0.5s\right] \times \frac{1}{s}$$

$$\Rightarrow \quad P(s) = \frac{10}{s} + \frac{10}{s^2} + 5$$

$$\Rightarrow \quad P(s) = \frac{10s + 10 + 5s^2}{s^2}$$

$$\Rightarrow \quad P(s) = \frac{10s + 10 + 5s^2}{s^2} = \frac{C_1}{s} + \frac{C_2}{s^2}$$

$$\Rightarrow \quad 10s + 10 + 5s^2 = C_1(s) + C_2$$

Comparing the coefficient of 's'

$$10 = C_1$$

Comparing the coefficient of constant term,

$$10 = C_2$$

$$\Rightarrow \quad P(s) = \frac{10}{s} + \frac{10}{s^2}$$

$$\Rightarrow \quad P(t) = 10 + 10t$$

$P(t)$	t
10	0
20	1
30	2
40	3
50	4

■ **Example 11.7** A PID controller is at steady state with an output pressure of 10 Psi. The set point and pen point are initially together. At time $t = 0$, the set point is moved away from the pen point at a rate of 0.5 inch/min. The motion of the set point is in the direction of lower readings. If the knob settings are :

$$k_c = 2 \text{ Psi/inch of pen travel}$$
$$\tau_I = 1.5 \text{ min}$$
$$\tau_D = 0.5 \text{ min}$$

Plot the output pressure versus time.

Solution : The control action by PID controller is given by the expression

$$p = p_s + k_c\varepsilon + \frac{k_c}{\tau_I}\int_0^t \varepsilon\, dt + k_c\tau_D \frac{d\varepsilon}{dt} \qquad \text{...(11.15)}$$

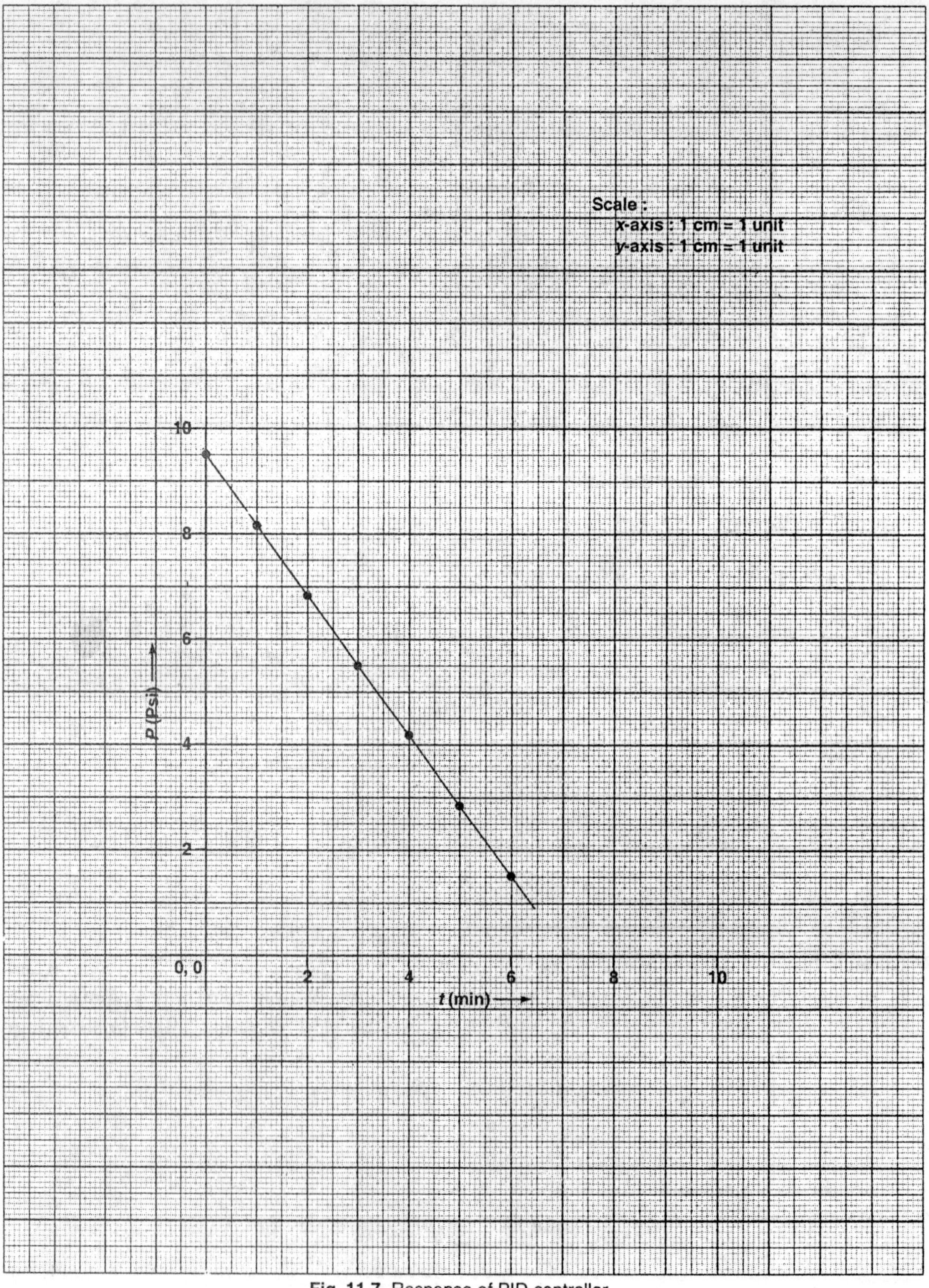

Fig. 11.7 Response of PID controller.

From the given data, we have

$$p_s = 10 \text{ Psi}$$
$$\varepsilon = -0.5t \text{ inch}$$
$$k_c = 2 \text{ Psi/inch of pen travel}$$
$$\tau_I = 1.5 \text{ min}$$
$$\tau_D = 0.5 \text{ min}$$

$$\Rightarrow \quad p = 10 + 2\times(-0.5t) + \frac{2}{1.5}\int_0^t (-0.5t)\,dt + 2\times 0.5\times\frac{d(-0.5t)}{dt}$$

$$\Rightarrow \quad p = 10 - t - \frac{0.6t^2}{2} - 0.5$$

$$\Rightarrow \quad p = 9.5 - t - 0.33t^2$$

p (Psi)	t (min)
9.50	0
8.17	1
6.84	2
5.51	3
4.18	4
2.85	5
1.52	6

11.4 FINAL CONTROL ELEMENT

The final control element is the mechanism which alters the value of the manipulated variable in response to the output signal from an automatic controller or direct manual manipulation. The automatic control installations normally consists of two parts :

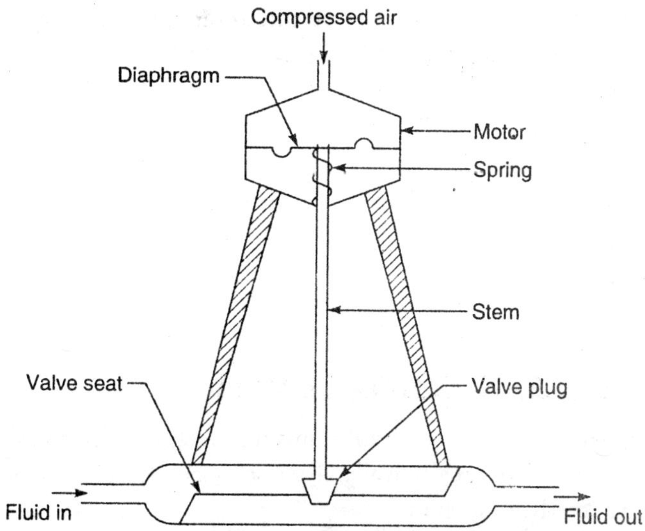

Fig. 11.8 Pneumatic control valve.

(i) An actuator which translate the output signal of the controlling device into an action involving a large force or the manipulation of large power.

(ii) A manipulated variable.

For example, the actuator may be used to change the position of a valve plug in an orifice, the velocity of a rotating device or the amount of power being delivered to an electrical load.

In automatic process control, the most frequently used final control element is a pneumatic control valve. It consists of a diaphragm, motor, actuator and a process fluid control valve. A pneumatic control valve is shown in Fig. 11.8.

The pneumatic control valve work on the principle of compressed air. Generally, the pneumatic control valve are of the following type :

(i) Air to close

(ii) Air to open

The most common body materials are cast iron and carbon steel. Low alloy steel such as chromium molybdenum are used for high temperature. Various grades of stainless steels, bronze, monel nickel. Inconel and many castable alloys are available for corrosive chemical service or for high or low temperature applications. When it is not practical to cast valve bodies, they may be machined from solid bar or forced material particularly in small sizes.

The range of pressure used in air to close pneumatic control valve is 3 Psi to 15 Psi. At 3 Psi it is fully open and 15 Psi it is fully closed. Initially, when the pressure is less the position of the valve plug is at higher level from the valve seat. This make the valve fully open and at this condition the flow rate is maximum. As the compressed air added in the chamber, the pressure increases which make the diaphragm to move downward along with stem and valve plug. At 15 Psi the valve plug sit on the valve seat completely and the flow is minimized to zero. Again, when the pressure releases slowly the spring bring back the valve plug to its original condition.

According to the definition stated for a final control element, the actuator is a transducer. It translates the control signal from one form or level of energy or power to another form. For example, from a pneumatic signal into a mechanical action which is used to manipulate a process variable. In general, a valve actuator may be termed as linear position actuator. It provides an output position that is proportional to its input signal. The motion of the actuator output element is generally translational, although the translational motion may be transformed into rotary motion to operate some valves. The following are the actuators may be used in automatic process control.

(i) Pneumatic actuator

(ii) Hydraulic actuator

(iii) Electric actuator

The other final control elements includes :

(i) Variable speed drive

(ii) Variable electric power actuator

(iii) Variable output pumps.

11.5 CHARACTERISTICS OF CONTROL VALVE

The valve plug and seat geometry is the most significant single factor in determining the characteristics of the control valve. A variety of plug types is available and may be used to provide a control valve with any one of several flow characteristics such as :

(i) Inherent flow characteristics
(ii) Installed flow characteristics

There are three basic types of valve characteristics as observed with constant pressure drop across the valve. This classification is based on the sensitivity of the rate of change of flow through the valve to the valve stem position.

(i) Decreasing sensitivity
(ii) Linear sensitivity
(iii) Increasing sensitivity

11.5.1 Decreasing Sensitivity

It is represented by the equation

$$\left.\frac{dQ}{dL}\right|_{Q=0} > n > \left.\frac{dQ}{dL}\right|_{Q=\max} \qquad \ldots(11.21)$$

where, L = Percentage of maximum valve stem travel
Q = Percentage of maximum flow
n = Valve flow coefficient

The basic characteristics for decreasing sensitivity is schematically illustrated in Fig. 11.9.

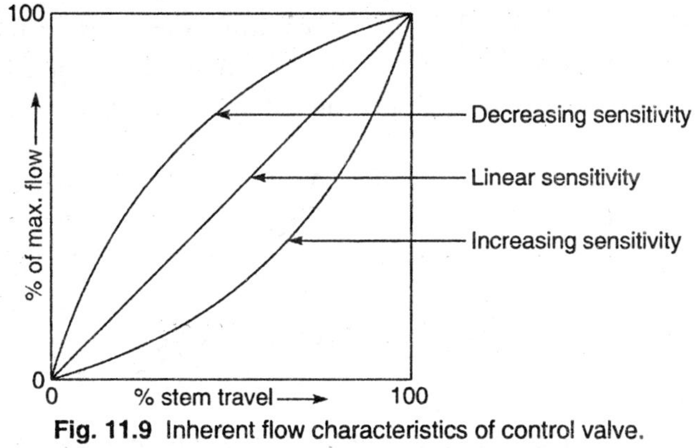

Fig. 11.9 Inherent flow characteristics of control valve.

Example of the decreasing sensitivity type are quick opening control valve, global valve and weirs. The inherent flow characteristics are determined by the valve orifice and plug geometry. They describe the flow rate through the valve as a function of the stem position with pressure drop across the valve kept constant.

11.5.2 Linear Sensitivity

It is represented by the equation

$$\frac{dQ}{dL} = n \qquad \ldots(11.22)$$

The basic characteristics for linear sensitivity is schematically illustrated in Fig. 11.9.

11.5.3 Increasing Sensitivity

It is represented by the equation

$$\left.\frac{dQ}{dL}\right|_{Q=0} < n < \left.\frac{dQ}{dL}\right|_{Q=\max} \qquad \text{...(11.23)}$$

Besides sensitivity two other valve properties are important. They are rangeability and turn down ratio.

The rangeability is defined as the ratio of maximum controllable flow to minimum controllable flow. It is represented by the symbol R.

$$R = \frac{\text{Maximum controllable flow}}{\text{Minimum controllable flow}}$$

The turndown is defined as the ratio of normal maximum flow to minimum controllable flow. It is represented by the symbol T.

$$T = \frac{\text{Normal maximum flow}}{\text{Minimum controllable flow}}$$

Generally, valve are sized so that $T = 0.7R$, where R lies, depending on the type of valve between 20 to 50.

The usefulness of rangeability and turndown may be indicated by a specific control valve application. For example, if the design of a tank level control system calls for a 25 : 1 change in inlet flow rate to maintain the level because of a change in outlet flow rate, then the indicated control valve turndown ratio is atleast 25 and the rangeability ratio at least 35. Typically, increasing sensitivity type valves have higher rangeability.

When control valves are combined with other fluid handling equipment in processing systems, the composite flow rate characteristic differ from the characteristics of any single component of the system. Flow rates through the valve are no longer determined solely by the geometry of the valve body and plug.

REVIEW QUESTIONS

11.1 What are the basic modes of control action?

11.2 What do you understand by two position control? Explain. Give its advantages and disadvantages.

11.3 What is proportional controller?

11.4 Define proportional band.

11.5 What is derivative control? Explain.

11.6 What is integral control? Explain.

11.7 Discuss on combination of control mode.

11.8 With a neat diagram explain the working of pneumatic control valve.

11.9 Discuss on final control element.

11.10 With a neat diagram explain the working of air to close pneumatic control valve.

11.11 With a neat diagram explain the working of air to open pneumatic control valve.

11.12 Other than pneumatic control valve name few more final control element.

11.13 Discuss on characteristics of control valve.

11.14 Define rangeability.

11.15 Define turndown of a control valve.

EXERCISES

11.1 Starting from the basic derive the transfer function for proportional controller.

11.2 Starting from the basic derive the transfer function for derivative controller.

11.3 Starting from the basic derive the transfer function for integral controller.

11.4 Starting from the basic derive the transfer function for proportional integral controller.

11.5 For a step change in input of a PI controller show that

$$P(t) = k_c A + \frac{k_c}{\tau_I} At$$

11.6 For a linear change in input of a PD controller show that

$$P(t) = k_c At + k_c \tau_D A$$

11.7 For a linear change in input of a PID controller show that

$$P(t) = k_c At + \frac{k_c At^2}{2\tau_I} + k_c \tau_D A$$

11.8 Starting from the basic derive the transfer function for PD controller.

11.9 Starting from the basic derive the transfer function for PID controller.

11.10 Suppose the proportional band of a controller is changed to 75%. Find the gain k_c. Temperature change is necessary to make a valve go from fully open to fully closed position.

11.11 A pneumatic PI controller has an output pressure of 10 Psi when the set point and pen point are together. The set point and pen point are suddenly displaced by 0.5 inch and the following data are obtained.

t (sec)	P (Psi)
0	10
0	8
20	7
60	5
90	3.5

Determine the actual gain and integral time.

11.12 A PD controller has a sensitivity of 1 and $\tau_D = 1$ min. If the deviation is sinusoidal, calculate the phase of oscillation of manipulated variable.

11.13 An ideal PD controller has the transfer function

$$\frac{P(s)}{\varepsilon(s)} = k_c(\tau_D s + 1)$$

where as an actual PD controller has the transfer function

$$\frac{P(s)}{\varepsilon(s)} = k_c \left[\frac{(\tau_D s + 1)}{((\tau_D/\beta)\, s + 1)} \right]$$

where, β = Large constant in an industrial controller.

If a unit step change in error is introduced into a controller having the second transfer function, show that

$$P(t) = k_c\,[1 + Ae^{-t\beta/\tau_D}]$$

Determine A which is function of β. For $\beta = 4$ and $k_c = 0.5$, plot $p(t)$ vs t/τ_D. As $\beta \to \infty$, show that the unit step response approaches that of the ideal controller.

11.14 An electrode temperature controller has an output range of 0 to 15 mA and a chart calibration of 5 to 200°C. The output is 5 mA at 80°C. The sensitivity is 0.5 mA/°C. Determine :

(i) The proportional band

(ii) The output at 90°C.

11.15 A unit step change in error is introduced into a PID controller. If $k_c = 15$, $\tau_I = 1$ and $\tau_D = 0.5$. Plot the response of the controller against time.

□□□

Chapter 12

Block Diagram and Transient Response of Closed Loop Control System

12.1 INTRODUCTION

A system represented by a diagram consisting of blocks, summing junction and arrows is called block diagram. We can find the overall transfer function of the control system from these block diagram by writing different algebraic equation and the process is known as block diagram algebra. The block diagram of a typical closed loop control system is shown in Fig. 12.1.

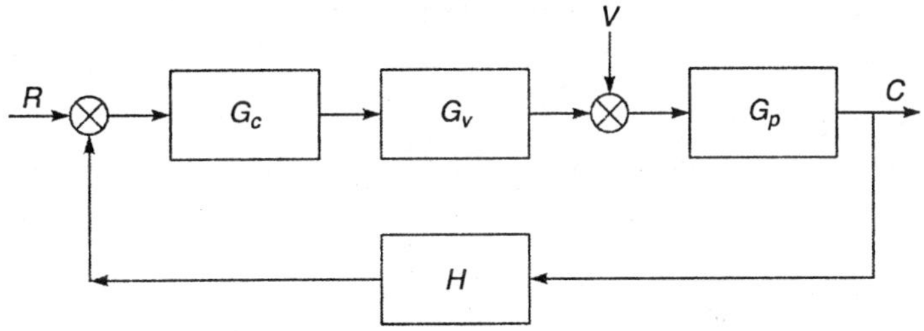

Fig. 12.1 Block diagram of closed loop control system.

where, C = Controlled variable
R = Setpoint or reference point
V = Load variable
G_C = Transfer function of the controller
G_V = Transfer function of the final control element
G_P = Transfer function of the process
H = Transfer function of the measuring element

When, $H = 1$ measuring instrument has negligible lag and is known as unity feed back control system.

12.2 BLOCK DIAGRAM REDUCTION PROCESS

To reduce the block diagram into block diagram algebra the following step by step procedure to be adopted :

(i) In a block diagram each element of a control system is represented by means of rectangular block. The transfer function of the element will be written inside the block.

(ii) The output signal from an element can be found by multiplying the input signal by the transfer function.

(iii) A circle in a block diagram represents the summing junction. The input signals are algebraically added to produce the output signal. A summing junction is shown in Fig. 12.2.

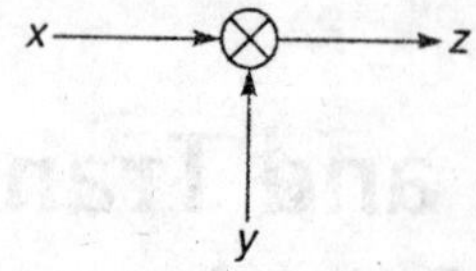

Fig. 12.2 Summing junction.

From the Fig. 12.2, we have

$$z = x + y \qquad ...(12.1)$$

(iv) Two or more blocks in series is equivalent to a single block whose transfer function is the product of the individual transfer function. A block diagram in series is shown in Fig. 12.3.

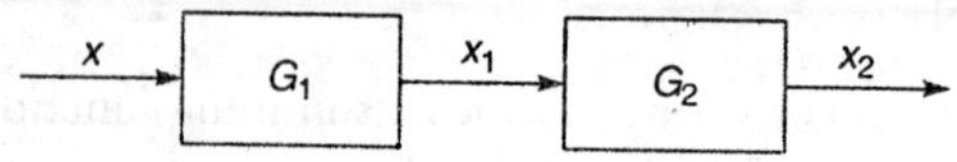

Fig. 12.3 Block diagram in series.

From the Fig. 12.3, we have

$$x_1 = xG_1 \qquad ...(12.2)$$

$$x_2 = x_1 G_2 \qquad ...(12.3)$$

$$\Rightarrow \quad x_2 = xG_1 G_2 \qquad ...(12.4)$$

The equation (12.4) may be represented by the single block diagram as shown in Fig. 12.4.

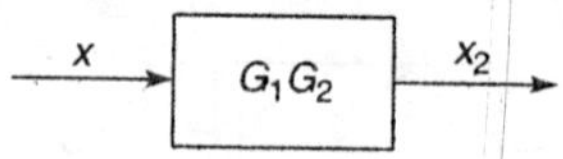

Fig. 12.4 Block diagram after reduction.

(v) Two or more blocks in parallel can be reduced to a single block diagram whose transfer function is equal to the sum of the individual transfer function. A block diagram in parallel is shown in Fig 12.5.

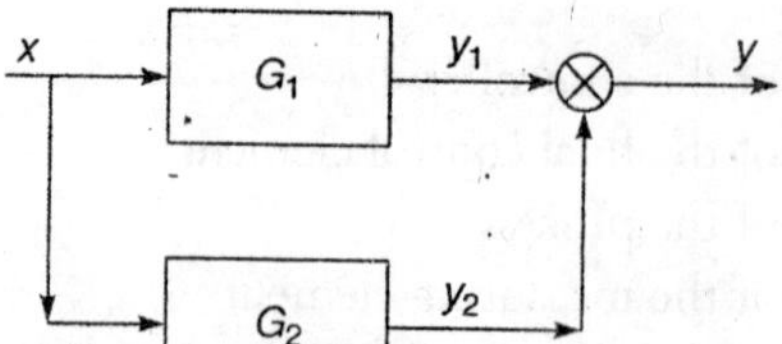

Fig. 12.5 Block diagram in parallel.

From the Fig. 12.5, we have

$$y = y_1 + y_2 \qquad ...(12.5)$$

$$y_1 = xG_1$$

$$y_2 = xG_2$$

$$\Rightarrow \quad y = xG_1 + xG_2$$

$$\Rightarrow \quad y = x(G_1 + G_2) \qquad ...(12.6)$$

The equation (12.6) may be represented by the single block diagram as shown in Fig. 12.6.

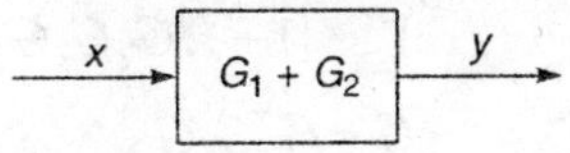

Fig. 12.6 Block diagram after reduction.

Using these principle, we can find the closed loop transfer function.

12.3 CONTROL PROBLEM

The control problem normally arises in the system either due to change in load variable or due to change in set point. They can be classified as :

(i) Regulator problem
(ii) Servo problem

12.3.1 Regulator Problem

The control problem arises because of the change in load variable where the set point is held constant is called regulator problem. This type of problems are very common in chemical industry. The examples of regulator problems are :

(i) Controlling the temperature of a chemical reaction which is taking place at a fixed temperature.
(ii) Controlling the level of a storage tank at a fixed point.

A schematic block diagram of regulator problem is shown in Fig. 12.7.

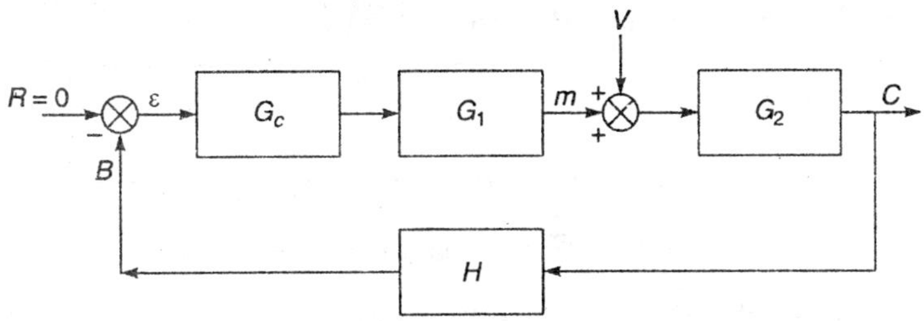

Fig. 12.7 Schematic block diagram of regulator problem.

We can write the following algebraic equation between the different variables of regulator problem control system as

$$\varepsilon = R - B$$

$$m = \varepsilon G_c G_1$$

$$B = CH$$

$$C = (V + m) G_2$$

$$C = (V + \varepsilon G_c G_1) G_2$$

$$\Rightarrow \quad C = VG_2 + \varepsilon G_c G_1 G_2$$

$$\Rightarrow \quad C = VG_2 + (R - B) G_c G_1 G_2$$

$$\Rightarrow \quad C = VG_2 + (O - B) G_c G_1 G_2$$

$$\Rightarrow \quad C = VG_2 - CHG_c G_1 G_2$$

$$\Rightarrow \qquad \frac{C}{V} = \frac{G_2}{[1 + HG_c G_1 G_2]}$$

Let,
$$G = G_c G_1 G_2$$

$$\Rightarrow \qquad \frac{C}{V} = \frac{G_2}{1 + GH} \qquad \ldots(12.7)$$

This is the transfer function of regulator problems.

12.3.2 Servo Problem

The control problem arises because of the change in set point where the load variable is kept constant is called servo problem. This type of problems are rare in chemical industry. The examples of servo problems are :

(i) Controlling the temperature of a chemical reaction which is taking place at different stages at different temperature.

(ii) Tracking of aircraft.

However, in a system both set point and load may be changing simultaneously. In such cases for convincing of analysis, response of the system is found out for the change in load or set point respectively assuming the other to be constant. Then, these responses are added linearly to get the overall response. A schematic block diagram of servo problem is shown in Fig. 12.8.

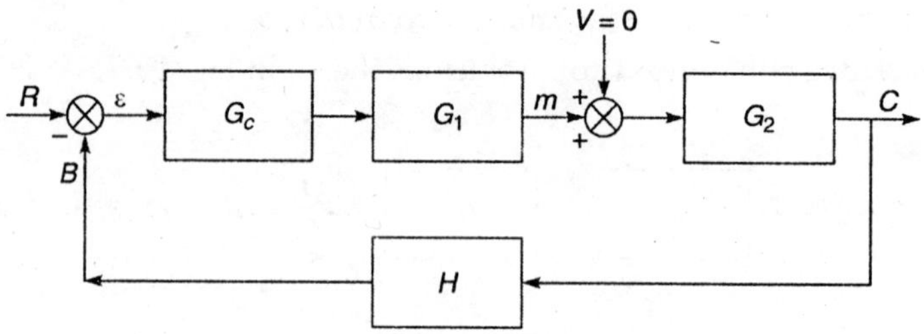

Fig. 12.8 Schematic block diagram of servo problem.

We can write the following algebraic equation between the different variables of servo problem control system as :

$$\varepsilon = R - B$$
$$m = \varepsilon G_c G_1$$
$$B = CH$$
$$C = (V + m) G_2$$
$$C = (0 + m) G_2$$
$$C = mG_2$$
$$C = \varepsilon G_c G_1 G_2$$
$$C = (R - B) G_c G_1 G_2$$
$$C = (R - CH) G_c G_1 G_2$$
$$C = RG_c G_1 G_2 - CHG_c G_1 G_2$$
$$C[1 + HG_c G_1 G_2] = RG_c G_1 G_2$$

$$\Rightarrow \qquad \frac{C}{R} = \frac{G_c G_1 G_2}{[1 + HG_c G_1 G_2]}$$

Let,

$$G = G_c G_1 G_2$$

$$\Rightarrow \qquad \frac{C}{R} = \frac{G}{1+GH} \qquad \text{...(12.8)}$$

This is the transfer function of servo problems.

■ **Example 12.1** Determine the transfer function $Y(s)/X(s)$ for the block diagram shown in Fig. 12.9.

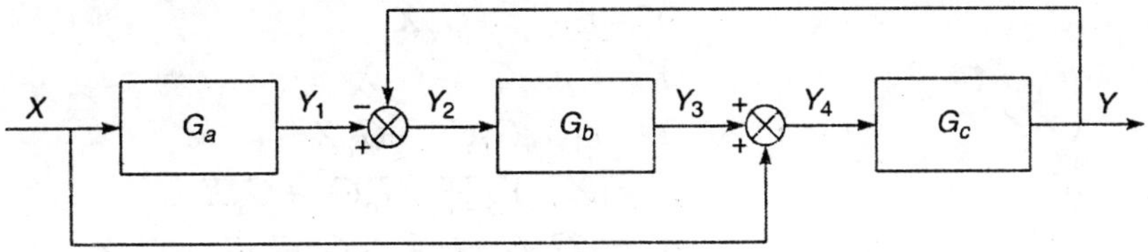

Fig. 12.9 Block diagram.

Solution : We can write the following algebraic equation from the block diagram

$$Y_1 = XG_a$$

$$Y_2 = Y_1 - Y$$

$$Y_3 = Y_2 G_b$$

$$Y_4 = Y_3 + X$$

$$Y = Y_4 G_c$$

$$\Rightarrow \qquad Y = (Y_3 + X) G_c$$

$$Y = Y_3 G_c + XG_c$$

$$\Rightarrow \qquad Y = Y_2 G_b G_c + XG_c$$

$$\Rightarrow \qquad Y = (Y_1 - Y) G_b G_c + XG_c$$

$$Y = Y_1 G_b G_c - YG_b G_c + XG_c$$

$$\Rightarrow \qquad Y = XG_a G_b G_c - YG_b G_c + XG_c$$

$$\Rightarrow \qquad Y[1 + G_b G_c] = X[G_a G_b G_c + G_c]$$

$$\Rightarrow \qquad \frac{Y}{X} = \frac{G_c + G_a G_b G_c}{1 + G_b G_c}$$

■ **Example 12.2** Determine the transfer function Y/X for the block diagram shown in Fig. 12.10.

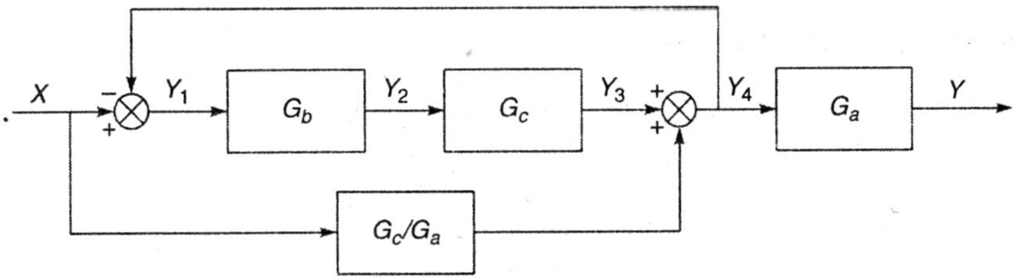

Fig. 12.10 Block diagram.

Solution : We can write the following algebraic equation from the block diagram

$$Y_1 = X - Y_4$$

$$Y_2 = Y_1 G_b$$
$$Y_3 = Y_2 G_c$$
$$Y_4 = Y_3 + \frac{G_c}{G_b} X$$
$$Y = Y_4 G_a$$
$$\Rightarrow \quad Y = \left[Y_3 + \frac{G_c}{G_a} X \right] G_a$$
$$Y = Y_3 G_a + G_c X$$
$$\Rightarrow \quad Y = Y_2 G_c G_a + G_c X$$
$$\Rightarrow \quad Y = Y_1 G_b G_c G_a + G_c X$$
$$\Rightarrow \quad Y = [X - Y_4] G_a G_b G_c + G_c X$$
$$Y = X G_a G_b G_c - Y_4 G_a G_b G_c + G_c X$$
$$\Rightarrow \quad Y = X G_a G_b G_c - \left[\frac{Y}{G_a} \right] G_a G_b G_c + G_c X$$
$$Y = X G_a G_b G_c - Y G_b G_c + X G_c$$
$$Y[1 + G_b G_c] = X[G_c + G_a G_b G_c]$$
$$\Rightarrow \quad \frac{Y}{X} = \frac{G_c + G_a G_b G_c}{1 + G_b G_c}$$

■ **Example 12.3** Reduce the multiloop block diagram shown in Fig. 12.11 to a single loop block diagram. And also, determine the transfer function.

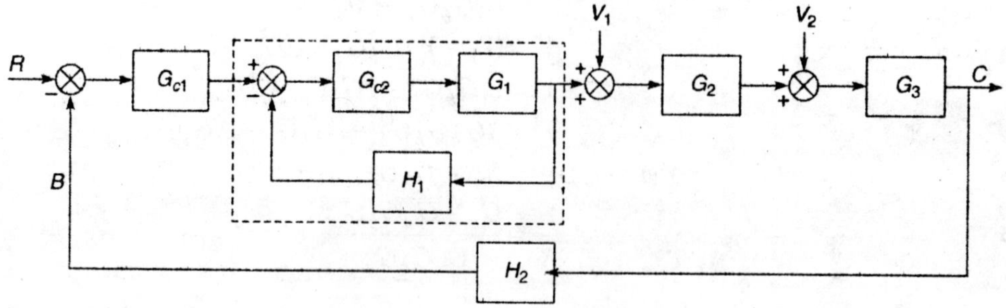

Fig. 12.11 Multiloop block diagram.

Solution : From the inner loop block diagram, we have

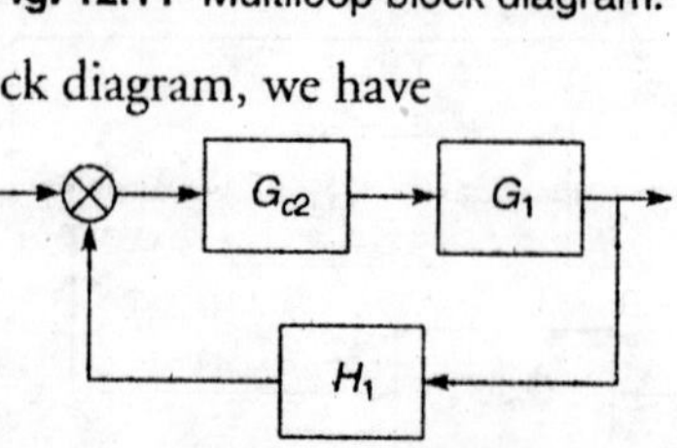

Fig. 12.12 Inner loop block diagram.

$$\Rightarrow \quad G_a = \frac{G_{c2} G_1}{1 + G_{c2} G_1 H_1} \qquad \text{...(12.9)}$$

The equation (12.9) may be represented by the single block as shown in Fig. 12.13.

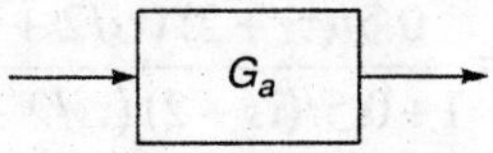

Fig. 12.13 Single block.

Now, replace the inner loop block diagram presented inside the dotted line by a single block as shown in Fig. 12.13. This will give a single loop block diagram shown in Fig. 12.14.

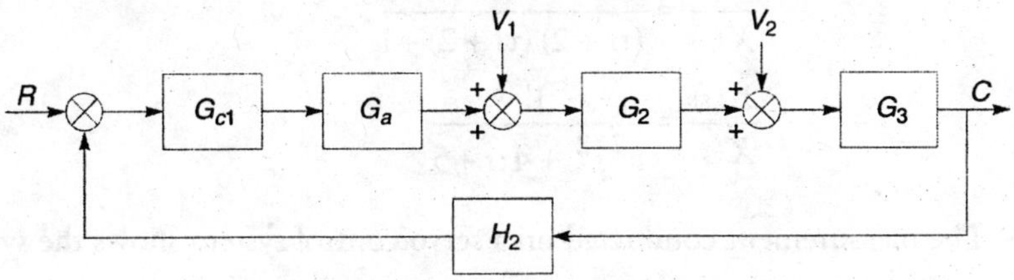

Fig. 12.14 Reduced block diagram.

From the reduced block diagram, we get the transfer function

$$\frac{C}{R} = \frac{G_{c1}G_aG_2G_3}{1 + H_2G_{c1}G_aG_2G_3} \qquad \text{...(12.10)}$$

■ **Example 12.4** Find the transfer function $Y(s)/X(s)$ for the block diagram shown in Fig. 12.15.

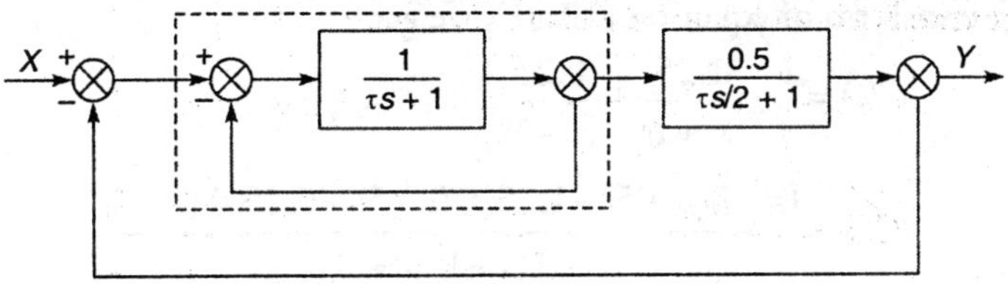

Fig. 12.15 Multiloop block diagram.

Solution : From the inner loop block diagram, we have

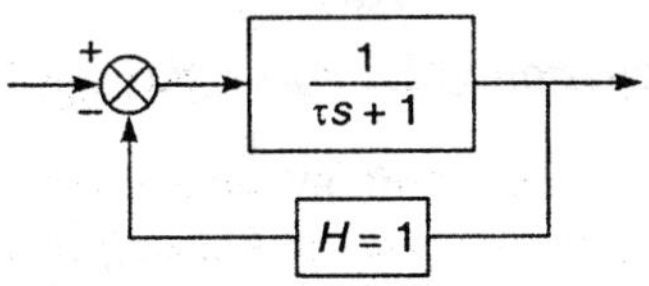

Fig. 12.16 Inner loop block diagram.

$$G_a = \frac{1/(\tau s + 1)}{1 + 1/(\tau s + 1)}$$

$$G_a = \frac{1/(\tau s + 1)}{[(\tau s + 1) + 1]/(\tau s + 1)}$$

$$G_a = \frac{1}{(\tau s + 2)}$$

Therefore, the multiloop block diagram reduce to

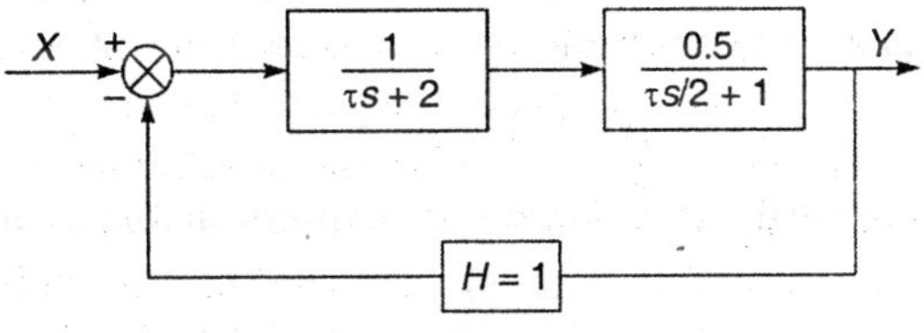

Fig. 12.17 Reduced block diagram.

From the reduced block diagram, we get the transfer function

$$\frac{Y(s)}{X(s)} = \frac{0.5/(\tau s+2)(\tau s/2+1)}{1+0.5/(\tau s+2)(\tau s/2+1)}$$

$$\frac{Y(s)}{X(s)} = \frac{0.5}{(\tau s+2)(\tau s/2+1)+0.5}$$

$$\frac{Y(s)}{X(s)} = \frac{1}{(\tau s+2)(\tau s+2)+1}$$

$$\frac{Y(s)}{X(s)} = \frac{1}{\tau^2 s^2+4\tau s+5}$$

■ **Example 12.5** The measurement conducted on a servo control system shows the system response

$$C(t) = 1-1.25e^{-t}+0.25e^{-5t}$$

when subjected to a unit step change in input. Obtain the expression for the closed loop transfer function. [VTU, Dec. 2012]

Solution : From the given data, we have

$$C(t) = 1-1.25e^{-t}+0.25e^{-5t} \qquad \text{...(12.11)}$$

Taking Laplace transform of equation (12.11), we get

$$C(s) = \frac{1}{s}-\frac{1.25}{(s+1)}+\frac{0.25}{(s+5)} \qquad \text{...(12.12)}$$

$$C(s) = \frac{(s+1)(s+5)-1.25(s)(s+5)+0.25(s)(s+1)}{(s)(s+1)(s+5)} \qquad \text{...(12.13)}$$

From the unit step change in input, we have

$$R(t) = 1 \qquad \text{...(12.14)}$$

Taking Laplace transform of equation (12.14), we get

$$R(s) = \frac{1}{s} \qquad \text{...(12.15)}$$

Dividing equation (12.13) by equation (12.15), we get

$$\Rightarrow \quad \frac{C(s)}{R(s)} = \frac{(s^2+6s+5)-1.25(s^2+5s)+0.25(s^2+s)}{(s+1)(s+5)}$$

$$\Rightarrow \quad \frac{C(s)}{R(s)} = \frac{s^2+6s+5-1.25s^2-6.25s+0.25s^2+0.25s}{(s+1)(s+5)}$$

$$\Rightarrow \quad \frac{C(s)}{R(s)} = \frac{5}{(s+1)(s+5)}$$

■ **Example 12.6** The measurement conducted on a servo control system shows the system response

$$C(t) = 1-3e^{-2t}+2e^{-3t}$$

when subjected to unit step change in input. Obtain the expression for the open loop transfer function.

Solution : From the given data, we have

$$C(t) = 1 - 3e^{-2t} + 2e^{-3t} \qquad \text{...(12.16)}$$

Taking Laplace transform of equation (12.16), we get

$$C(s) = \frac{1}{s} - \frac{3}{(s+2)} + \frac{2}{(s+3)} \qquad \text{...(12.17)}$$

$$C(s) = \frac{(s+2)(s+3) - 3(s)(s+3) + 2(s)(s+2)}{(s)(s+2)(s+3)}$$

$$C(s) = \frac{s^2 + 5s + 6 - 3s^2 - 9s + 2s^2 + 4s}{(s)(s+2)(s+3)}$$

$$C(s) = \frac{6}{(s)(s+2)(s+3)} \qquad \text{...(12.18)}$$

From the unit step change in input, we have

$$R(t) = 1 \qquad \text{...(12.19)}$$

Taking Laplace transform of equation (12.19), we get

$$R(s) = \frac{1}{s} \qquad \text{...(12.20)}$$

Dividing equation (12.18) by equation (12.20), we get

$$T(s) = \frac{C(s)}{R(s)} = \frac{6}{(s+2)(s+3)} \qquad \text{...(12.21)}$$

For closed loop transfer function, we have

$$T(s) = \frac{G(s)}{1+G(s)} \qquad \text{...(12.22)}$$

$$\Rightarrow \qquad T(s)[1+G(s)] = G(s)$$

$$\Rightarrow \qquad G(s) = \frac{T(s)}{1-T(s)} \qquad \text{...(12.23)}$$

where, $G(s)$ = Open loop transfer function
$T(s)$ = Closed loop transfer function

Substituting equation (12.21) in equation (12.23), we get

$$G(s) = \frac{6/(s+2)(s+3)}{1-6/(s+2)(s+3)}$$

$$G(s) = \frac{6}{(s+2)(s+3)-6}$$

$$G(s) = \frac{6}{s^2 + 5s + 6 - 6}$$

$$G(s) = \frac{6}{(s)(s+5)} \qquad \text{...(12.24)}$$

■ **Example 12.7** A simple feed back control system consists of a proportional plus integral controller coupled to a process characterized by time constants of 2 and 5 sec. The controlled variable is measured by an instrument with a time constant of 4 secs and as time delay of 1 sec. The reference signal is fed to the controller summing junction through a first order pneumatic transmitter with a time constant of 1 sec.

(i) Draw a neat block diagram for this system. Insert transfer function representing each block.

(ii) Write the transfer function relating the controlled variable '*C*' to the reference '*R*'.

(iii) Write the transfer function relating the measuring instrument '*M*' to the reference '*R*'.

Solution :

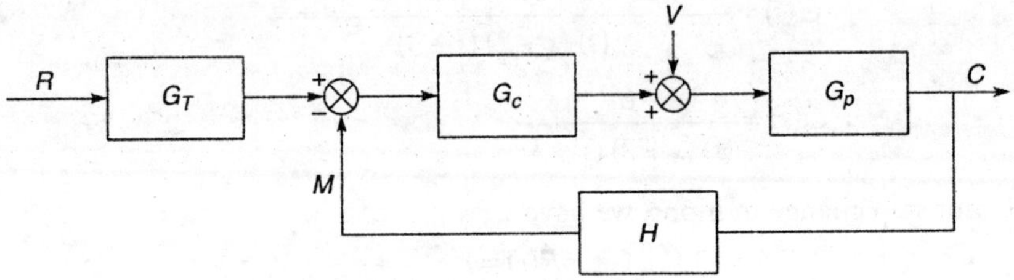

Fig. 12.18 Block diagram.

(i) Transfer function of final control element is assumed as one, since it is not given in the problem.

(ii) $$\frac{C}{R} = \frac{G_T G_C G_P}{1 + G_C G_P H}$$

(iii) $$\frac{M}{R} = \frac{G_T G_C G_P H}{1 + G_C G_P H}$$

Note : G_T would not come in denominator, because it is not inside the loop.

■ **Example 12.8** Draw the block diagram for the following servo problem. Controller is proportional, valve is first order, measuring instrument has no lag. Find proportional gain k_c of the controller for a critically damped system. The gain of the controlled valve and process are both 1 and time constants are 2 and 10 sec respectively.

Solution :

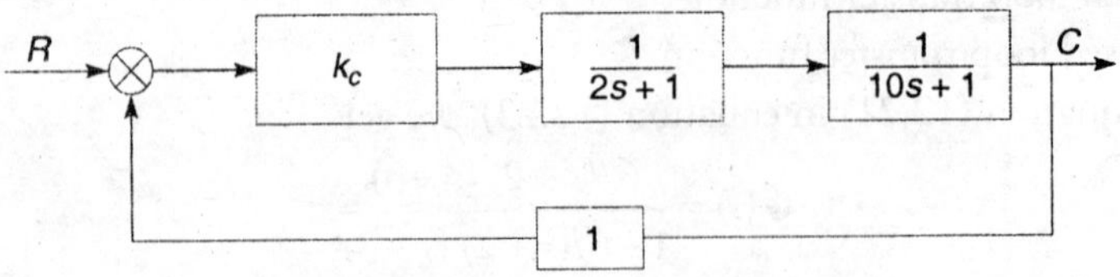

Fig. 12.19 Block diagram.

From the servo problem transfer function, we have

$$\frac{C(s)}{R(s)} = \frac{G}{1 + GH}$$

$$\Rightarrow \quad \frac{C(s)}{R(s)} = \frac{k_c/(2s+1)(10s+1)}{1 + k_c/(2s+1)(10s+1)}$$

$$\Rightarrow \quad \frac{C(s)}{R(s)} = \frac{k_c}{(2s+1)(10s+1)+k_c}$$

$$\Rightarrow \quad \frac{C(s)}{R(s)} = \frac{k_c}{20s^2+12s+(1+k_c)} \qquad \text{...(12.25)}$$

Dividing numerator and denominator of equation (12.25) by $(1+k_c)$, we get

$$\frac{C(s)}{R(s)} = \frac{k_c/(1+k_c)}{\left(\frac{20}{1+k_c}\right)s^2+\left(\frac{12}{1+k_c}\right)s+1} \qquad \text{...(12.26)}$$

From the standard second order transfer function, we have

$$\frac{C(s)}{R(s)} = \frac{k_c}{\tau^2 s^2+2\tau\xi s+1} \qquad \text{...(12.27)}$$

Comparing the equation (12.26) and (12.27), we get

$$\tau^2 = \frac{20}{(1+k_c)}$$

$$2\tau\xi = \frac{12}{(1+k_c)}$$

From the given data, we have

$$\xi = 1$$

$$\Rightarrow \quad 2\tau = \frac{12}{1+k_c}$$

$$\Rightarrow \quad \tau = \frac{6}{1+k_c}$$

$$\Rightarrow \quad \left(\frac{6}{1+k_c}\right)^2 = \frac{20}{1+k_c}$$

$$\Rightarrow \quad \frac{36}{1+k_c} = 20$$

$$\Rightarrow \quad 1+k_c = 1.8$$

$$\Rightarrow \quad k_c - 0.8$$

■ **Example 12.9** A system has two major process and is controlled with proportional controller. The transfer function of the process are

$$G_1 = \frac{1}{2s+1} \quad ; \quad G_2 = \frac{1}{0.4s+1}$$

where, G_1 and G_2 are connected in series and time constants are in minutes. Find the recommended controller gain for a damping coefficient of 0.5.

Solution :

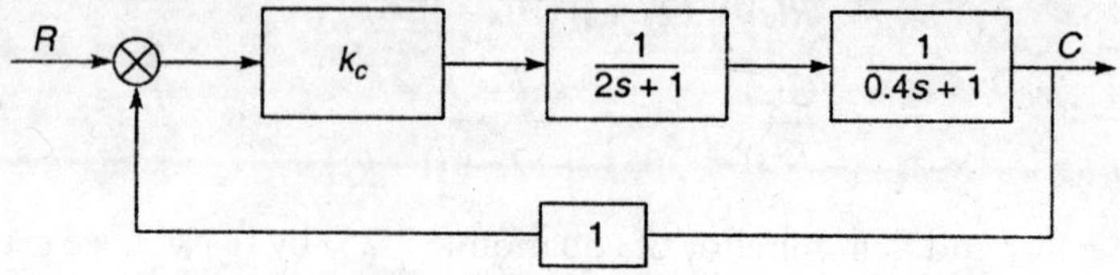

Fig. 12.20 Block diagram.

From the servo problem transfer function, we have

$$\frac{C(s)}{R(s)} = \frac{G}{1+GH}$$

$$\Rightarrow \quad \frac{C(s)}{R(s)} = \frac{k_c/(2s+1)(0.4s+1)}{1+k_c/(2s+1)(0.4s+1)}$$

$$\Rightarrow \quad \frac{C(s)}{R(s)} = \frac{k_c}{(2s+1)(0.4s+1)+k_c}$$

$$\Rightarrow \quad \frac{C(s)}{R(s)} = \frac{k_c}{0.8s^2+2.4s+(1+k_c)} \quad \text{...(12.28)}$$

Dividing numerator and denominator of equation (12.28) by $(1+k_c)$, we get

$$\frac{C(s)}{R(s)} = \frac{k_c/(1+k_c)}{\left(\frac{0.8}{1+k_c}\right)s^2+\left(\frac{2.4}{1+k_c}\right)s+1} \quad \text{...(12.29)}$$

From the standard second order transfer function, we have

$$\frac{C(s)}{R(s)} = \frac{k_c}{\tau^2 s^2 + 2\tau\xi s + 1} \quad \text{...(12.30)}$$

Comparing the equation (12.29) and (12.30), we get

$$\tau^2 = \frac{0.8}{1+k_c}$$

$$2\tau\xi = \frac{2.4}{1+k_c}$$

From the given data, we have

$$\xi = 0.5$$

$$\Rightarrow \quad \tau = \frac{2.4}{1+k_c}$$

$$\Rightarrow \quad \left(\frac{2.4}{1+k_c}\right)^2 = \frac{0.8}{1+k_c}$$

$$\Rightarrow \quad \frac{5.76}{1+k_c} = 0.8$$

$$\Rightarrow \qquad 1+k_c = 7.2$$

$$\Rightarrow \qquad k_c = 6.2$$

■ **Example 12.10** A system has two major process and is controlled with proportional controller. The transfer function of the process are :

$$G_1 = \frac{0.1}{2s+1} \quad ; \quad G_2 = \frac{1}{5s+1}$$

where, G_1 and G_2 are connected in series and time constants are in minutes. Find the recommended controller gain for a damping coefficient of 0.5. **[VUT, Jan. 2013]**

Solution :

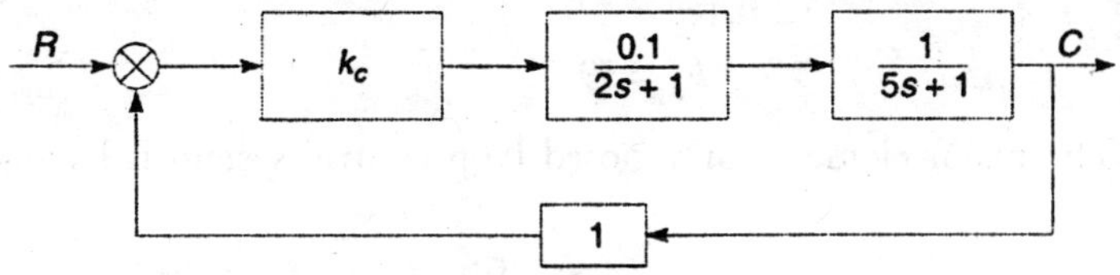

Fig. 12.21 Block diagram.

From the servo problem transfer function, we have

$$\frac{C(s)}{R(s)} = \frac{G}{1+GH}$$

$$\Rightarrow \qquad \frac{C(s)}{R(s)} = \frac{0.1k_c/(2s+1)(5s+1)}{1+0.1k_c/(2s+1)(5s+1)}$$

$$\Rightarrow \qquad \frac{C(s)}{R(s)} = \frac{0.1k_c}{(2s+1)(5s+1)+0.1k_c}$$

$$\Rightarrow \qquad \frac{C(s)}{R(s)} = \frac{0.1k_c}{10s^2+7s+(1+0.1k_c)} \qquad \text{...(12.31)}$$

Dividing numerator and denominator of equation (12.31) by $(1+0.1k_c)$, we get

$$\frac{C(s)}{R(s)} = \frac{0.1k_c/(1+0.1k_c)}{\left(\frac{10}{1+0.1k_c}\right)s^2+\left(\frac{7}{1+0.1k_c}\right)s+1} \qquad \text{...(12.32)}$$

From the standard second order transfer function, we have

$$\frac{C(s)}{R(s)} = \frac{k_c}{\tau^2 s^2 + 2\tau\xi s + 1} \qquad \text{...(12.33)}$$

Comparing the equation (12.32) and (12.33), we get

$$\tau^2 = \frac{10}{1+0.1k_c}$$

$$2\tau\xi = \frac{7}{1+0.1k_c}$$

From the given data, we have

$$\xi = 0.5$$

$$\Rightarrow \quad \tau = \frac{7}{1+0.1k_c}$$

$$\Rightarrow \quad \left(\frac{7}{1+0.1k_c}\right)^2 = \frac{10}{1+0.1k_c}$$

$$\Rightarrow \quad \frac{49}{1+0.1k_c} = 10$$

$$\Rightarrow \quad 1+0.1k_c = 4.9$$

$$\Rightarrow \quad k_c = 39$$

■ **Example 12.11** The major elements of a closed loop control system is having following transfer functions

$$G_P(s) = \frac{1.5}{(s+1)(2s+1)}$$

$$G_c(s) = k_c$$

Transfer function for rest of the components in the loop are unity. All time constants are in minutes. Determine the transient response for a unit impulse input in set point at a controller gain that made the system critically damped. [VTU, July 2012]

Solution :

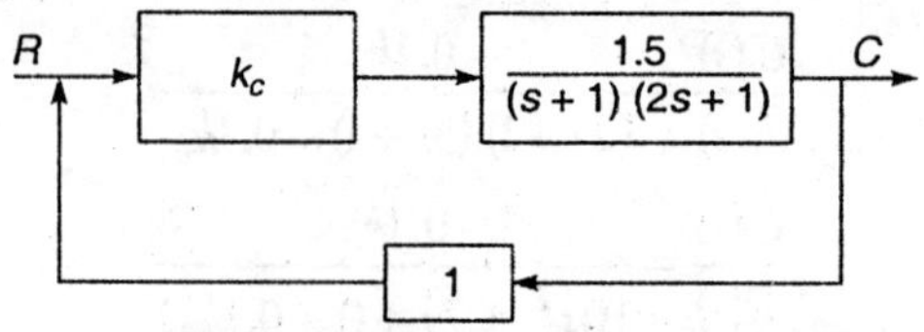

Fig. 12.22 Block diagram.

From the servo problem transfer function, we have

$$\frac{C(s)}{R(s)} = \frac{G}{1+GH}$$

$$\Rightarrow \quad \frac{C(s)}{R(s)} = \frac{1.5k_c/(s+1)(2s+1)}{1+1.5k_c/(s+1)(2s+1)}$$

$$\Rightarrow \quad \frac{C(s)}{R(s)} = \frac{1.5k_c}{(s+1)(2s+1)+1.5k_c}$$

$$\Rightarrow \quad \frac{C(s)}{R(s)} = \frac{1.5k_c}{2s^2+3s+(1+1.5k_c)} \quad \text{...(12.34)}$$

Dividing numerator and denominator of equation (12.34) by $(1+1.5k_c)$, we get

$$\frac{C(s)}{R(s)} = \frac{1.5k_c/(1+1.5k_c)}{\left(\frac{2}{1+1.5k_c}\right)s^2+\left(\frac{3}{1+1.5k_c}\right)s+1} \quad \text{...(12.35)}$$

From the standard second order transfer function, we have

$$\frac{C(s)}{R(s)} = \frac{k_c}{\tau^2 s^2 + 2\tau\xi s + 1} \qquad ...(12.36)$$

Comparing the equation (12.35) and (12.36), we get

$$\tau^2 = \frac{2}{1+1.5k_c}$$

$$2\tau\xi = \frac{3}{1+1.5k_c}$$

From the given data, we have

$$\xi = 1$$

$$\Rightarrow \quad \tau = \frac{3}{2(1+1.5k_c)}$$

$$\Rightarrow \quad \frac{9}{4(1+1.5k_c)^2} = \frac{2}{(1+1.5k_c)}$$

$$\Rightarrow \quad 1+1.5k_c = 1.125$$

$$\Rightarrow \quad k_c = 0.083$$

From the definition of transfer function, we have

$$\frac{C(s)}{R(s)} = G(s)$$

$$\Rightarrow \quad \frac{C(s)}{R(s)} = \frac{1.5k_c}{(s+1)(2s+1)}$$

From the given data, we have

$$R(t) = 1\partial(t)$$

$$R(s) = 1$$

$$\Rightarrow \quad C(s) = \frac{1.5 \times 0.083}{(s+1)\,2\,(s+0.5)}$$

$$\Rightarrow \quad C(s) = \frac{0.062}{(s+1)(s+0.5)}$$

$$\Rightarrow \quad C(s) - \frac{0.062}{(s+1)(s+0.5)} - \frac{C_1}{(s+1)} + \frac{C_2}{(s+0.5)}$$

$$\Rightarrow \quad 0.062 = C_1(s+0.5) + C_2(s+1)$$

Put, $s = -0.5$; $C_2 = 0.124$

Put, $s = -1$; $C_1 = -0.124$

$$\Rightarrow \quad C(s) = -\frac{0.124}{(s+1)} + \frac{0.124}{(s+0.5)}$$

$$\Rightarrow \quad C(t) = -0.124e^{-t} + 0.124e^{-0.5t}$$

$$\Rightarrow \quad C(t) = 0.124\,[e^{-0.5t} - e^{-t}]$$

■ **Example** 12.12 The major elements of a closed loop control system is having following transfer functions

$$G_P(s) = \frac{2.5}{(s+1)(2s+1)}$$

$$G_c(s) = k_c$$

Transfer function for rest of the components in the loop are unity. All time constants are in minutes. Determine the transient response for a unit step input in set point at a controller gain that made the system critically damped.

Solution :

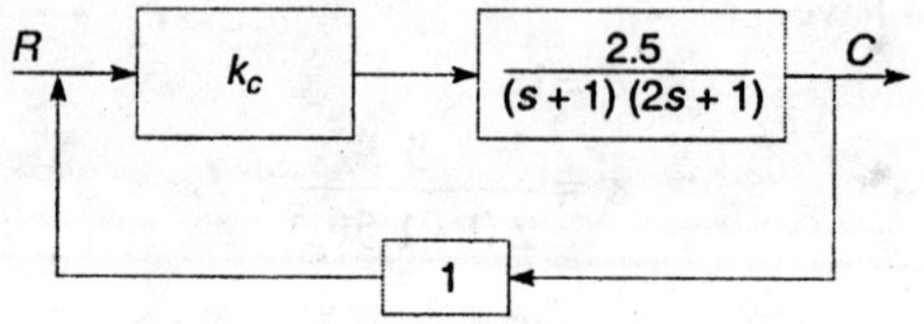

Fig. 12.23 Block diagram.

From the servo problem transfer function, we have

$$\frac{C(s)}{R(s)} = \frac{G}{1+GH}$$

$$\Rightarrow \quad \frac{C(s)}{R(s)} = \frac{2.5k_c/(s+1)(2s+1)}{1+2.5k_c/(s+1)(2s+1)}$$

$$\Rightarrow \quad \frac{C(s)}{R(s)} = \frac{2.5k_c}{(s+1)(2s+1)+2.5k_c}$$

$$\Rightarrow \quad \frac{C(s)}{R(s)} = \frac{2.5k_c}{2s^2+3s+(1+2.5k_c)} \quad ...(12.37)$$

Dividing numerator and denominator of equation (12.37) by $(1+2.5k_c)$, we get

$$\frac{C(s)}{R(s)} = \frac{2.5k_c/(1+2.5k_c)}{\left(\frac{2}{1+2.5k_c}\right)s^2+\left(\frac{3}{1+2.5k_c}\right)s+1} \quad ...(12.38)$$

From the standard second order transfer function, we have

$$\frac{C(s)}{R(s)} = \frac{k_c}{\tau^2 s^2 + 2\tau\xi s + 1} \quad ...(12.39)$$

Comparing the equation (12.38) and (12.39), we get

$$\tau^2 = \frac{2}{1+2.5k_c}$$

$$2\tau\xi = \frac{3}{1+2.5k_c}$$

From the given data critically damped system, we have

$$\xi = 1$$

$$\Rightarrow \quad \tau = \frac{3}{2\,(1+2.5k_c)}$$

$$\Rightarrow \quad \frac{9}{4\,(1+2.5k_c)^2} = \frac{2}{(1+2.5k_c)}$$

$$\Rightarrow \quad 1+2.5k_c = 1.125$$

$$\Rightarrow \quad k_c = 0.05$$

From the definition of transfer function, we have

$$\frac{C(s)}{R(s)} = G(s)$$

$$\Rightarrow \quad \frac{C(s)}{R(s)} = \frac{2.5k_c}{(s+1)\,(2s+1)}$$

From the given data, we have

$$R(t) = 1$$

$$\Rightarrow \quad R(s) = 1/s$$

$$\Rightarrow \quad C(s) = \frac{2.5 \times 0.05}{(s)\,(s+1)\,2\,(s+0.5)}$$

$$\Rightarrow \quad C(s) = \frac{0.063}{(s)\,(s+1)\,(s+0.5)}$$

$$\Rightarrow \quad C(s) = \frac{0.063}{(s)\,(s+1)\,(s+0.5)} = \frac{C_1}{s} + \frac{C_2}{(s+1)} + \frac{C_3}{(s+0.5)}$$

$$\Rightarrow \quad 0.063 = C_1\,(s+1)\,(s+0.5) + C_2\,(s)\,(s+0.5) + C_3\,(s)\,(s+1)$$

Put, $s = -1$; $C_2 = 0.126$

Put, $s = -0.05$; $C_3 = -0.126$

Put, $s = 0$; $C_1 = 0.126$

$$\Rightarrow \quad C(s) = \frac{0.126}{s} + \frac{0.126}{(s+1)} - \frac{0.126}{(s+0.5)}$$

$$\Rightarrow \quad C(t) = 0.126\,[1 + e^{-t} - e^{-0.5t}]$$

■ **Example 12.13** A system has two major process and is controlled with proportional integral controller. The transfer function of the process are

$$G_1 = \frac{1}{(s+1)} \quad ; \quad G_2 = \frac{1}{(5s+1)}$$

where, G_1 and G_2 are connected in series and time constants are in minutes. Find the recommended controller gain for $\tau_1 = 1$ and dumping coefficient of 0.5.

Solution :

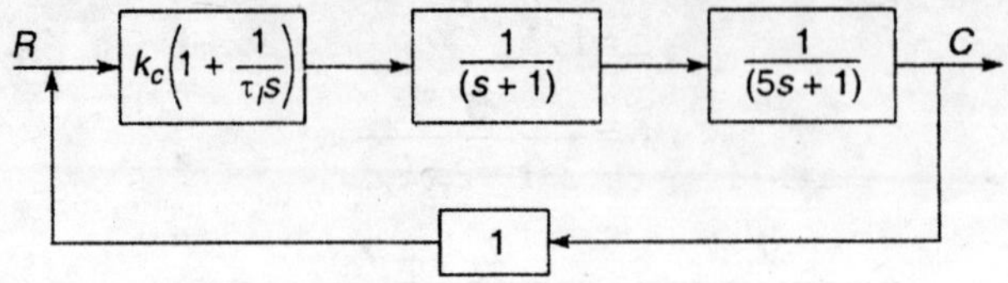

Fig. 12.24 Block diagram.

From the servo problem transfer function, we have

$$\frac{C(s)}{R(s)} = \frac{G}{1+GH}$$

$$\Rightarrow \quad \frac{C(s)}{R(s)} = \frac{k_c/(s)\,(5s+1)}{1+k_c/(s)\,(5s+1)}$$

$$\Rightarrow \quad \frac{C(s)}{R(s)} = \frac{k_c}{(s)\,(5s+1)+k_c}$$

$$\Rightarrow \quad \frac{C(s)}{R(s)} = \frac{k_c}{5s^2+s+k_c} \qquad \ldots(12.40)$$

Dividing numerator and denominator of equation (12.40) by k_c, we get

$$\frac{C(s)}{R(s)} = \frac{k_c/k_c}{\left(\frac{5}{k_c}\right)s^2+\left(\frac{1}{k_c}\right)s+1} \qquad \ldots(12.41)$$

From the standard second order transfer function, we have

$$\frac{C(s)}{R(s)} = \frac{k_c}{\tau^2 s^2+2\tau\xi s+1} \qquad \ldots(12.42)$$

Comparing the equation (12.41) and (12.42), we get

$$\tau^2 = \frac{5}{k_c}$$

$$2\tau\xi = \frac{1}{k_c}$$

From the given data, we have

$$\xi = 0.5$$

$$\Rightarrow \quad \tau = \frac{1}{k_c}$$

$$\Rightarrow \quad \left(\frac{1}{k_c}\right)^2 = \frac{5}{k_c}$$

$$\Rightarrow \quad k_c = 0.2$$

12.4 LUMPED AND DISTRIBUTED PARAMETER SYSTEM

A system in which a resistance and a capacitance are present only in particular place or region is called lumped parameter system. If they are present throughout the system, then it is known as distributed parameter system. For example, consider a mercury thermometer kept in a constant temperature water at 60°C. Here, all the resistance for heat transfer is in the film surrounding the bulb and also the capacitance is in the mercury. Therefore, the thermometer system is to be considered as lumped parameter.

For another example, consider a continuous stirred tank reactor (CSTR), in which a first order reaction is being carried out. If the resistance is present throughout the batch of liquid level in a tank or the glass wall of the thermometer expands or contacts due to heat, it becomes a distributed parameter system.

A lumped parameter system is described by ordinary differential equation where as a distributed parameter system is described by a partial differential equation.

REVIEW QUESTIONS

12.1 What is block diagram?

12.2 Explain the method of reduction of multiloop block diagram into a single loop block diagram.

12.3 Define regulator problem control system. Give suitable example.

12.4 Define servo problem control system. Give suitable example.

12.5 Define lumped parameter. Give suitable example.

12.6 Define distributed parameter. Give suitable example.

EXERCISES

12.1 Draw neat block diagram of a regulator control problem and derive the transfer function.

12.2 Draw neat block diagram of a servo control problem and derive the transfer function.

12.3 Using the block diagram reduction technique, find the overall transfer function of the multiloop block diagram shown in Fig. 12.25.

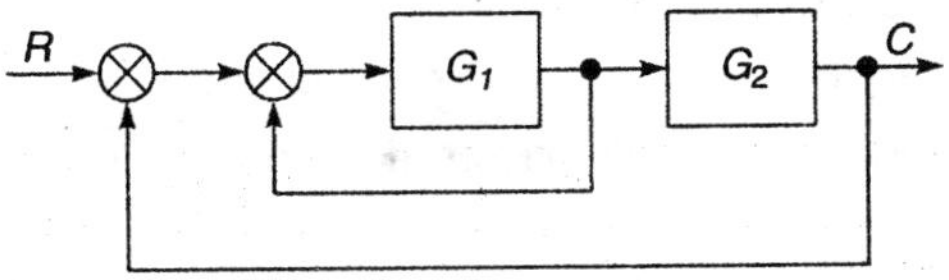

Fig. 12.25 Multiloop block diagram.

12.4 Using the block diagram reduction technique, find the overall transfer function $C(s)/R(s)$ of the multiloop block diagram shown in Fig. 12.26. [VTU, Jan. 2015]

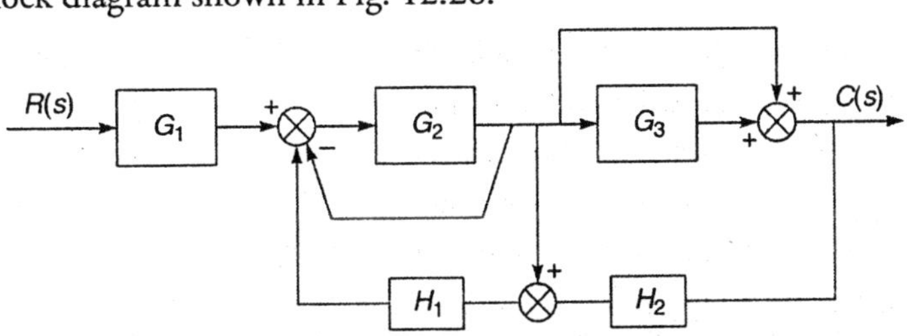

Fig. 12.26 Multiloop block diagram.

12.5 The measurement conducted on a servo control system shows the system response

$$C(t) = 1 + 0.2e^{-6t} - 1.2e^{-t}$$

when subjected to unit step change in input. Obtain the expression for closed loop transfer function.

12.6 The unit step response of single loop, unit feed back control system is given by

$$C(t) = 1 - 1.25e^{-2t} + 0.25e^{-10t}$$

Determine its closed loop transfer function.

12.7 The unit step response of single loop unit feed back control system is given by

$$C(t) = 1 - 1.25e^{-2t} + 0.25e^{-10t}$$

Determine its open loop transfer function.

12.8 A closed loop system has time constant of 1 min and 2 min and a proportional controller. Obtain the response of a closed loop system to a unit step change in set point at a controller gain that gives the damping factor of 0.5.

12.9 The major elements of a closed loop control system is having the following transfer function

$$G_P(s) = \frac{1.2}{(s+1)(s+2)}$$

$$G_c(s) = k_c$$

Transfer function for rest of the components in the loop are unity. All time constants are in minutes. Determine the transient response for a unit impulse input in set point at a controller gain that made the system critically damped.

12.10 A system has two major process and is controlled with a proportional controller. The transfer function of the process are

$$G_1 = \frac{1}{(2s+1)} \quad ; \quad G_2 = \frac{1}{(5s+1)}$$

where, G_1 and G_2 are connected in series and time constants are in minutes. Find out the recommended controller gain for a damping coefficient of 0.5.

12.11 Determine C/L, the overall transfer function of a control system having the following transfer function for its elements in a servo type negative feed back system.

Controller, $G_c = k_c$

Final control element, $G_v = k_v$

Process, $G_P = \dfrac{k_P}{(\tau_1 s+1)(\tau_2 s+1)}$

Measuring element, $H = \dfrac{1}{(\tau_m s+1)}$

Control variable is C and set point is L.

12.12 The major elements of a closed loop control system is having the following transfer function.

$$G_P(s) = \frac{2.0}{(s+1)(s+2)}$$

$$G_c(s) = k_c$$

Transfer function for rest of the components in the loop are unity. All time constants are in minutes. Determine the transient response for a unit step input in set point at a controller gain that made the system critically damped.

12.13 A system has two major process and is controlled with a proportional controller. The transfer function of the process are

$$G_1(s) = \frac{1}{(0.2s+1)} \quad ; \quad G_2(s) = \frac{1}{(0.5s+1)}$$

where, $G_1(s)$ and $G_2(s)$ are connected in series and time constants are in minutes. Find the recommended controller gain for a damping coefficient of 1.

12.14 A proportional derivative controller is used in a system with a first order process and a first order measuring lag. Find the expression for characteristic time τ_c and damping coefficient ξ, for closed loop system. If process time constant is 1 min and measuring time constant is 10 sec, find the controller gain by using $\xi = 0.5$ and $\tau_D = 3$ seconds.

□□□

Chapter 13

Stability

13.1 INTRODUCTION

The linear control system is said to be stable if its output response is bounded for bounded inputs. The term bounded means being within certain finite limits always. For example, a step function, an impulse function etc.

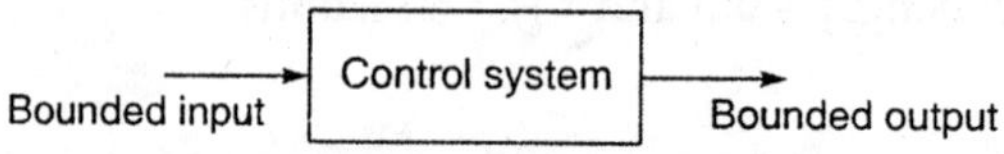

Fig. 13.1 Stable system.

The linear function is unbounded since there is no finite limit for the function. Also a function of the type e^{mt} is unbounded. For example, consider the closed loop transfer function of servo control

$$\frac{C}{R} = \frac{G}{1+GH} \qquad \ldots(13.1)$$

where, G = Open loop transfer function.

The response C for any output change depends on the factors of the denominator. In other words, the nature of response is depend upon the roots of the characteristics equation $1+GH=0$ and also the controller parameter k_c.

The relative stability of a control system depends upon the settling time. A system is said to be more stable if the settling time is less. If the settling time is more then the system is said to be less stable.

A linear system is said to be conditionally stable, if a particular parameter of a system at certain condition, its output is bounded one.

A linear system is said to be critically or marginally stable if its output oscillates with constant frequency and amplitude for a bounded input. Such oscillations of output are called sustained (undamped) oscillation.

13.2 STABILITY CRITERION

The stability of a control system depends upon the characteristic equation of its roots. The roots can be real imaginary or complex and can lie anywhere on the complex plane as shown in the Fig. 13.2.

Hence, with respect to the location of the roots in the complex plane, the following criteria can be stated.

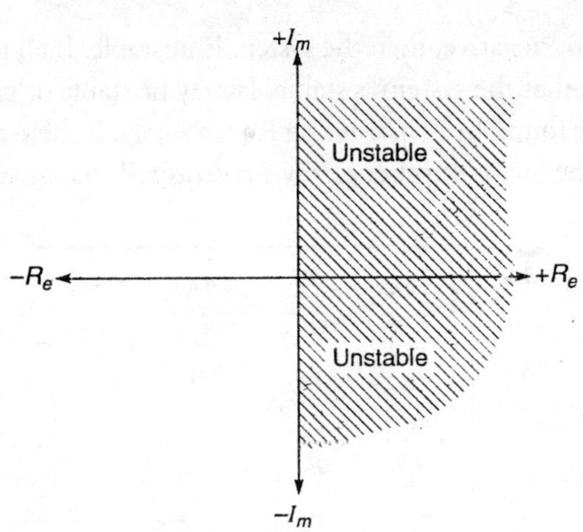

Fig. 13.2 Complex plane.

A control system will be unstable if any roots of its characteristics equation lie either in the imaginary axis or through to its right slashed portion. Otherwise the system will be stable. It means that for stability, all the roots should have –ve real parts.

General conclusions regarding system stability :

(i) If all the roots of the characteristics equation has negative real parts, then the response is bounded and eventually decreases to zero.

(ii) If any root of the characteristics equation has a positive real part, the response is unbounded. The system is therefore unstable.

(iii) If the characteristics equation has repeated roots on the *S*-plane, the response is unbounded. The system is therefore unstable.

(iv) If one or more non-repeated roots of the characteristics equations are on the *S*-plane, then the response is unbounded. The system is therefore unstable.

13.3 METHODS OF ANALYSIS OF STABILITY OF CONTROL SYSTEM

The different methods for analysing the stability of control system are :

(i) Routh Hurwitz criteria
(ii) Root locus technique
(iii) Freequency response technique
  (a) Bode stability criteria
  (b) Nyquist stability criteria

13.3.1 Routh Hurwitz Criteria

This is an algebraic method for determining the stability of control system. The procedure is as follows :

The characteristics equation of control system is expanded and written in the polynomial form as given in equation (13.1)

$$a_0 s^n + a_1 s^{n-1} + a_2 s^{n-2} + \ldots + a_n = 0 \qquad \ldots(13.1)$$

where, $a_0, a_1, a_2, \ldots, a_n$ are coefficient of '*s*'.

If any coefficient is zero or negative, then the system is unstable. If all the coefficients are non-zero and positive, it does not mean that the system is stable. It may be stable or unstable. The determination of stability of the system can be found by constructing Routh Array. If there are 'n' number of coefficient in the polynomial equation, the '$n+1$' number of rows is required to construct in Routh Array.

Routh Array :

$$
\begin{array}{c|cccc}
1 & a_0 & a_2 & a_4 & a_6 \\
2 & a_1 & a_3 & a_5 & a_7 \\
3 & b_1 & b_2 & b_3 & \\
4 & c_1 & c_2 & c_3 & \\
5 & d_1 & d_2 & & \\
6 & e_1 & e_2 & & \\
7 & f_1 & & & \\
8 & g_1 & & &
\end{array}
$$

The calculations of each element of the Routh Array are explained.

$$b_1 = \frac{a_1 a_2 - a_3 a_0}{a_1} \quad ; \quad b_2 = \frac{a_1 a_4 - a_5 a_0}{a_1}$$

$$b_3 = \frac{a_1 a_6 - a_7 a_0}{a_1} \quad ; \quad c_1 = \frac{b_1 a_3 - b_2 a_1}{b_1}$$

$$c_2 = \frac{b_1 a_5 - b_3 a_1}{b_1} \quad ; \quad c_3 = \frac{b_1 a_7 - (o) a_1}{b_1}$$

$$d_1 = \frac{c_1 b_2 - c_2 b_1}{c_1} \quad ; \quad d_2 = \frac{c_1 b_3 - c_3 b_1}{c_1}$$

$$e_1 = \frac{d_1 c_2 - d_2 c_1}{d_1} \quad ; \quad e_2 = \frac{d_1 c_3 - (o) c_1}{d_1}$$

$$f_1 = \frac{e_1 d_2 - e_2 d_1}{e_1} \quad ; \quad g_1 = \frac{f_1 e_2 - (o) e_1}{f_1}$$

Theorem of Routh Hurwitz Test :

(i) Necessary and sufficient test condition for stability is that all the elements in the first column of the Routh Array must be non-zero and positive.

(ii) The number of roots with +ve real parts will be equal to the number of sign changes in the first column.

(iii) If one pair of roots lie on the imaginary axis and all other to its left, then the nth row elements vanishes and the previous row elements remains as it is. The location of the roots on imaginary axis can be found by solving the equation

$$Cs^2 + D = 0 \qquad \ldots(13.2)$$

where, coefficient C and D are the elements in the array $(n-1)^{\text{th}}$ row.

■ **Example 13.1** For the control system described by the following characteristics equation.

$$s^4 + 3s^3 + 5s^2 + 4s + 2 = 0$$

Check the stability. Find the number of roots having +ve real parts.

Solution : Characteristics equation :

$$s^4 + 3s^3 + 5s^2 + 4s + 2 = 0$$

$$n = 4$$

No. of rows, $n + 1 = 4 + 1 = 5$

Routh Array :

1	1	5	2
2	3	4	0
3	11/3	2	
4	26/3	0	
5	2		

From the Routh Array it is observed that all the elements in the first column are non-zero and +ve. Therefore, the system is stable. There are no sign change, hence no root is having +ve real parts, *i.e.*, all the roots have –ve real parts.

■ **Example 13.2** For the control system described by the following characteristic equation

$$s^4 + 2s^3 + s^2 + 4s + 2 = 0$$

Check the stability. Find the number of roots having +ve real parts.

Solution : Characteristic equation :

$$s^4 + 2s^3 + s^2 + 4s + 2 = 0$$

$$n = 4$$

No. of rows = 4 + 1 = 5

Routh Array :

1	1	1	2
2	2	4	0
3	–1	2	
4	8	0	
5	2		

From the Routh Array it is observed that one of the element of the first column is –ve. Therefore, the system is unstable. There are two sign changes +ve to –ve and –ve to +ve. Hence, the number of roots having +ve real parts is 2.

■ **Example 13.3** Determine the stability of a control system whose characteristic equation is

$$s^5 + s^4 + 2s^3 + 2s^2 + 3s + 5 = 0$$

Find the number of roots with +ve real parts.

Solution : Characteristics equation :

$$s^5 + s^4 + 2s^3 + 2s^2 + 3s + 5 = 0$$

$$n = 5$$

$$\text{No. of rows} = 5 + 1 = 6$$

Routh Array :

1	1	2	3
2	1	2	5
3	0	–2	

Since there is a zero in the first column of third row. The system is unstable. To determine the number of roots with +ve real parts substitute $s = 1/z$ in the characteristics equation and get new polynomial equation in terms of 'z'. Then, construct Routh Array.

$$\frac{1}{z^5} + \frac{1}{z^4} + \frac{2}{z^3} + \frac{2}{z^2} + \frac{3}{z} + 5 = 0$$

$$\Rightarrow \quad 1 + z + 2z^2 + 2z^3 + 3z^4 + 5z^5 = 0$$

$$\Rightarrow \quad 5z^5 + 3z^4 + 2z^3 + 2z^2 + z + 1 = 0$$

Routh Array :

1	5	2	1
2	3	2	1
3	–4/3	–2/3	0
4	1/2	1	
5	2	0	
6	1		

It is observed from the Routh Array that there are two sign changes +ve to –ve and –ve to +ve. Hence, the number of roots with +ve real parts are 2.

■ **Example 13.4** For the characteristics equation given below determine the number of roots with positive real part.

$$s^6 + s^5 + 3s^4 + 2s^2 + 5s^2 + 3s + 1 = 0$$

Solution : Characteristics equation :

$$s^6 + s^5 + 3s^4 + 2s^2 + 5s^2 + 3s + 1 = 0$$

$$n = 6$$

$$\text{No. of rows} = 6 + 1 = 7$$

Routh Array :

1	1	3	5	1
2	1	2	3	0
3	1	2	1	
4	0	2	0	

Since there is a zero in the first column of fourth row. The system is unstable. To determine the number of roots with positive real parts replace '0' by ε and construct the Routh Array again.

Routh Array :

1	1	3	5	1
2	1	2	3	0
3	1	2	1	
4	ε	2	0	
5	$\frac{2\varepsilon-2}{\varepsilon}$	1	0	
6	$\frac{\left(\frac{2\varepsilon-2}{\varepsilon}\right)2-\varepsilon}{\left(\frac{2\varepsilon-2}{\varepsilon}\right)}$	0		
7	1			

On letting $\varepsilon \to 0$, we get

1	1	3	5	1
2	1	2	3	0
3	1	2	1	
4	0	2	0	
5	$-\infty$	1	0	
6	2	0		
7	1			

There are two sign changes in the first column of Routh Array. Hence 2 roots +ve real parts.

■ **Example 13.5** Discuss the stability of control system represented by the characteristic equation

$$s^6 + 2s^5 + 8s^4 + 12s^3 + 20s^2 + 16s + 16 = 0$$

Solution : Characteristics equation :

$$s^6 + 2s^5 + 8s^4 + 12s^3 + 20s^2 + 16s + 16 = 0$$

$$n = 6$$

$$\text{No. of rows} = 6 + 1 = 7$$

Routh Array :

1	1	8	20	16
2	2	12	16	0
3	2	12	16	
4	0	0	0	

From the Routh Array it is observed that all the elements of the fourth row is zero. It indicates that the system is not stable. It can have sustained oscillation.

The procedure to construct the Routh Array for the system having sustained oscillations are as follows :

(i) Select the auxiliary polynomial which is generally above the zero rows.
(ii) Find the differential of this polynomial.
(iii) Replace the row of zeros with the coefficients of the differential polynomial and complete the Routh Array.
(iv) Find the roots of the auxiliary polynomial. These will be the roots of original characteristic equation.

Therefore, the auxiliary polynomial is

$$2s^4 + 12s^2 + 16 = 0 \quad \text{...(13.3)}$$

Differentiating equation (13.3) w.r.t. 's', we get

$$8s^3 + 24s = 0 \quad \text{...(13.4)}$$

Now, replace the zero rows with the coefficient of equation (13.4) and reconstruct the Routh Array.

1	1	8	20	16
2	2	12	16	0
3	2	12	16	
4	8	24	0	
5	6	16		
6	8/3	0		
7	16			

There are no sign change in the first column of the Routh Array. Hence, no root has +ve real parts. From the auxiliary polynomial equation, we have

$$2s^4 + 12s^2 + 16 = 0 \quad \text{...(13.3)}$$

From the formula of quadratic equation, we have

$$x = \frac{-b \pm \sqrt{b^2 - 4ac}}{2a}$$

$$\Rightarrow \quad s^2 = \frac{-12 \pm \sqrt{144 - 4 \times 2 \times 16}}{2 \times 2}$$

$$\Rightarrow \quad s^2 = \frac{-12 \pm \sqrt{16}}{4}$$

$$\Rightarrow \quad s^2 = \frac{-12 \pm 4}{4}$$

$$\Rightarrow \quad s^2 = \frac{-12 + 4}{4}, \quad \frac{-12 - 4}{4}$$

$$\Rightarrow \quad s^2 = -2, \ -4$$

$$\Rightarrow \quad s = \pm j\sqrt{2}; \quad \pm j2$$

There are four roots in the imaginary axis. Therefore, the system will be under sustained oscillation or it will have limited stable response.

■ **Example 13.6** For the characteristic equation given below determine the number of roots with positive real part.

$$s^8+s^7+4s^6+3s^5+14s^4+11s^3+20s^2+9s+9=0$$

Solution : Characteristic equation :

$$s^8+s^7+4s^6+3s^5+14s^4+11s^3+20s^2+9s+9=0$$

$$n=8$$

$$\text{No. of rows}=8+1=9$$

Routh Array :

1	1	3	14	20	9
2	1	3	11	9	0
3	1	3	11	9	
4	0	0	0	0	

All the element of fourth row is zero. From the auxiliary polynomial, we have

$$s^6+3s^4+11s^2+9=0 \qquad \text{...(13.5)}$$

Differentiating equation (13.5) w.r.t. 's', we get

$$6s^5+12s^3+22s=0 \qquad \text{...(13.6)}$$

Now replace the zeroth rows with the coefficient of equation (13.6) and reconstruct the Routh Array.

1	1	4	14	20	9
2	1	3	11	9	0
3	1	3	11	9	
4	6	12	22	0	
5	1	22/3	9		
6	–32	–32	0		
7	19/3	9			
8	256/19	0			
9	9				

There are two sign changes in the first column of the Routh Array. Hence, two roots are having positive real part.

■ **Example 13.7** The characteristic equation of a control system is given by

$$1+\frac{k}{(s)(s+1)(s+2)(s+3)}=0$$

Determine the value of k for which the system is stable.

Solution : Characteristic equation :

$$1+\frac{k}{(s)(s+1)(s+2)(s+3)}=0$$

$$\Rightarrow \quad (s)(s+1)(s+2)(s+3)+k=0$$

$$\Rightarrow \quad s^4+6s^3+11s^2+6s+k=0$$

Routh Arrays :

1	1	11	k
2	6	6	0
3	10	k	
4	$\frac{60-6k}{10}$	0	
5	k		

On letting,

$$\frac{60-6k}{6}=0$$

$$\Rightarrow \quad 60-6k=0$$

$$\Rightarrow \quad 6k=60$$

$$\Rightarrow \quad k=10$$

The value of k for which the system is stable is $0<k<10$.

■ **Example 13.8** A feedback control system has a characteristic equaion

$$s^6+2s^5+9s^4+16s^3+24s^2+32s+16=0$$

Determine the number of poles :

(i) On the right half of S-plane

(ii) On imaginary axis

(iii) On the left half of S-plane.

Solution : Characteristic equation :

$$s^6+2s^5+9s^4+16s^3+24s^2+32s+16=0$$

$$n=6$$

$$\text{No. of rows}=6+1=7$$

Routh Array :

1	1	9	24	16
2	2	16	32	0
3	1	8	16	
4	0	0	0	

Since all the elements of fourth row is zero, the auxiliary polynomial is exist. From the auxiliary polynomial, we have

$$s^4+8s^2+16=0 \qquad \ldots(13.7)$$

Differentiating equation (13.7) w.r.t. '*s*', we get

$$4s^3 + 16s = 0 \qquad \ldots(13.8)$$

Now, replace the zeroth rows with the coefficient of equation (13.8) and reconstruct the Routh Array.

Routh Array :

1	1	9	24	16
2	2	16	32	0
3	1	8	16	
4	4	16	0	
5	4	16		
6	0	0		

Since, all the element of sixth row is zero, the second auxiliary polynomial is exist. From the second auxiliary polynomial, we have

$$4s^2 + 16 = 0 \qquad \ldots(13.9)$$

Differentiating equation (13.9) w.r.t. '*s*', we get

$$8s = 0 \qquad \ldots(13.10)$$

Now, replace the zero rows with the coefficient of equation (13.10) and reconstruct the Routh Array.

Routh Array :

1	1	9	24	16
2	2	16	32	0
3	1	8	16	
4	4	16	0	
5	4	16		
6	8	0		
7	16			

It is observed from the Routh Array that there is no sign change in the first column. Hence, no roots lie on the right half of *S*-plane.

From the first auxiliary polynomial equation, we have

$$s^4 + 8s^2 + 16 = 0 \qquad \ldots(13.7)$$

From the formula of quadratic equation, we have

$$x = \frac{-b \pm \sqrt{b^2 - 4ac}}{2a}$$

$$\Rightarrow \qquad s^2 = \frac{-8 \pm \sqrt{64 - 4 \times 1 \times 16}}{2 \times 1}$$

$$\Rightarrow \qquad s^2 = \frac{-8 \pm 0}{2}$$

$$\Rightarrow \qquad s^2 = -4$$

$$\Rightarrow \qquad s = \pm j2 \ ; \ \pm j2$$

There are four roots in the imaginary axis. Remaining 2 roots lie in the left half of S-plane.

■ **Example 13.9** For the control system described by the characteristic equation

$$s^3 + 7s^2 + 25s + 39 = 0$$

Find the number of rots having and real parts, more negative than –1.

Solution : Characteristic equation :

$$s^3 + 7s^2 + 25s + 39 = 0$$

To determine the no. of roots having more negative than –1, substitute $s = z - 1$ in the characteristic equation and get new equation in terms of 'z'.

$$(z-1)^3 + 7(z-1)^2 + 25(z-1) + 39 = 0$$

$$z^3 - 3z^2 + 3z - 1 + 7(z^2 - 2z + 1) + 25(z-1) + 39 = 0$$

$$z^3 - 3z^2 + 3z - 1 + 7z^2 - 14z + 7 + 25z - 25 + 39 = 0$$

$$z^3 + 4z^2 + 14z + 20 = 0$$

Thus, we obtained new characteristic equation in terms of z. Now, construct the Routh Array.

New characteristic equation

$$z^3 + 4z^2 + 14z + 20 = 0$$

$$n = 3$$

$$\text{No. of rows} = 3 + 1 = 4$$

Routh Arrray :

1	1	14
2	4	20
3	9	0
4	20	

There are no sign changes in the first column. Hence, no roots lie on the right half of S-plane. Therefore, all the 3 roots have real parts more negative than –1.

■ **Example 13.10** The open loop transfer function of a single loop unity feed back control system is given by

$$G(s) = \frac{k(s+13)}{(s)(s+3)(s+7)}$$

Find the value of k for which the closed loop system is stable.

Solution : The characteristic equation of closed loop system is given by the equation

$$1 + G(s) = 0$$

$$\Rightarrow \qquad 1 + \frac{k(s+13)}{(s)(s+3)(s+7)} = 0$$

$$\Rightarrow \qquad (s)(s+3)(s+7)+k(s+13)=0$$

$$\Rightarrow \qquad s^3+10s^2+(21+k)s+13k=0$$

Routh Array :

$$\begin{array}{c|cc} 1 & 1 & (21+k) \\ 2 & 10 & 13k \\ 3 & \dfrac{10(21+k)-13k}{10} & 0 \\ 4 & 13k & \end{array}$$

On letting,

$$\frac{10(21+k)-13k}{10}=0$$

$$\Rightarrow \qquad 10(21+k)-13k=0$$

$$\Rightarrow \qquad 210+10k-13k=0$$

$$\Rightarrow \qquad 3k=210$$

$$\Rightarrow \qquad k=70$$

The range of k for which the system is stable is $0<k<70$.

■ **Example 13.11** The open loop transfer function of chemical process is given by

$$G(s)=\frac{k}{(s+1)(50s^2+12s+0.5)}$$

Find the range of k for which this system is stable. What will be the frequency of sustained oscillation?

Solution : The characteristic equation of closed loop transfer function is given by the equation

$$1+G(s)=0$$

$$\Rightarrow \qquad 1+\frac{k}{(s+1)(50s^2+12s+0.5)}=0$$

$$\Rightarrow \qquad (s+1)(50s^2+12s+0.5)+k=0$$

$$\Rightarrow \qquad 50s^3+62s^2+12.5s+(0.5+k)=0$$

Routh Array :

$$\begin{array}{c|cc} 1 & 50 & 12.5 \\ 2 & 62 & (0.5+k) \\ 3 & \dfrac{(62\times 12.5)-50(0.5+k)}{62} & 0 \\ 4 & (0.5+k) & \end{array}$$

On letting,

$$\frac{(62\times 12.5)-50(0.5+k)}{62}=0$$

$$\Rightarrow \qquad (62\times 12.5)-50(0.5+k)=0$$

$$\Rightarrow \quad 50(0.5+k)=775$$
$$\Rightarrow \quad 0.5+k=15.5$$
$$\Rightarrow \quad k=15$$

On letting, $0.5+k=0$

$$\Rightarrow \quad k=-0.5$$

The range of k for which the system stable is $-0.5<k<15$.

To determine the frequency of sustained oscillation consider the equation of the form $Cs^2+D=0$ from the Routh Array.

$$\Rightarrow \quad 62s^2+(0.5+k)=0$$
$$\Rightarrow \quad 62s^2+(0.5+15)=0$$
$$\Rightarrow \quad 62s^2=-15.5$$
$$\Rightarrow \quad s^2=-0.25$$
$$\Rightarrow \quad s^2=j^2(0.25)$$
$$\Rightarrow \quad s=j(0.5) \qquad \ldots(13.11)$$

On letting, $s=j(\omega)$ ...(13.12)

Comparing equations (13.11) and (13.12), we get

$$\omega=0.5 \text{ radians/s}$$

This is the frequency of sustained oscillation.

From the equation of cyclical frequency, we have

$$\omega=2\pi f$$
$$\Rightarrow \quad 0.5=2\pi f$$
$$\Rightarrow \quad f=0.5/(2\pi)$$
$$\Rightarrow \quad f=0.0796 \text{ cycle/s}$$

■ **Example 13.12** The open loop transfer function of a single loop, unity feed back control system is given by

$$G(s)=\frac{k(s+13)}{(s)(s+3)(s+7)}$$

Find the value of k for which the closed loop poles are more negative than -1.

Solution : The characteristic equation of closed loop system is given by the equation

$$1+G(s)=0$$
$$\Rightarrow \quad 1+\frac{k(s+13)}{(s)(s+3)(s+7)}=0$$
$$\Rightarrow \quad (s)(s+3)(s+7)+k(s+13)=0$$
$$\Rightarrow \quad s^3+10^2+(21+k)s+13k=0 \qquad \ldots(13.13)$$

On letting, $s=z-1$ in equation (13.13), we get

$$(z-1)^3+10(z-1)^2+(21+k)(z-1)+13k=0$$

$$z^3 - 3z^2 + 3z - 1 + 10(z^2 - 2z + 1) + 21z + kz - 21 - k + 13k = 0$$

$$z^3 + 7z^2 + (4+k)z + 12(k-1) = 0$$

Routh Array :

1	1	$(4 + k)$
2	7	$12(k - 1)$
3	$\dfrac{7(4+k)-12(k-1)}{7}$	0
4	$12(k + 1)$	

On letting, $$\frac{7(4+k)-12(k-1)}{7} = 0$$

$\Rightarrow$ $$7(4+k) - 12(k-1) = 0$$

$\Rightarrow$ $$28 + 7k - 12k + 12 = 0$$

$\Rightarrow$ $$40 - 5k = 0$$

$\Rightarrow$ $$5k = 40$$

$\Rightarrow$ $$k = 8$$

On letting, $$12(k-1) = 0$$

$\Rightarrow$ $$k - 1 = 0$$

$\Rightarrow$ $$k = 1$$

The range of k for which the closed loop poles are more negative than –1 is $1 < k < 8$.

■ **Example 13.13** A unity feed back control system has transfer function

$$G(s) = \frac{k}{(s)(s+2)(s+4)(s+6)}$$

Using Routh Hurwitz criterion, find the range of k for stability. Also find k_{max} and ω_{max}.

Solution : The characteristic equation of feed back control system is given by the equation

$$1 + G(s)\,H(s) = 0$$

$$1 + \frac{k}{(s)(s+2)(s+4)(s+6)} = 0$$

$\Rightarrow$ $$(s)(s+2)(s+4)(s+6) + k = 0$$

$\Rightarrow$ $$s^4 + 12s^3 + 44s^2 + 48s + k = 0 \quad \ldots(13.13)$$

Constructing Routh Array using characteristic equation (13.13), we get

Routh Array :

1	1	44	k
2	12	48	0
3	40	k	
4	$\dfrac{1920-12k}{40}$	0	
5	k		

It is observed from the Routh Array that for stability $k > 0$.

On letting, $$\frac{1920-12k}{40} > 0$$

$$\Rightarrow \quad 1920-12k > 0$$

$$\Rightarrow \quad 1920 > 12k$$

$$\Rightarrow \quad k < 160$$

The range of k for system to be stable is $0 < k < 160$.

Therefore, $$k_{max} = 160$$

To determine the frequency of sustained oscillation, consider the equation of the form $Cs^2 + D = 0$ from the Routh Array.

$$\Rightarrow \quad 40s^2 + 160 = 0$$

$$\Rightarrow \quad 40s^2 = -160$$

$$\Rightarrow \quad 40s^2 = j^2 160$$

$$\Rightarrow \quad s^2 = j^2 4$$

$$\Rightarrow \quad s = \pm j2$$

On letting, $$s = j\omega$$

$$\omega_{max} = 2 \text{ radian/s}$$

■ **Example 13.14** For the system represented by open loop transfer function

$$G(s) = \frac{k}{(s+1)(s+2)(s+3)}$$

Using Routh Hurwitz criteria :

(i) Determine the range of k for which the system is stable.

(ii) The value of k for which the system develops a sustained oscillations and the corresponding frequencies.

Solution : The characteristic equation of feed back system is given by the equation

$$1 + G(s) = 0$$

$$\Rightarrow \quad 1 + \frac{k}{(s+1)(s+2)(s+3)} = 0$$

$$\Rightarrow \quad (s+1)(s+2)(s+3) + k = 0$$

$$\Rightarrow \quad s^3 + 6s^2 + 11s + (6+k) = 0$$

Routh Array :

1	1	11
2	6	$(6+k)$
3	$\frac{66-(6+k)}{6}$	0
4	$(6+k)$	

On letting, $$\frac{66-(6+k)}{6}=0$$

$\Rightarrow$ $$66-(6+k)=0$$

$\Rightarrow$ $$6+k=66$$

$\Rightarrow$ $$k=60$$

On letting, $$6+k=0$$

$\Rightarrow$ $$k=-6$$

The range of k for system stability is $-6<k<60$.

Therefore, $$k_{max}=60$$

To determine the frequency of sustained oscillation, consider the equation of the form $Cs^2+D=0$ from the Routh Array.

$\Rightarrow$ $$6s^2+66=0$$

$\Rightarrow$ $$6s^2=-66$$

$\Rightarrow$ $$s^2=-11$$

$\Rightarrow$ $$s^2=j^2 11$$

$\Rightarrow$ $$s=\pm j\sqrt{11}$$

On letting, $$s=j\omega$$

$\Rightarrow$ $$\omega_{max}=3.313 \text{ rad/s}$$

■ **Example 13.15** The characteristic equation of a feedback control system is given by

$$1+\frac{k}{(s)(s+1)(s+2)(s+3)}=0$$

Determine the value of k for which this system is stable.

Solution : The characteristic equation of the feedback control system is given by

$$1+\frac{k}{(s)(s+1)(s+2)(s+3)}=0$$

$\Rightarrow$ $$(s)(s+1)(s+2)(s+3)+k=0$$

$\Rightarrow$ $$s^4+6s^3+11s^2+6s+k=0$$

Routh Array :

1	1	11	k
2	6	6	0
3	10	k	
4	$\frac{60-6k}{10}$	0	
5	k		

On letting,
$$\frac{60-6k}{10}=0$$
$$\Rightarrow \quad 60-6k=0$$
$$\Rightarrow \quad 6k=60$$
$$\Rightarrow \quad k=10$$

The range of k for which this system stable is $0<k<10$.

■ **Example 13.16** The open loop transfer function of a unity feed back control system is given by

$$G(s)=\frac{k}{(s+2)(s+4)(s^2+6s+5)}$$

By applying Routh Hurwitz criteria discuss the stability of the closed loop system as a function of k. Determine the value of k which will cause sustained oscillation in the closed loop system. What are the corresponding oscillation frequencies?

Solution : The characteristic equation of unity feed back control system is given by

$$1+\frac{k}{(s+2)(s+4)(s^2+6s+5)}=0$$
$$\Rightarrow \quad (s+2)(s+4)(s^2+6s+5)+k=0$$
$$\Rightarrow \quad s^4+12s^3+49s^2+78s+(40+k)=0$$

Routh Array :

1	1	49	$(40+k)$
2	12	78	0
3	$\frac{85}{3}$	$(40+k)$	
4	$\frac{6630-36(40+k)}{85}$	0	
5	$(40+k)$		

On letting,
$$\frac{6630-36(40+k)}{85}=0$$
$$\Rightarrow \quad 6630-36(40+k)=0$$
$$\Rightarrow \quad 36(40+k)=6630$$
$$\Rightarrow \quad k=\frac{865}{6}$$

On letting,
$$40+k=0$$
$$\Rightarrow \quad k=-40$$

The range of k for which this system stable is $-40<k<865/6$ and also, $k_{max}=865/6$.

To determine the frequency of sustained oscillation, consider the equation of the form $Cs^2+D=0$ from the Routh Array.

$$\Rightarrow \quad (85/3)s^2+(40+k)=0$$

$$\Rightarrow \qquad (85/3)\,s^2 + [40 + (865/6)] = 0$$

$$\Rightarrow \qquad (85/3)\,s^2 = -(1105/6)$$

$$\Rightarrow \qquad s^2 = -6.5$$

$$\Rightarrow \qquad s^2 = j^2\,6.5$$

$$\Rightarrow \qquad s = \pm\, j\sqrt{6.5}$$

On letting, $s = j\omega$

$$\Rightarrow \qquad \omega = 2.55 \text{ rad/s}$$

■ **Example 13.17** The polynomial $s^4 + s^3 + 2s^2 + 3s + 4 = 0$ has all roots in LHS of S-plane. Use Routh Hurwitz criteria to determine the number of roots lying between $s = -1$ and $s = -2$.

Solution : The characteristic equation is given by

$$s^4 + s^3 + 2s^2 + 3s + 4 = 0 \qquad \text{...(13.14)}$$

Letting, $s = (s-1)$ in equation (13.14), we get

$$(s-1)^4 + (s-1)^3 + 2(s-1)^2 + 3(s-1) + 4 = 0$$

$$(s-1)^2(s-1)^2 + (s-1)(s-1)^2 + 2(s-1)^2 + 3(s-1) + 4 = 0$$

$$(s^4 - 4s^3 + 6s^2 - 4s + 1) + (s^3 - 3s^2 + 3s - 1) + (2s^2 - 4s + 2) + (3s - 3) + 4 = 0$$

$$s^4 - 3s^3 + 5s^2 - 2s + 3 = 0$$

Routh Array :

1	1	5	3
2	–3	–2	0
3	4.33	3	
4	0.078	0	
5	3		

There are two sign changes in the first column. Hence, there are two roots on RHS of $s = -1$.
On letting, $s = s - 2$ in equation (13.14), we get

$$(s-2)^4 + (s-2)^3 + 2(s-2)^2 + 3(s-2) + 4 = 0$$

$$(s-2)^2(s-2)^2 + (s-2)(s-2)^2 + 2(s-2)^2 + 3(s-2) + 4 = 0$$

$$(s^4 - 8s^3 + 24s^2 - 32s + 16) + (s^3 - 6s^2 + 12s - 8) + (2s^2 - 8s + 8) + (3s - 6) + 4 = 0$$

$$s^4 - 7s^3 + 20s^2 - 25s + 14 = 0$$

Routh Array :

1	1	20	14
2	–7	–25	0
3	16.43	14	
4	–19.04	0	
5	14		

There are four sign changes in the first column of Routh Array. Hence, there are four roots on RHS of $s = -2$. Therefore, number of roots lying between $s = -1$ and $s = -2$ are two.

■ **Example 13.18** A negative feed back control system has

$$G(s) = \frac{k}{(s)(s^2 + 2s + 1)}$$

$$H(s) = \frac{1}{(s + 2)}$$

Determine the range of k for the absolute stability of the system. Also, determine the frequency of sustained oscillations for the limiting value of k.

Solution : The characteristic equation is given by

$$1 + G(s)\,H(s) = 0$$

$$\Rightarrow \quad 1 + \frac{k}{(s)(s^2 + 2s + 1)} \times \frac{1}{(s + 2)} = 0$$

$$\Rightarrow \quad (s)(s^2 + 2s + 1)(s + 2) + k = 0$$

$$\Rightarrow \quad s^4 + 4s^3 + 5s^2 + 2s + k = 0$$

Routh Array :

1	1	5	k
2	4	2	0
3	9/2	k	
4	$(9 - 4k)/(9/2)$	0	
5	k		

On letting,
$$(9 - 4k)/(9/2) = 0$$

$$\Rightarrow \quad 9 - 4k = 0$$

$$\Rightarrow \quad 4k = 9$$

$$\Rightarrow \quad k = 2.25$$

Range of k for absolute stability is $0 < k < 2.25$.

To determine the frequency of sustained oscillation, consider the equation of the form $Cs^2 + D = 0$ from the Routh Array.

$$\Rightarrow \quad (9/2)\,s^2 + 2.25 = 0$$

$$\Rightarrow \quad (9/2)\,s^2 = -2.25$$

$$\Rightarrow \quad s^2 = -0.5$$

$$\Rightarrow \quad s^2 = j^2 0.5$$

$$\Rightarrow \quad s = \pm j\sqrt{0.5}$$

$$\Rightarrow \quad s = j(0.707)$$

On letting,
$$s = j\omega$$

$$\Rightarrow \quad \omega = 0.707 \text{ rad/s}$$

■ Example 13.19 The open loop transfer function of a unity feed back control system is given by

$$G(s) = \frac{30}{(s)(1+0.05s)(1+0.2s)}$$

Using Routh Hurwitz criteria show that the system is unstable.

Solution : The characteristic equation of unity feed back control system is given by

$$1+G(s)=0$$

$$\Rightarrow \quad 1+\frac{30}{(s)(1+0.05s)(1+0.2s)}=0$$

$$\Rightarrow \quad 1+\frac{30}{(s)(0.05)(s+20)(0.2)(s+5)}=0$$

$$\Rightarrow \quad 1+\frac{3000}{(s)(s+20)(s+5)}=0$$

$$\Rightarrow \quad (s)(s+20)(s+5)+3000=0$$

$$\Rightarrow \quad s^3+25s^2+100s+3000=0$$

Routh Array :

1	1	100
2	25	3000
3	–20	0
4	3000	

Since, there is a negative sign in the first column of the Routh Array, the system is thus unstable.

■ Example 13.20 A feed back control system has open loop transfer function

$$G(s) = \frac{ke^{-s}}{(s)(s^2+8s+9)}$$

Determine the maximum value of k for which the closed loop control system is stable.

Solution : The characteristic equation of closed loop control system is given by

$$1+G(s)H(s)=0$$

On letting,

$$H(s)=1$$

$$\Rightarrow \quad 1+G(s)=0$$

$$\Rightarrow \quad 1+\frac{ke^{-s}}{(s)(s^2+8s+9)}=0$$

From the mathematical interpretation, we have

$$e^{-s}=1-s$$

$$\Rightarrow \quad 1+\frac{k(1-s)}{(s)(s^2+8s+9)}=0$$

$$\Rightarrow \quad (s)(s^2+8s+9)+k(1-s)=0$$

$$\Rightarrow \qquad s^3 + 8s^2 + 9s + k - ks = 0$$

$$\Rightarrow \qquad s^3 + 8s^2 + (9 - k)\, s + k = 0$$

Routh Array :

1	1	$(9-k)$
2	8	k
3	$\dfrac{8(9-k)-k}{8}$	0
4	k	

On letting,

$$\frac{8(9-k)-k}{8} = 0$$

$$\Rightarrow \qquad 8(9-k) - k = 0$$

$$\Rightarrow \qquad 72 - 8k - k = 0$$

$$\Rightarrow \qquad 9k = 72$$

$$\Rightarrow \qquad k = 8$$

The range of k for which the closed loop control system stable is $0 < k < 8$.

■ **Example 13.21** The open loop transfer function of a unity feed back control system is given by

$$G(s) = \frac{1}{(s+3)(s+4)(s^2 + 6s + 25)}$$

Determine the range of values of k for system stability. What is the value of k which gives sustained oscillation? What is the oscillation frequency?

Solution : The characteristic equation of unity feed back control system is given by

$$1 + G(s) = 0$$

$$\Rightarrow \qquad 1 + \frac{k}{(s+3)(s+4)(s^2 + 6s + 25)} = 0$$

$$\Rightarrow \qquad (s+3)(s+4)(s^2 + 6s + 25) + k = 0$$

$$\Rightarrow \qquad s^4 + 13s^3 + 79s^2 + 247s + (300 + k) = 0$$

Routh Array :

1	1	79	$(300 + k)$
2	13	247	0
3	60	$(300 + k)$	
4	$\dfrac{10920 - 13k}{60}$	0	
5	$(300 + k)$		

On letting,

$$\frac{10920 - 13k}{60} = 0$$

$$\Rightarrow \qquad 10920 - 13k = 0$$

$$\Rightarrow \quad 13k = 10920$$

$$\Rightarrow \quad k = 840$$

The range of values of k for which the system stable is $-300 < k < 840$.

To determine the frequency of sustained oscillation, consider the equation of the form $Cs^2 + D = 0$ from the Routh Array.

$$\Rightarrow \quad 60s^2 + 1140 = 0$$

$$\Rightarrow \quad 60s^2 = -1140$$

$$\Rightarrow \quad s^2 = -19$$

$$\Rightarrow \quad s^2 = j^2(19)$$

$$\Rightarrow \quad s = \pm j(4.36)$$

On letting,

$$s = j\omega$$

$$\omega = 4.36 \text{ rad/s}$$

Hence, the oscillation frequency is 4.36 rad/s.

Merits and Demerits

Following are the merits and demerits of Routh Hurwitz criteria.

Merits :

(i) This is only an algebraic method to determine the stability. It does not involve any plot or diagram.

(ii) It requires only the characteristic equation and no other data. Hence, it is a simplest method.

(iii) The number of roots with +ve real parts can be found.

(iv) The frequency of sustained oscillation can also be determined.

Demerits :

(i) It does not give any information on actual location of the roots in the complex plane.

(ii) It does not give any information on degree of stability.

(iii) It can be used for the system having characteristic equation in polynomial form only.

(iv) Applicable to linear system only.

13.3.2 Root Locus Technique

It is a graphical presentation of the roots of the control system. It simplifies the problems involve in root solving by providing an effective graphical procedure.

The first step in applying the root locus technique to determine the roots of the characteristic equation of the closed loop control system is to write open loop transfer function in the standard form

$$G(s) = \frac{kN}{D} \qquad \text{...(13.15)}$$

where, k = Constant

$$N = (s + z_1)(s + z_2)\ldots(s + z_i)\ldots(s + z_m)$$

$$D = (s + p_1)(s + p_2)\ldots(s + p_i)\ldots(s + p_n)$$

The term z_i is called zero of the open loop transfer function and p_j is called pole of the open loop transfer function.

From the characteristic equation of closed loop control system, we have

$$1+G(s)=0 \qquad ...(13.16)$$

Substituting equation (13.15) in equation (13.16), we get

$$1+\frac{kN}{D}=0$$

It is assumed that $n>m$, which is true for all physical systems. This being the case, the characteristic equation will be nth order and have 'n' roots. To develop the graphical method for determining the root locus the characteristic equation may be written as

$$\frac{kN}{D}=-1$$

$$\Rightarrow \quad \frac{k\,(s+z_1)\,(s+z_2)\ldots(s+z_i)\ldots(s+z_m)}{(s+p_1)\,(s+p_2)\ldots(s+p_j)\ldots(s+p_n)}=-1 \qquad ...(13.17)$$

Since the left hand side of equation (13.17) is in the form of general complex, we may write the above equation in equivalent form involving magnitude and phase angle criteria.

Magnitude Criteria :

$$\frac{k\,|s+z_1|\,|s+z_2|\ldots|s+z_i|\ldots|s+z_m|}{|s+p_1|\,|s+p_2|\ldots|s+p_j|\ldots|s+p_n|}=1 \qquad ...(13.18)$$

Phase Angle Criteria :

$$|[\{L(s+z_1)+L(s+z_2)+\ldots+L(s+z_i)+\ldots+L(s+z_m)\}$$
$$-\{L(s+p_1)+L(s+p_2)+\ldots+L(s+p_j)+\ldots+L(s+p_m)\}]|=(2i+1)\,\pi \qquad ...(13.19)$$

where, 'i' is an positive or negative integer or zero.

The equation (13.18) and (13.19) may be used to find the root locus by trial and error method. The path of the locus is found generally using the "angle criterion" which is independent of k. After the locus is established, the gain k for any point on it may be obtained by the equation

$$k=\frac{|(s_c+p_1)\,(s_c+p_2)\ldots(s_c+p_n)|}{|(s_c+z_1)\,(s_c+z_2)\ldots(s_c+z_m)|} \qquad ...(13.20)$$

This method of finding k is referred as "magnitude criterion".

Rules for Plotting the Root-Locus Diagram :

(i) The number of loci or branches is equal to the number of open-loop poles. All the open loop poles and zeros are first plotted in the complex plane. Poles are marked as '×' and zeros are marked as 'O'.

(ii) The root-loci begin at open loop poles and terminates at open loop zeros. The termination of the loci will occur at the zero at infinity along the asymptotes.

(iii) **Locus on Real Axis :** The real axis is part of the root locus when the sum of the number of poles and zeros to the right of a point on the real axis is odd. It is necessary to consider only the real poles and zeros in applying this rule. The complex poles and zeros always occur in conjugate pairs and their effects cancel in checking the angle criterion for points on the real axis.

(iv) **Asymptotes :** There are $(n-m)$ loci which approach as $k \to \infty$ asymptotically $(n-m)$ straight lines, radiating from the "centre of gravity" of the poles and zeros of the open loop transfer function. It is also called centroid. It is given by the equation

$$r = \frac{\text{Sum of real parts of the pole} - \text{Sum of real parts of the zeros}}{(n-m)} \quad \ldots(13.21)$$

$$r = \frac{\sum_{j=1}^{n} p_j - \sum_{i=1}^{m} z_i}{(n-m)} \quad \ldots(13.22)$$

(v) **Asymptote Angle :** The angles made by asymptotic lines along with real axis are given by the equation

$$\alpha = \frac{\pi(2k+1)}{(n-m)} \quad \ldots(13.23)$$

where, $k = 0, 1, 2, 3, \ldots (n-m)$.

(vi) **Break Away Point :** The point at which two root loci emerging from adjacent poles or moving towards adjacent zeros on the real axis, intersect and then leave or enter the real axis is determined by the solution of the equation

$$\sum_{j=1}^{n} \frac{1}{(s+p_j)} = \sum_{i=1}^{m} \frac{1}{(s+z_i)} \quad \ldots(13.24)$$

The equation (13.24) is solved by trial and checking it for various test point at $s = s_c$, on the real axis between the poles or zeros of intersect. For real poles or zeros, the terms in the denominator are obtained by simply measuring the distance along real axis between the test point and the poles or zeros.

■ **Example 13.22** Plot the root locus diagram for the open loop transfer function

$$G(s) = \frac{k}{(s+1)(s+2)(s+3)}$$

Solution :

(i) *Calculate the no. of zeros and poles :*

$\Rightarrow \quad m = 0$

$\Rightarrow \quad n = 3 \quad ; \quad p_1 = -1$

$p_2 = -2$

$p_3 = -3$

(ii) *Calculate the no. of asymptotes :*

$$n - m = 3 - 0 = 3$$

(iii) *Calculate the angle of asymptotes :*

$$\alpha = \frac{\pi(2k+1)}{n-m}$$

$$\Rightarrow \quad \alpha_1 = \frac{\pi(2 \times 0 + 1)}{3}$$

$\Rightarrow \qquad \alpha_1 = 60°$

$\Rightarrow \qquad \alpha_2 = 180°$

$\Rightarrow \qquad \alpha_3 = 300°$

(iii) *Calculate the centre of gravity* :

$$r = \frac{\sum_{j=1}^{n} p_j - \sum_{i=1}^{m} z_i}{(n-m)}$$

$$\Rightarrow \qquad r = \frac{[(-1) + (-2 + (-3))] - (0)}{3}$$

$$\Rightarrow \qquad r = \frac{(-6)}{3}$$

$$\Rightarrow \qquad r = -2$$

(v) *Calculate the breakaway point* :

$$\sum_{j=1}^{n} \frac{1}{(s+p_j)} = \sum_{i=1}^{m} \frac{1}{(s+z_i)}$$

$$\Rightarrow \qquad \frac{1}{(s+1)} + \frac{1}{(s+2)} + \frac{1}{(s+3)} = 0$$

$$\Rightarrow \qquad \frac{(s+2)(s+3) + (s+1)(s+3) + (s+1)(s+2)}{(s+1)(s+2)(s+3)} = 0$$

$$\Rightarrow \qquad (s^2 + 5s + 6) + (s^2 + 4s + 3) + (s^2 + 3s + 2) = 0$$

$$\Rightarrow \qquad 3s^2 + 12s + 11 = 0 \qquad \ldots(13.25)$$

The equation (13.25) is of the form of quadratic equation. It can be solve using formula

$$x = \frac{-b \pm \sqrt{b^2 - 4ac}}{2a}$$

$$\Rightarrow \qquad s = \frac{-12 \pm \sqrt{(12)^2 - (4 \times 3 \times 11)}}{(2 \times 3)}$$

$$\Rightarrow \qquad s = \frac{-12 \pm \sqrt{144 - 132}}{6}$$

$$\Rightarrow \qquad s = \frac{-12 \pm \sqrt{12}}{6}$$

$$\Rightarrow \qquad s = \frac{-12 \pm 2\sqrt{3}}{6}$$

$$\Rightarrow \qquad s = \frac{-6 \pm \sqrt{3}}{3}$$

$$\Rightarrow \qquad s = \frac{-6+\sqrt{3}}{3}, \quad \frac{-6-\sqrt{3}}{3}$$

$$\Rightarrow \qquad s = -1.42, \quad -2.57$$

The centre of gravity is found to be –2.

Therefore, appropriate value of breakaway point is $s = -1.42$.

In this problem there are 3 poles and no zero. Hence, there are 3 asymptote which move towards the infinity with angles 60°, 180° and 300° respectively. The pole p_3 goes to infinity with an angle 180°. The poles p_2 and p_1 move each other in the opposite direction and meet at breakaway point $s = -1.42$. These two poles after meeting, break and goes to infinity with the asymptotic locus.

Take a ordinary graph sheet. Hold it vertically upward. Draw a horizontal line from the centre of the graph and a vertical line from the right end of the graph. Select the suitable scale for *x*-axis and *y*-axis. Locate the poles p_1, p_2, p_3 and zeros on *x*-axis. Also, locate the centre of gravity '*r*' and breakaway point '*B*' on the *x*-axis.

Draw a horizontal line from the point p_3 to its left towards the infinity. Also, draw horizontal lines from the points p_2 and p_1 towards the breakaway point. These two horizontal line meets at breakaway point '*B*'. From this point draw a asymptotic locus which move towards the infinity with an angle 60°. Consider a point '*L*' on the asymptotic locus. Draw the lines Lp_1, Lp_2 and Lp_3. Name the angle formed as α_1, α_2 and α_3 respectively.

Then, apply the phase angle criterion to verify the assumed point '*L*' is on the locus of asymptote or not.

From the phase angle criterion, we have

$$|\{(L\beta_1 + L\beta_2 + \ldots) - (L\alpha_1 + L\alpha_2 + L\alpha_3)\}| = (2\pi + 1)\,\pi \qquad \ldots(13.19)$$

$$|\{(0) - (90 + 57 + 36)\}| = (2i + 1)\,\pi \qquad \ldots(13.20)$$

On letting $i = 0$ in equation (13.20), we get

$$183° = 180°$$

It is observed from the phase angle criterion that the point '*L*' is within the ±5% of the final value. Hence, the assumed locus point '*L*' is correct. If the locus point by phase angle criterion is less than 170°, *i.e.*, the assumed locus point is not correct and the point to be shifted towards the right of assumed point '*L*'. If the locus point by phase angle criterion is more than 190°, *i.e.*, the assumed locus point is not correct and point to be shifted towards the left of assumed point '*L*'. This procedure to be continued till it satisfy the phase angle criterion.

From the magnitude criterion, we have

$$k = \frac{|(s_c + p_1)(s_c + p_2)\ldots(s_c + p_n)|}{|(s_c + z_1)(s_c + z_2)\ldots(s_c + z_m)|} \qquad \ldots(13.20)$$

$$\Rightarrow \qquad k = \frac{|(-1)(-2)(-3)|}{1}$$

$$\Rightarrow \qquad k = 6$$

From the polynomial form of characteristic equation, we have

$$s^3 + 6s^2 + 11s + 6 + k = 0$$

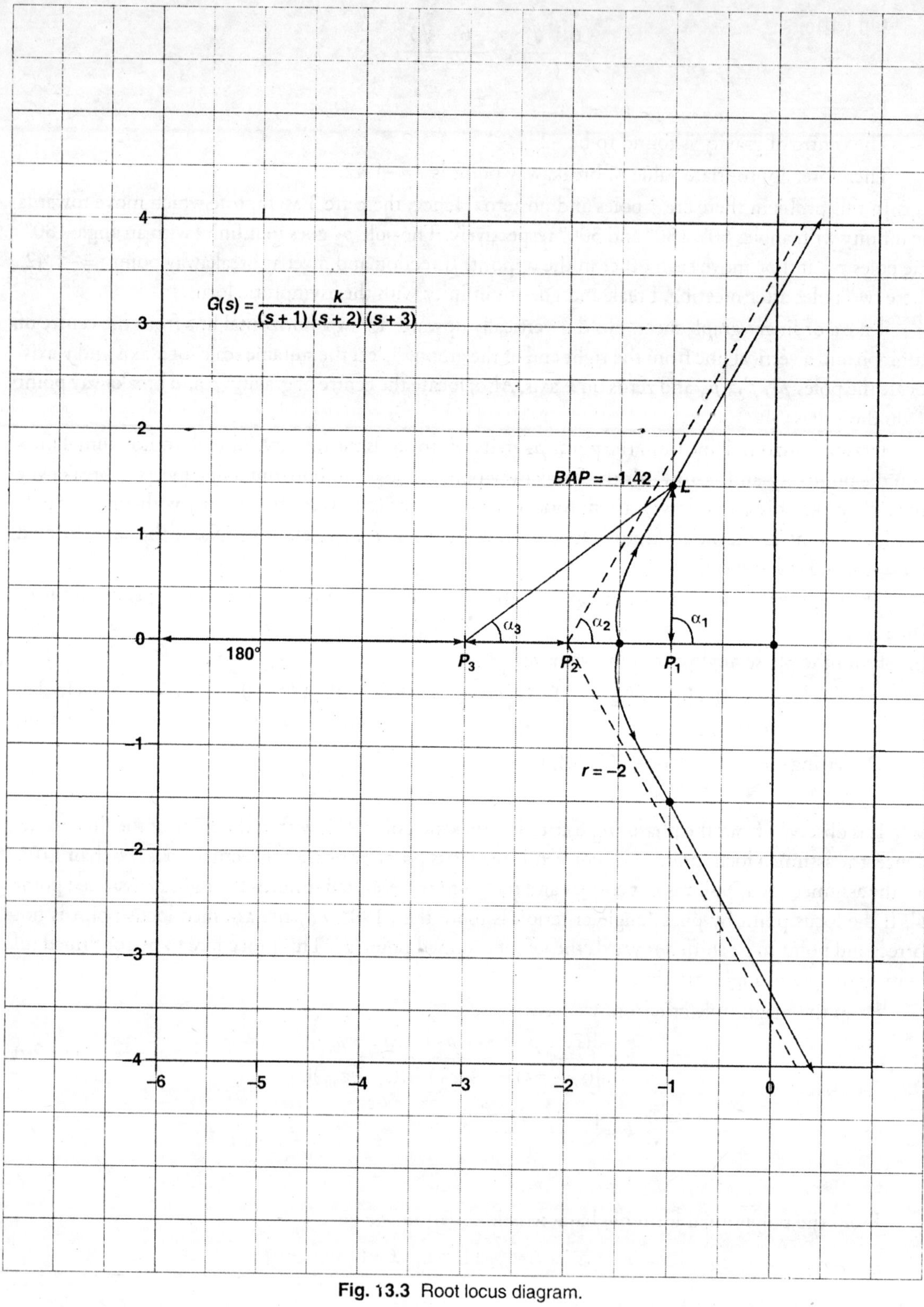

Fig. 13.3 Root locus diagram.

On letting, $k = 6$

$$\Rightarrow \quad s^3 + 6s^2 + 11s + 12 = 0$$

Routh Array :

1	1	11
2	6	12
3	9	0
4	12	

From the Routh Array it is observed that all the elements in the first column are non-zero and +ve. Therefore, the control system is stable.

■ **Example 13.23** Plot the root locus diagram for the open loop transfer function

$$G(s) = \frac{k}{(s)\,(s+1)\,(s+2)\,(s+3)}$$

Solution :

(i) *Calculate the no. of zeros and poles :*

$\Rightarrow \quad m = 0$

$\Rightarrow \quad n = 4 \quad ; \quad p_1 = 0$

$p_2 = -1$

$p_3 = -2$

$p_4 = -3$

(ii) *Calculate the no. of asymptotes :*

$$n - m = 4 - 0 = 4$$

(iii) *Calculate the angle of asymptotes :*

$$\alpha = \frac{\pi(2k+1)}{n-m}$$

$$\Rightarrow \quad \alpha_1 = 45^\circ$$

$$\Rightarrow \quad \alpha_2 = 135^\circ$$

$$\Rightarrow \quad \alpha_3 = 225^\circ$$

$$\Rightarrow \quad \alpha_4 = 315^\circ$$

(iv) *Calculate the centre of gravity :*

$$r = \frac{\sum_{j=1}^{n} p_j - \sum_{i=1}^{m} z_i}{n-m}$$

$$\Rightarrow \quad r = \frac{[(0) + (-1) + (-2) + (-3)] - 0}{4}$$

$$\Rightarrow \quad r = \frac{(-6)}{4}$$

$$\Rightarrow \quad r = -1.5$$

(v) *Calculate the breakaway point* :

From the characteristic equation, we have

$$1+G(s)\,H(s)=0$$

$$\Rightarrow \quad 1+\frac{k}{(s)\,(s+1)\,(s+3)\,(s+3)}=0$$

$$\Rightarrow \quad (s)\,(s+1)\,(s+2)\,(s+3)+k=0$$

$$\Rightarrow \quad s^4+6s^3+11s^2+6s+k=0 \qquad \ldots(13.26)$$

$$\Rightarrow \quad k=-(s^4+6s^3+11s^2+6s) \qquad \ldots(13.27)$$

Differentiating equation (13.27) w.r.t. 's' and equalizing to zero, we get

$$\frac{dk}{ds}=0$$

$$\Rightarrow \quad (4s^3+18s^2+22s+6)=0 \qquad \ldots(13.28)$$

On letting, $s=-1.5$, we get

$$4(-1.5)^3+18(-1.5)^2+22(-1.5)+6=0$$

$$-13.5+40.5-33+6=0$$

$$\Rightarrow \quad 0=0$$

Therefore, $(s+1.5)$ is a factor of equation (13.28).

Now, dividing equation (13.28) by $(s+1.5)$, we get

$$
\begin{array}{r|l|l}
(s+1.5) & 4s^3+18s^2+22s+5 & 4s^2+12s+4 \\
 & 4s^3+6s^2 & \\
 & -\quad - & \\
\hline
 & 12s^2+22s & \\
 & 12s^2+18s & \\
 & -\quad - & \\
\hline
 & 4s+6 & \\
 & 4s+6 & \\
 & -\quad - & \\
\hline
 & 0 \quad 0 &
\end{array}
$$

Therefore, $\quad (s+1.5)\,(4s^2+12s+4)=0$

The equation $4s^2+12s+4=0$ is of the form of quadratic equation. It can be solve using the formula

$$s=\frac{-b\pm\sqrt{b^2-4ac}}{2a}$$

$$\Rightarrow \quad s=\frac{-12\pm\sqrt{144-(4\times4\times4)}}{2\times4}$$

$$\Rightarrow \qquad s = \frac{-12 \pm \sqrt{80}}{2 \times 4}$$

$$\Rightarrow \qquad s = \frac{-12 \pm \sqrt{5 \times 16}}{2 \times 4}$$

$$\Rightarrow \qquad s = \frac{-3 \times 4 \pm 4\sqrt{5}}{2 \times 4}$$

$$\Rightarrow \qquad s = \frac{-3 \pm \sqrt{5}}{2}$$

$$\Rightarrow \qquad s = \frac{-3 + \sqrt{5}}{2}, \quad \frac{-3 - \sqrt{5}}{2}$$

$$\Rightarrow \qquad s = -0.382, \quad -2.618$$

Therefore, $s = -0.382, -2.618$ and -1.5.

Substituting the value of 's' in equation (13.27), we get

When, $s = -0.382$; $k = 1$

$s = 2.618$; $k = 1$

$s = -1.5$; $k = -0.563$

Since k is +ve and real for $s = -0.382$ and $s = -2.618$, they are valid breakaway points. Where as k is –ve and real for $s = -1.5$. Hence, it is invalid breakaway point.

(vi) *Calculate the intersection point with imaginary axis* :

From the polynomial form of the characteristic equation, we have

$$s^4 + 6s^3 - 11s^2 + 6s + k = 0 \qquad \ldots(13.26)$$

Letting $s = j\omega$ in equation (13.26), we get

$$\Rightarrow \qquad (j\omega)^4 + 6(j\omega)^3 + 11(j\omega)^2 + 6(j\omega) + k = 0$$

$$\Rightarrow \qquad \omega^4 - 6j\omega^3 - 11\omega^2 + 6j\omega + k = 0 \qquad \ldots(13.29)$$

On equating imaginary parts to zero, we get

$$-6j\omega^3 + 6j\omega = 0$$

$$\Rightarrow \qquad -6\omega^3 + 6\omega = 0$$

$$\Rightarrow \qquad -6\omega(\omega^2 - 1) = 0$$

$$\Rightarrow \qquad \omega^2 - 1 = 0$$

$$\Rightarrow \qquad \omega^2 = 1$$

$$\Rightarrow \qquad \omega = \pm 1$$

On equating real parts to zero, we get

$$\omega^4 - 11\omega^2 + k = 0$$

$$\Rightarrow \qquad k = 11\omega^2 - \omega^4$$

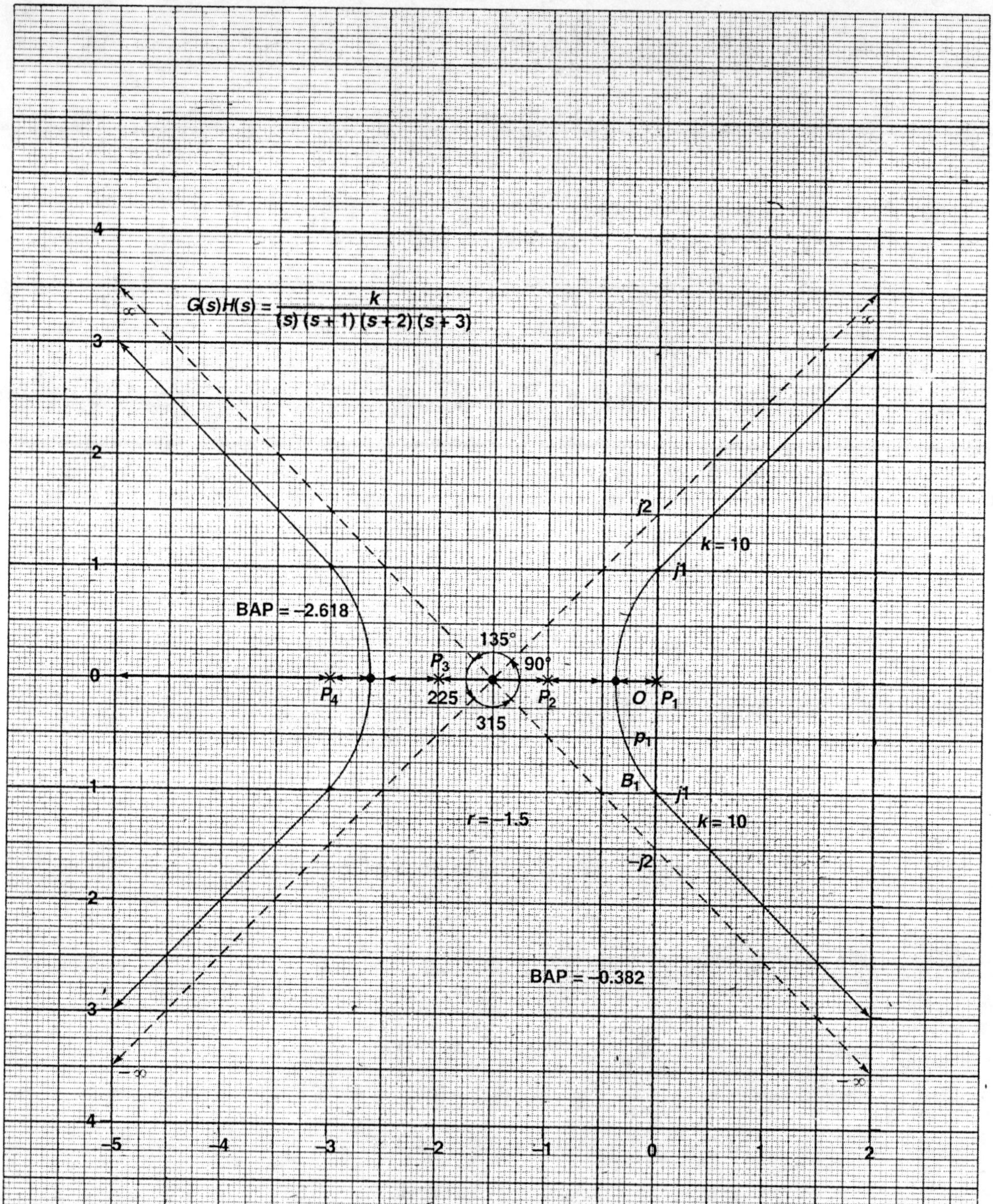

Fig. 13.4 Root locus diagram.

$$\Rightarrow \qquad k = 11 \times (1)^2 - (4)^4$$

$$\Rightarrow \qquad k = 10$$

Since k is +ve and real the imaginary axis intersected at $s = \pm j1$.

From the polynomial form of the characteristic equation, we have

$$s^4 + 6s^3 + 11s^2 + 6s + k = 0$$

$$\Rightarrow \qquad s^4 + 6s^3 + 11s^2 + 6s + 10 = 0$$

Routh Array :

1	1	11	10
2	6	6	0
3	10	10	
4	0	0	

Since all the elements of fourth row is zero, the auxiliary polynomial is exists. From the auxiliary polynomial, we have

$$10s^2 + 10 = 0 \qquad \text{...(13.30)}$$

Differentiating equation (13.30) w.r.t. 's', we get

$$20s + 0 = 0 \qquad \text{...(13.31)}$$

Now, replace the zeroth rows with coefficient of equation (13.31) and reconstruct the Routh Array.

Routh Array :

1	1	11	10
2	6	6	0
3	10	10	
4	20	0	
5	10		

It is observed from the Routh Array that there are no sign change in the first column. Hence, the control system is stable.

■ **Example 13.24** A feed back control system has operating transfer function

$$G(s) = \frac{k}{(s)(s^2 + 4s + 13)}$$

Sketch the root locus by determining the following :

(i) No. of asymptotes and asymptotes angle

(ii) Centre of gravity

(iii) Breakaway point, if any

(iv) Angle of departure of root loci from the poles

(v) The value of k and the frequency at which the root loci crosses imaginary axis.

Solution : Transfer function of the control system is given by

$$G(s) = \frac{k}{(s)(s^2 + 4s + 13)}$$

$$\Rightarrow \qquad G(s) = \frac{k}{(s)[s - (-2 + 3j)][s - (-2 - 3j)]}$$

No. of poles, $n = 3$; $\quad p_1 = 0$

$p_2 = -2 + 3j$

$p_3 = -2 - 3j$

No. of zeros, $m = 0$

(i) *No. of asymptotes :*

$$n - m = 3 - 0 = 3$$

Angle of asymptotes :

$$\alpha = \frac{\pi(2k + 1)}{(n - m)}$$

where, $\quad k = 0, 1, 2$

$\Rightarrow \quad \alpha_1 = 60°$

$\Rightarrow \quad \alpha_2 = 180°$

$\Rightarrow \quad \alpha_3 = 300°$

(ii) *Centre of gravity :*

$$r = \frac{\sum_{j=1}^{n} p_j - \sum_{i=1}^{m} z_i}{(n - m)}$$

$$\Rightarrow \qquad r = \frac{[(0) + (-2 + 3i) + (-2 - 3i)] - (0)}{3}$$

$$\Rightarrow \qquad r = -1.33$$

(iii) *Breakaway point :*

$$\sum_{j=0}^{n} \frac{1}{(s + j)} = \sum_{i=0}^{m} \frac{1}{(s + z_i)}$$

$$\frac{1}{s} + \frac{1}{(s + 2 - 3i)} + \frac{1}{(s + 2 + 3i)} = 0$$

$$\Rightarrow \qquad \frac{(s^2 + 4s + 13) + s(s + 2 + 3i) + s(s + 2 - 3i)}{(s)(s + 2 - 3i)(s + 2 + 3i)} = 0$$

$$\Rightarrow \qquad (s^2 + 4s + 13) + (s^2 + 2s + 3is) + (s^2 + 2s - 3is) = 0$$

$$\Rightarrow \qquad 3s^2 + 8s + 13 = 0$$

$$\Rightarrow \qquad s = \frac{-b \pm \sqrt{b^2 - 4ac}}{2a}$$

$$\Rightarrow \qquad s = \frac{-8 \pm \sqrt{(8)^2 - 4 \times 3 \times 13}}{2 \times 3}$$

$$\Rightarrow \qquad s = \frac{-8 \pm \sqrt{64 - 156}}{2 \times 3}$$

$$\Rightarrow \qquad s = \frac{-8 \pm \sqrt{-92}}{2 \times 3}$$

$$\Rightarrow \qquad s = \frac{-8 \pm 2\sqrt{-23}}{2 \times 3}$$

$$\Rightarrow \qquad s = \frac{-4 \pm j\sqrt{23}}{3}$$

$$\Rightarrow \qquad s = -1.33 \pm 1.6j$$

The breakaway point at $s = -1.33 \pm 1.6j$ is not possible. Hence, it is neglected.

(iv) *Angle of departure* : From the angle of departure, we have

$$\theta_1 = 180 - \tan^{-1}(y/x)$$

$$= 180 - \tan^{-1}(3/2) = 123.7°$$

$$\theta_2 = 90°$$

Angle of departure from $s = -2 + 3j$, we get

$$\theta_d = 180 - (\theta_1 + \theta_2)$$

$$= 180 - (123.7 + 90) = -33.7° = -34°$$

Angle of departure from $s = -2 - 3j$, we get

$$\theta_d^* = 34°$$

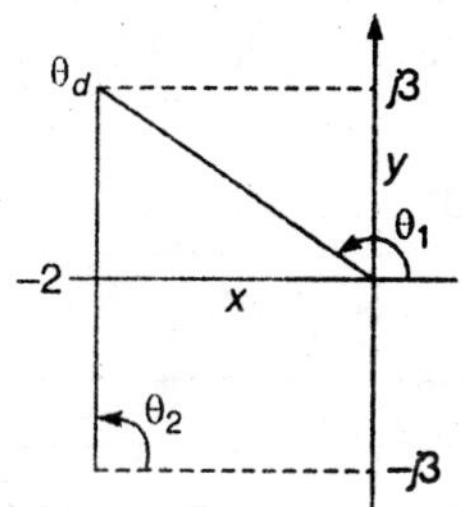

Fig. 13.5 Angle of departure.

(v) *Intersection point 'ω' on imaginary axis* : From the characteristic equation, we have

$$1 + G(s) = 0$$

$$1 + \frac{k}{(s)(s^2 + 4s + 13)} = 0$$

$$(s)(s^2 + 4s + 13) + k = 0$$

$$s^3 + 4s^2 + 13s + k = 0 \qquad \ldots(13.32)$$

Letting $s = j\omega$ in equation (13.32), we get

$$(j\omega)^3 + 4(j\omega)^2 + 13(j\omega) + k = 0$$

$$(j^3\omega^3) + 4j^2\omega^2 + 13j\omega + k = 0$$

$$-j\omega^3 - 4\omega^2 + 13j\omega + k = 0 \qquad \ldots(13.33)$$

On equating the imaginary part of equation (13.33) to zero, we get

$$-j\omega^3 + 13j\omega = 0$$

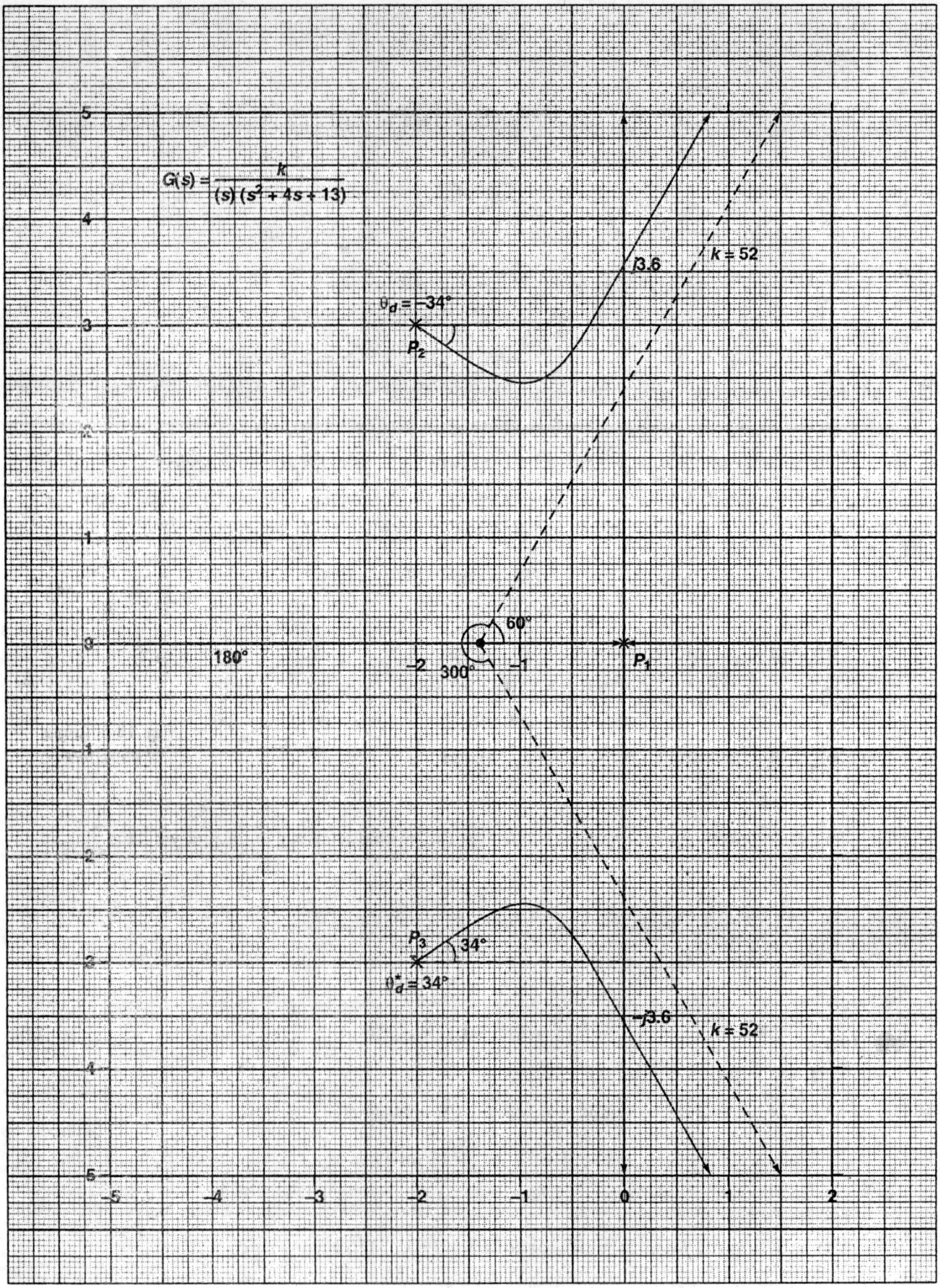

Fig. 13.6 Root locus diagram.

$$\Rightarrow \quad -\omega^2 + 13 = 0$$

$$\Rightarrow \quad \omega^2 = 13$$

$$\Rightarrow \quad \omega = \pm 3.60$$

On equating the real part of equation (13.33) to zero, we get

$$-4\omega^2 + k = 0$$

$$\Rightarrow \quad k = 4\omega^2$$

$$\Rightarrow \quad k = 4(3.60)^2$$

$$\Rightarrow \quad k = 51.84$$

$$\Rightarrow \quad k = 52$$

■ **Example 13.25** The open loop transfer function of a unity feed back control system is given by

$$G(s) = \frac{k}{(s)(s+4)(s^2+4s+20)}$$

Sketch the root locus diagram.

Solution : Transfer function of the control system is given by

$$G(s) = \frac{k}{(s)(s+4)(s^2+4s+20)}$$

$$\Rightarrow \quad G(s) = \frac{k}{(s)(s+4)(s+2-j4)(s+2+j4)}$$

(i) *No. of poles and zeros* :

$n = 4$; $p_1 = 0$

$p_2 = -4$

$p_3 = -2 + j4$

$p_4 = -2 - j4$

$m = 0$

(ii) *No. of asymptotes* :

$$n - m = 4 - 0 = 4$$

(iii) *Angle of asymptotes* :

$$\alpha = \frac{\pi(2k+1)}{(n-m)}$$

where, $k = 0, 1, 2, 3$

$$\Rightarrow \quad \alpha_1 = 45°$$

$$\Rightarrow \quad \alpha_2 = 135°$$

$$\Rightarrow \quad \alpha_3 = 225°$$

$$\Rightarrow \quad \alpha_4 = 315°$$

(iv) *Centre of gravity* :

$$r = \frac{\sum_{j=1}^{n} p_j - \sum_{i=1}^{m} z_i}{(n-m)}$$

$$r = \frac{[(-0)+(-4)+(-2+j4)+(-2-j4)]-(0)}{(4-0)}$$

$$r = \frac{-8}{4}$$

$$r = -2$$

(v) *Breakaway point* : From the characteristic equation, we have

$$1 + G(s)\,H(s) = 0$$

$$1 + \frac{k}{(s)(s+4)(s^2+4s+20)} = 0$$

$$(s)(s+4)(s^2+4s+20) + k = 0$$

$$s^4 + 8s^3 + 36s^2 + 80s + k = 0 \qquad \ldots(13.34)$$

$$\Rightarrow \qquad k = -(s^4 + 8s^3 + 36s^2 + 80s) \qquad \ldots(13.35)$$

Differentiating equation (13.35) w.r.t. 's' and equalizing to zero, we get

$$\frac{dk}{ds} = 0$$

$$\Rightarrow \qquad 4s^3 + 24s^2 + 72s + 80 = 0$$

$$\Rightarrow \qquad s^3 + 6s^2 + 18s + 20 = 0 \qquad \ldots(13.36)$$

On letting, $s = -2$, we get

$$\Rightarrow \qquad (-2)^3 + 6(-2)^2 + 18(-2) + 20 = 0$$

$$\Rightarrow \qquad -8 + 24 - 36 + 20 = 0$$

$$\Rightarrow \qquad 0 = 0$$

Therefore, $s + 2$ is a factor of equation (13.35).

Now, dividing equation (13.25) by $(s + 2)$, we get

$$
\begin{array}{r|l|l}
(s+2) & s^3 + 6s^2 + 18s + 20 & s^2 + 4s + 10 \\
 & s^3 + 2s^2 & \\
 & - \quad - & \\
\hline
 & 4s^2 + 18s & \\
 & 4s^2 + 8s & \\
 & - \quad - & \\
\hline
 & 10s + 20 & \\
 & 10s + 20 & \\
 & - \quad - & \\
\hline
 & 0 \quad 0 & \\
\hline
\end{array}
$$

Therefore, $(s+2)(s^2+4s+10)=0$

The equation $s^2+4s+10=0$ is of the form of quadratic equation. If can be solve using the formula

$$\Rightarrow \quad s=\frac{-b\pm\sqrt{b^2-4ac}}{2a}$$

$$\Rightarrow \quad s=\frac{-4\pm\sqrt{16-(4\times1\times10)}}{2\times1}$$

$$\Rightarrow \quad s=\frac{-4\pm\sqrt{-24}}{2}$$

$$\Rightarrow \quad s=\frac{-4\pm\sqrt{j^2(4\times6)}}{2}$$

$$\Rightarrow \quad s=\frac{-4\pm2j\sqrt{6}}{2}$$

$$\Rightarrow \quad s=-2\pm j\sqrt{6}$$

$$\Rightarrow \quad s=-2+j2.45;\quad -2-j2.45$$

Therefore, $s=0,-2,-2+j2.45$ and $-2-j2.45$

Substituting the value of 's' in equation (13.35), we get

When $s=-2$; $k=64$

$s=-2+j2.45$; $k=100$

$s=-2-j2.45$; $k=100$

Since, k is positive and real, all the roots are valid breakaway points.

(vi) *Angle of departure* : From the angle of departure, we have

$$\theta_1=180-\tan^{-1}(y/x)$$

$$=180-\tan^{-1}(4/2)=116.56°$$

$$\theta_2=90°$$

$$\theta_3=\tan^{-1}(4/2)=63.43°$$

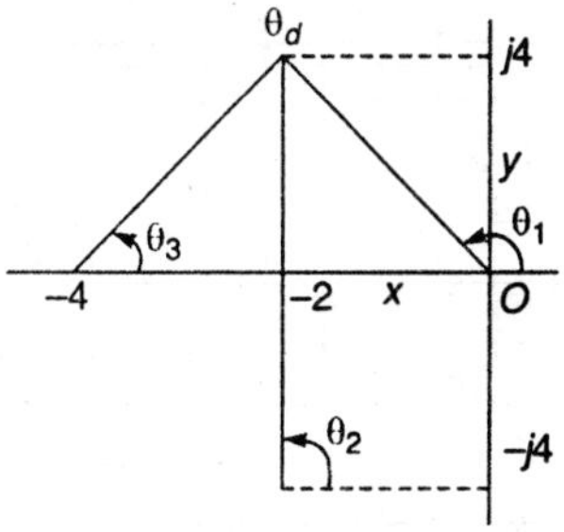

Fig. 13.7 Angle of departure.

Angle of departure from $s=-2+j4$, we get

$$\theta_d=180-(\theta_1+\theta_2+\theta_3)$$

$$=180-(116.56+90+53.43)$$

$$=-90°$$

Angle of departure from $s=-2-j4$, we get

$$\theta_d^*=90°$$

(vii) *Intersection point 'ω' on imaginary axis* : From the polynomial characteristic equation, we have

$$s^4+8s^3+36s^2+80s+k=0 \qquad \ldots(13.34)$$

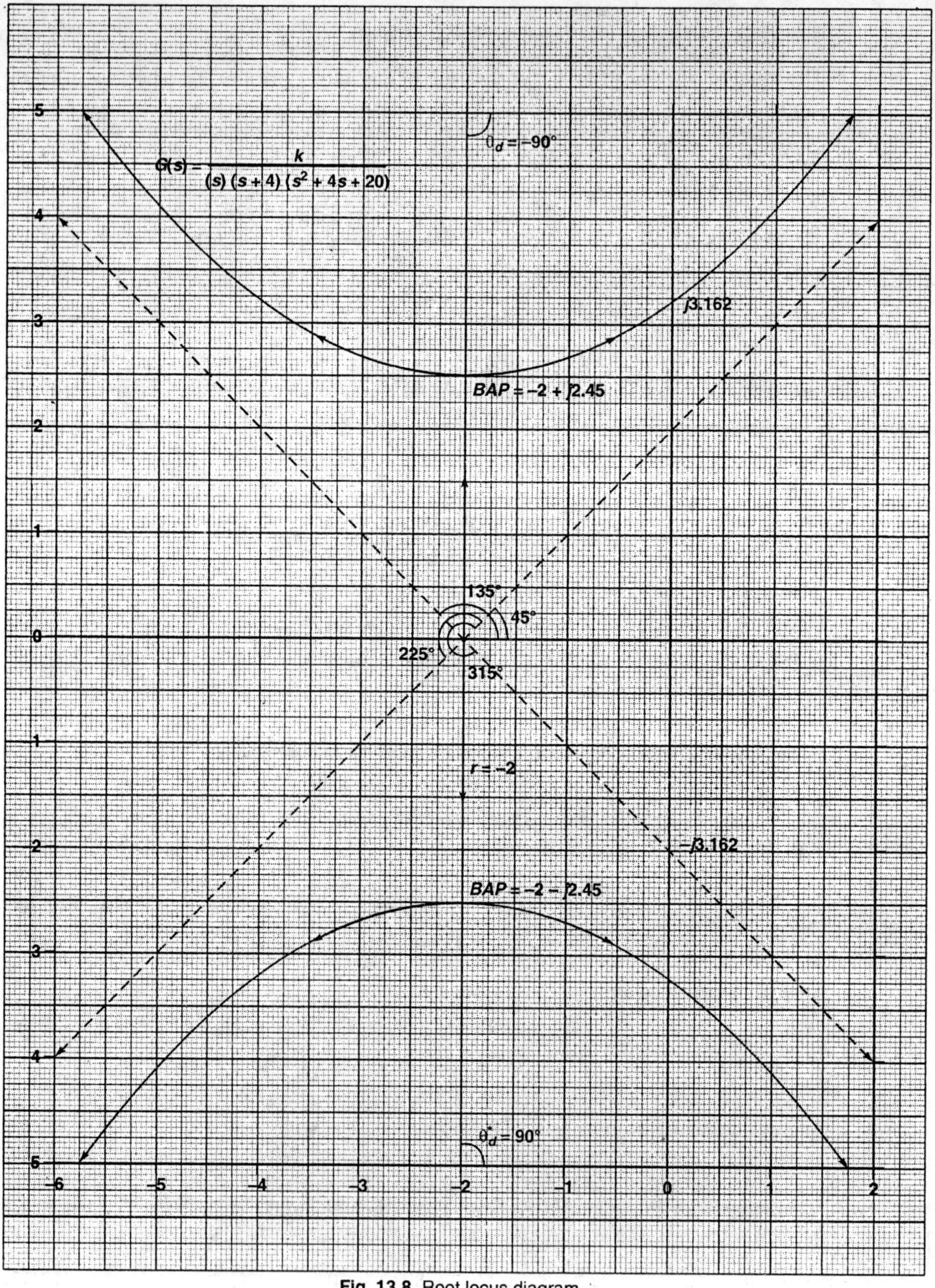

Fig. 13.8 Root locus diagram.

Letting, $s = j\omega$ in equation (13.34), we get

$$(j\omega)^4 + 8(j\omega)^3 + 36(j\omega)^2 + 80(j\omega) + k = 0$$

$$\Rightarrow \quad (j^4\omega^4) + 8(j^3\omega^3) + 36(j^2\omega^2) + 80(j\omega) + k = 0$$

$$\Rightarrow \quad \omega^4 - 8j\omega^3 - 36\omega^2 + 80j\omega + k = 0 \qquad \text{...(13.37)}$$

On equating the imaginary part of equation (13.37) to zero, we get

$$-8j\omega^3 + 80j\omega = 0$$

$$\Rightarrow \quad -\omega^2 + 10 = 0$$

$$\Rightarrow \quad \omega^2 = 10$$

$$\Rightarrow \quad \omega = \pm 3.162$$

On equating the real part of equation (13.37) to zero, we get

$$\omega^4 - 36\omega^2 + k = 0$$

$$\Rightarrow \quad k = -\omega^4 + 36\omega^2$$

$$\Rightarrow \quad k = -(3.162)^4 + 36(3.162)^2$$

$$\Rightarrow \quad k = -99.96 + 359.94$$

$$\Rightarrow \quad k = 260$$

The value of k is real and positive. Therefore, the imaginary axis part is $s = j3.162$. The entire root locus is shown in Fig. 13.8.

■ **Example 13.26** Draw the root locus diagram for the loop transfer function

$$G(s)\,H(s) = \frac{k}{(s)(s^2 + 8s + 17)}$$

and also evaluate the value of k.

Solution : Transfer function of the control system is given by

$$G(s)\,H(s) = \frac{k}{(s)(s^2 + 8s + 17)}$$

$$\Rightarrow \quad G(s)\,H(s) = \frac{k}{(s)(s + 4 - j)(s + 4 + j)}$$

(i) *No. of poles and zeros* :

$n = 3$; $p_1 = 0$

$p_2 = -4 + j$

$p_3 = -4 - j$

$m = 0$

(ii) *No. of asymptotes* :

$$n - m = 3 - 0 = 3$$

(iii) *Angle of asymptotes* :

$$\alpha = \frac{\pi(2k+1)}{(n-m)}$$

where, $k = 0, 1, 2, 3$

$\Rightarrow$ $$\alpha_1 = 60°$$

$\Rightarrow$ $$\alpha_2 = 180°$$

$\Rightarrow$ $$\alpha_3 = 300°$$

(iv) *Centre of gravity* :

$$r = \frac{\sum_{j=1}^{n} p_j - \sum_{i=1}^{m} z_i}{(n-m)}$$

$$r = \frac{[(-0) + (-4 + j) + (-4 - j)] - (0)}{3}$$

$$r = \frac{-8}{3}$$

$$r = -2.67$$

(v) *Breakaway point* : From the characteristic equation, we have

$$1 + G(s)\, H(s) = 0$$

$$1 + \frac{k}{(s)\,(s^2 + 8s + 17)} = 0$$

$$s^3 + 8s^2 + 17s + k = 0 \qquad \text{...(13.38)}$$

$\Rightarrow$ $$k = -(s^3 + 8s^2 + 17s) \qquad \text{...(13.39)}$$

Differentiating equation (13.39) w.r.t. '*s*' and equalizing to zero, we get

$$\frac{dk}{ds} = 0$$

$\Rightarrow$ $$3s^2 + 16 + 17 = 0$$

The equation $3s^2 + 16s + 17 = 0$ is of the form of quadratic equation. It can be solve using the formula

$$s = \frac{-b \pm \sqrt{b^2 - 4ac}}{2a}$$

$\Rightarrow$ $$s = \frac{-16 \pm \sqrt{256 - (4 \times 3 \times 17)}}{2 \times 3}$$

$\Rightarrow$ $$s = \frac{-16 \pm \sqrt{52}}{6}$$

$\Rightarrow$ $$s = -1.465; \quad -3.869$$

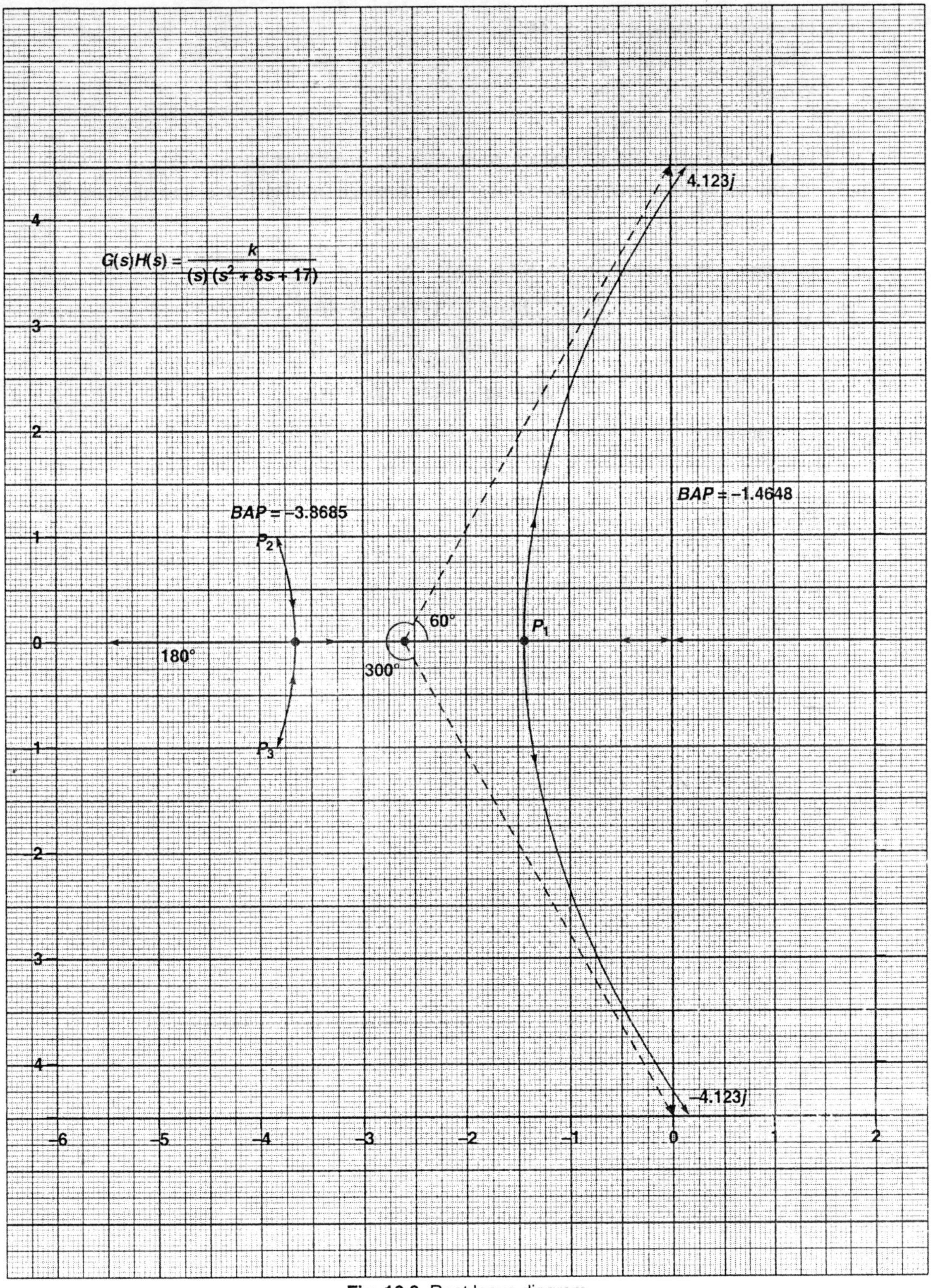

Fig. 13.9 Root locus diagram.

Therefore, $s = 0, -1.465, -3.869$

Substituting the value of 's' in equation (13.39), wet get

When, $s = -1.465$; $k = 10.879$

$s = -3.869$; $k = 3.935$

Since, k is positive and real, all the roots are valid breakaway points.

(vi) *Intersection point 'ω' on imaginary axis* : From the polynomial characteristic equation, we have

$$s^3 + 8s^2 + 17s + k = 0 \quad \text{...(13.38)}$$

Letting $s = j\omega$ in equation (13.38), we get

$$(j\omega)^3 + 8(j\omega)^2 + 17(j\omega) + k = 0$$

$$j^3\omega^3 + 8j^2\omega^2 + 17j\omega + k = 0$$

$$-j\omega^3 - 8\omega^2 + 17j\omega + k = 0 \quad \text{...(13.40)}$$

On equalizing the imaginary part of equation (13.40) to zero, we get

$$-j\omega^3 + 17j\omega = 0$$

$$\Rightarrow \quad -\omega^2 + 17 = 0$$

$$\Rightarrow \quad \omega = \pm 4.123$$

On equalizing the real part of equation (13.40) to zero, we get

$$-8\omega^2 + k = 0$$

$$\Rightarrow \quad k = 8\omega^2$$

$$\Rightarrow \quad k = 8(4.123)^2$$

$$\Rightarrow \quad k = 136$$

(vii) *Angle of departure* : From the angle of departure, we have

$$\theta_1 = 180 - \tan^{-1}(y/x)$$

$$= 180 - \tan^{-1}(1/4)$$

$$= 180 - 14.04 = 165.96°$$

$$\theta_2 = 90°$$

Angle of departure from $s = -4 + j1$, we get

$$\theta_d = 180 - (\theta_1 + \theta_2)$$

$$= 180 - (165.96 + 90) = -75.96°$$

$$\theta_d^* = 75.96°$$

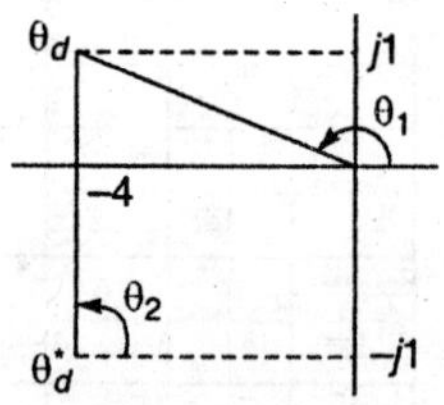

Fig. 13.10 Angle of departure.

■ **Example 13.27** Draw the root locus diagram for the loop transfer function

$$G(s) = \frac{k(s+1)}{s^2(s+9)}$$

Solution : Transfer function of the control system is given by

$$G(s) = \frac{k(s+1)}{s^2(s+9)}$$

(i) *No. of poles and zeros :*

$$n = 3 \quad ; \quad p_1 = 0$$
$$p_2 = 0$$
$$p_3 = -9$$
$$m = 1 \quad ; \quad z_1 = -1$$

(ii) *No. of asymptotes :*

$$n - m = 3 - 1 = 2$$

(iii) *Angle of asymptotes :*

$$\alpha = \frac{\pi(2k+1)}{n-m}$$

where, $k = 0, 1$

$$\Rightarrow \qquad \alpha_1 = 90°$$

$$\Rightarrow \qquad \alpha_2 = 270°$$

(iv) *Centre of gravity :*

$$r = \frac{\sum_{j=1}^{n} p_j - \sum_{i=1}^{m} z_i}{(n-m)} = \frac{[(0)+(0)+(-9)]-(-1)}{(3-1)} = -4$$

(v) *Breakaway point :* From the characteristic equation, we have

$$1 + G(s) = 0$$

$$\Rightarrow \qquad 1 + \frac{k(s+1)}{s^2(s+9)} = 0$$

$$\Rightarrow \qquad s^2(s+9) + k(s+1) = 0$$

$$\Rightarrow \qquad s^3 + 9s^2 + k(s+1) = 0 \qquad \text{...(13.41)}$$

$$\Rightarrow \qquad k = -\frac{(s^3 + 9s^2)}{(s+1)} \qquad \text{...(13.42)}$$

Differentiating equation (13.42) w.r.t. 's' and equalizing to zero, we get

$$\frac{dk}{ds} = 0$$

$$\Rightarrow \qquad -\left[\frac{(s+1)(3s^2+18s)-(s^3+9s^2).1}{(s+1)}\right] = 0$$

$$\Rightarrow \qquad (s+1)(3s^2+18s)-(s^3+9s^2) = 0$$

$$\Rightarrow \qquad 3s^3 + 18s^2 + 3s^2 + 18s - s^3 - 9s^2 = 0$$

$$\Rightarrow \qquad 2s^3 + 12s^2 + 18s = 0$$

$$\Rightarrow \qquad s^3 + 6s^2 + 9s = 0$$

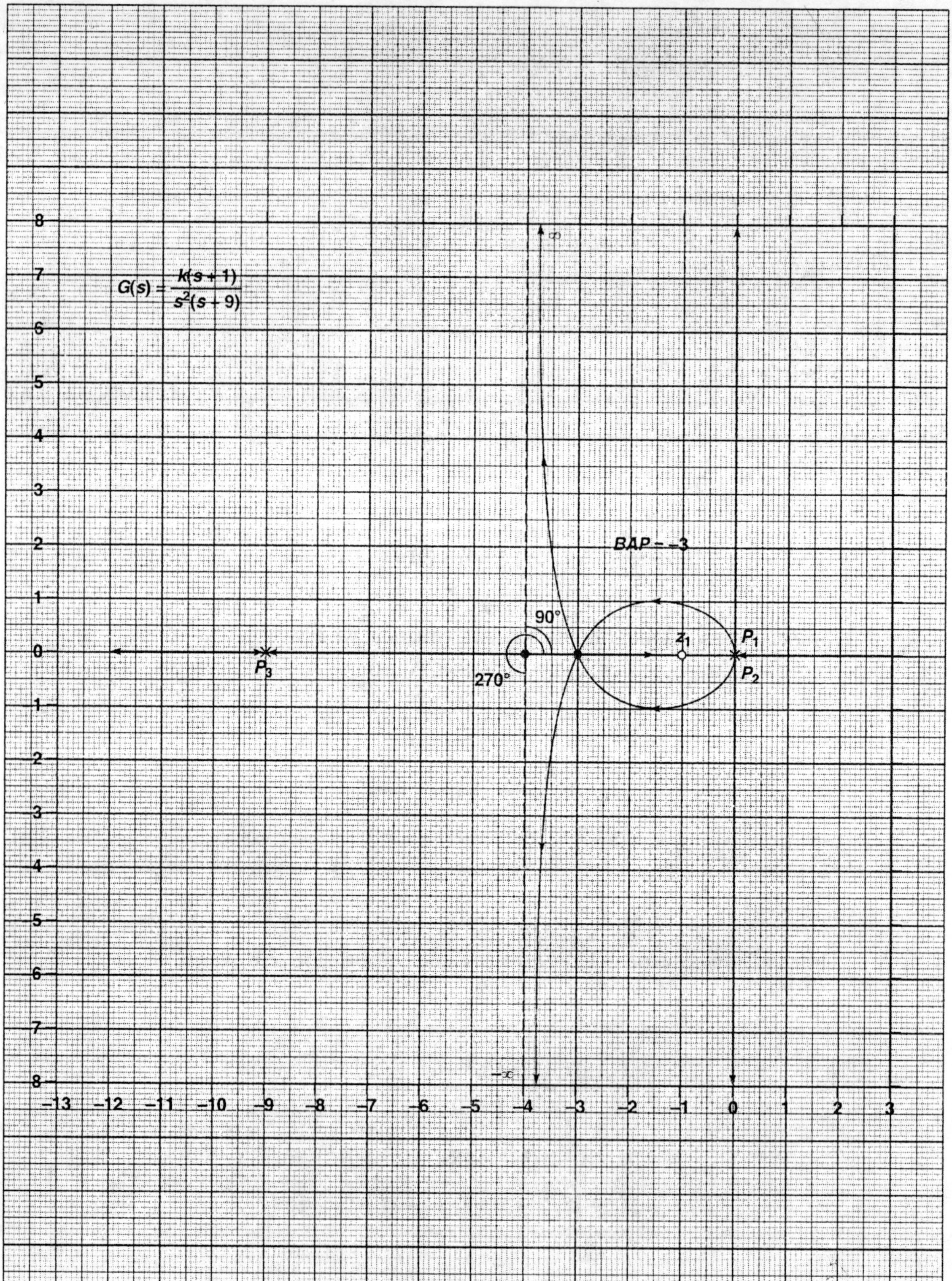

Fig. 13.11 Root locus diagram.

$$\Rightarrow \qquad (s)(s^2+6s+9)=0$$

$$\Rightarrow \qquad (s)(s+3)(s+3)=0$$

$$\Rightarrow \qquad s=0,-3,-3$$

The value of k at $s=-3$ is given by

$$k|_{s=-3}=-\left[\frac{(s^3+9s^2)}{(s+1)}\right]=27$$

k is positive. Hence, the breakaway point is $s=-3$.

■ **Example 13.28** Draw the root locus diagram for the loop transfer function

$$G(s)\,H(s)=\frac{k(s+4)}{(s)(s+2)}$$

Solution : Transfer function of the control system is given by

$$G(s)\,H(s)=\frac{k(s+4)}{(s)(s+2)}$$

(i) *No. of poles and zeros :*

$$n=2 \quad ; \quad p_1=0$$

$$p_2=-2$$

$$m=1 \quad ; \quad z_1=-4$$

(ii) *No. of asymptotes :*

$$n-m=2-1=1$$

(iii) *Angle of asymptotes :*

$$\alpha=\frac{\pi(2k+1)}{(n-m)}$$

where, $k=0$

$$\Rightarrow \qquad \alpha=180°$$

(iv) *Centre of gravity :*

$$r=\frac{\sum_{j=1}^{n} p_j-\sum_{i=1}^{m} z_i}{(n-m)}=\frac{[(0)+(-2)-(-4)]}{(2-1)}=2$$

(v) *Breakaway point :* From the characteristic equation, we have

$$1+G(s)\,H(s)=0$$

$$1+\frac{k(s+4)}{(s)(s+2)}=0$$

$$(s)(s+2)+k(s+4)=0$$

$$\Rightarrow \qquad k=-\frac{(s^2+2s)}{(s+4)} \qquad \ldots(13.43)$$

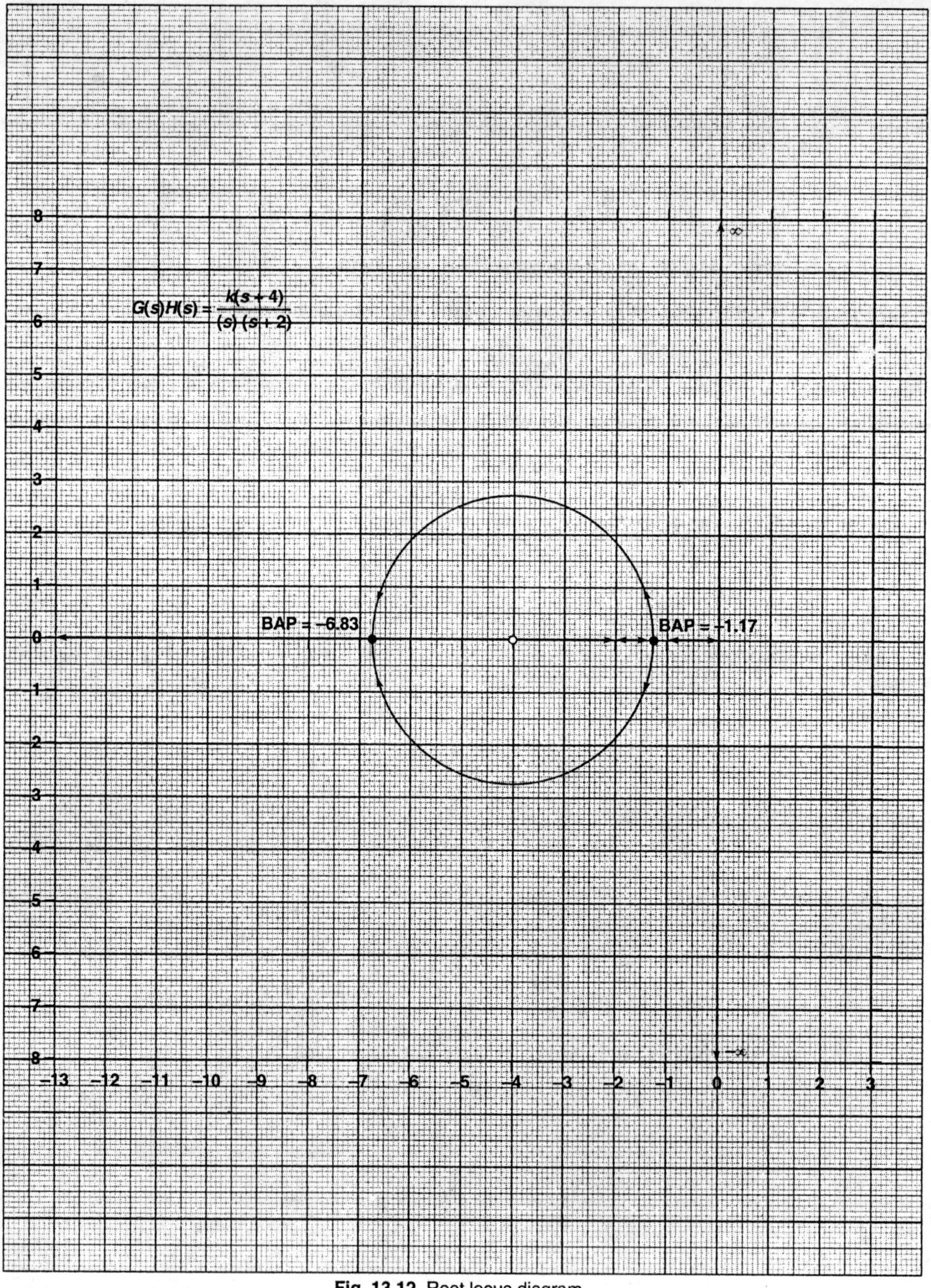

Fig. 13.12 Root locus diagram.

Differentiating equation (13.43) w.r.t. 's' and equalizing to zero, we get

$$\frac{dk}{ds} = 0$$

$$\Rightarrow \quad -\left[\frac{(s+4)(2s+2)-(s^2+2s).1}{(s+4)^2}\right] = 0$$

$$\Rightarrow \quad (s+4)(2s+2)-(s^2+2s) = 0$$

$$\Rightarrow \quad 2s^2+2s+8s+8-s^2-2s = 0$$

$$\Rightarrow \quad s^2+8s+8 = 0 \qquad \text{...(13.44),}$$

The equation $s^2+8s+8=0$ is of the form of quadratic equation. It can be solve using the formula

$$s = \frac{-b \pm \sqrt{b^2-4ac}}{2a}$$

$$\Rightarrow \quad s = \frac{-8 \pm \sqrt{64-4\times1\times8)}}{2\times1}$$

$$\Rightarrow \quad s = \frac{-8 \pm \sqrt{32}}{2}$$

$$\Rightarrow \quad s = -1.17; \quad -6.83$$

■ **Example 13.29** Draw the root locus diagram for the feed back control system having characteristic equation

$$1 + \frac{k}{(s)(s+1)(s+2)} = 0$$

Solution : From the open loop transfer function, we have

$$G(s)H(s) = \frac{k}{(s)(s+1)(s+2)}$$

(i) *No. of poles and zeros :*

$$n = 3 \quad ; \quad p_1 = 0, \quad p_2 = -1, \quad p_3 = -2$$

$$m = 0$$

(ii) *No. of asymptotes :*

$$n - m = 3 - 0 = 3$$

(iii) *Angle of asymptotes :*

$$\alpha = \frac{\pi(2k+1)}{(n-m)}$$

where, $k = 0, 1, 2$

$$\alpha_1 = 60°$$

$$\alpha_2 = 180°$$
$$\alpha_3 = 300°$$

(iv) *Centre of gravity* :

$$r = \frac{\sum_{j=1}^{n} p_j - \sum_{i=1}^{m} z_i}{(n-m)} = \frac{(0)+(-1)+(-2)-(0)}{(3-0)} = -1$$

(v) *Breakaway point* : From the feed back control system, we have

$$1 + \frac{k}{(s)(s+1)(s+2)} = 0$$

$$(s)(s+1)(s+2) + k = 0$$

$$s^3 + 3s^2 + 2s + k = 0 \quad \text{...(13.45)}$$

$$\Rightarrow \quad k = -(s^3 + 3s^2 + 2s) \quad \text{...(13.46)}$$

Differentiating equation (13.46) w.r.t. '*s*' and equalizing to zero, we get

$$\frac{dk}{ds} = 0$$

$$-(3s^2 + 6s + 2) = 0$$

$$3s^2 + 6s + 2 = 0 \quad \text{...(13.47)}$$

The equation $3s^2 + 6s + 2 = 0$ is of the form of quadratic equation. It can be solve using the formula

$$s = \frac{-b \pm \sqrt{b^2 - 4ac}}{2a}$$

$$\Rightarrow \quad s = \frac{-6 \pm \sqrt{36 - 4 \times 3 \times 2}}{2 \times 3}$$

$$\Rightarrow \quad s = \frac{-6 \pm \sqrt{12}}{6}$$

$$\Rightarrow \quad s = \frac{-6 \pm 2\sqrt{3}}{6}$$

$$\Rightarrow \quad s = \frac{-3 \pm \sqrt{3}}{3}$$

$$\Rightarrow \quad s = -0.42; \quad -1.58$$

Letting $s = j\omega$ in equation (13.45), we get

$$(j\omega)^3 + 3(j\omega)^2 + 2(j\omega) + k = 0$$

$$\Rightarrow \quad j^3\omega^3 + 3j^2\omega^2 + 2j\omega + k = 0$$

$$\Rightarrow \quad -j\omega^3 - 3\omega^2 + 2j\omega + k = 0 \quad \text{...(13.48)}$$

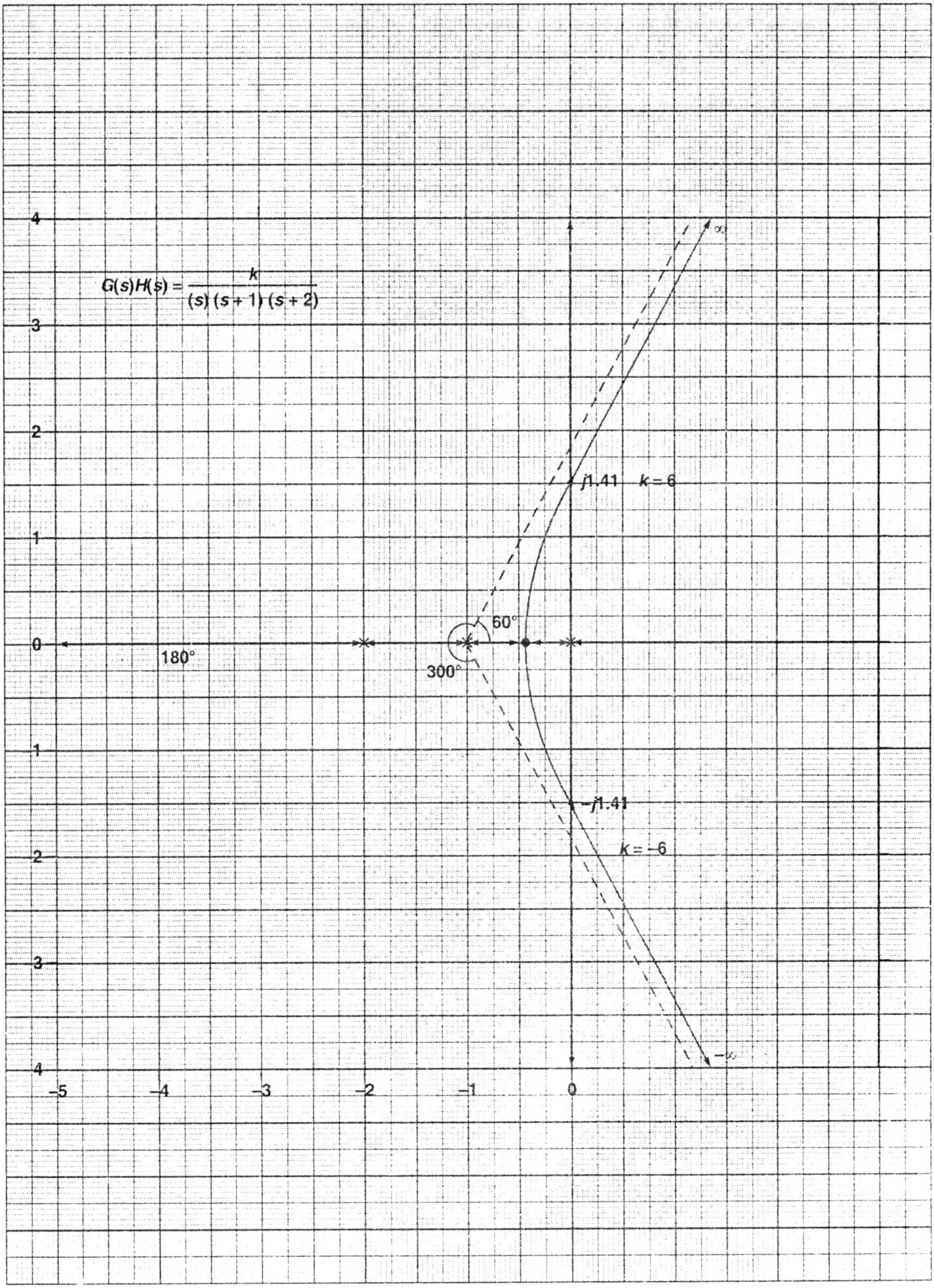

Fig. 13.13 Root locus diagram.

On equalizing the imaginary part of equation (13.47) to zero, we get

$$-j\omega^3 + 2j\omega = 0$$

$$\Rightarrow \quad -\omega^2 + 2 = 0$$

$$\Rightarrow \quad \omega^2 = 2$$

$$\Rightarrow \quad \omega = 1.414$$

On equalizing the real part of equation (13.48) to zero, we get

$$-3\omega^2 + k = 0$$

$$\Rightarrow \quad k = 3\omega^2$$

$$\Rightarrow \quad k = 3(1.414)^2$$

$$\Rightarrow \quad k = 6$$

■ **Example 13.30** Sketch the root locus diagram for all values at k ranging from 0 to ∞ for a negative feed back control system characterized by

$$G(s)H(s) = \frac{k(s+5)}{(s)(s+1)(s+2)}$$

Show all the salient points on the root locus.

Solution : Transfer function of the control system is given by

$$G(s)H(s) = \frac{k(s+5)}{(s)(s+1)(s+2)}$$

(i) *No. of poles and zeros :*

$n = 3$; $p_1 = 0$

$p_2 = -1$

$p_3 = -2$

$m = 1$; $z_1 = -5$

(ii) *No. of asymptotes :*

$$n - m = 3 - 1 = 2$$

(iii) *Angle of asymptotes :*

$$\alpha = \frac{\pi(2k+1)}{(n-m)}$$

where, $k = 0, 1$

$$\Rightarrow \quad \alpha_1 = 90^\circ$$

$$\Rightarrow \quad \alpha_2 = 270^\circ$$

(iv) *Centre of gravity :*

$$r = \frac{\sum_{j=1}^{n} p_j - \sum_{i=1}^{m} z_i}{(n-m)} = \frac{(0)+(-1)+(-2)-(-5)}{(3-1)} = 1$$

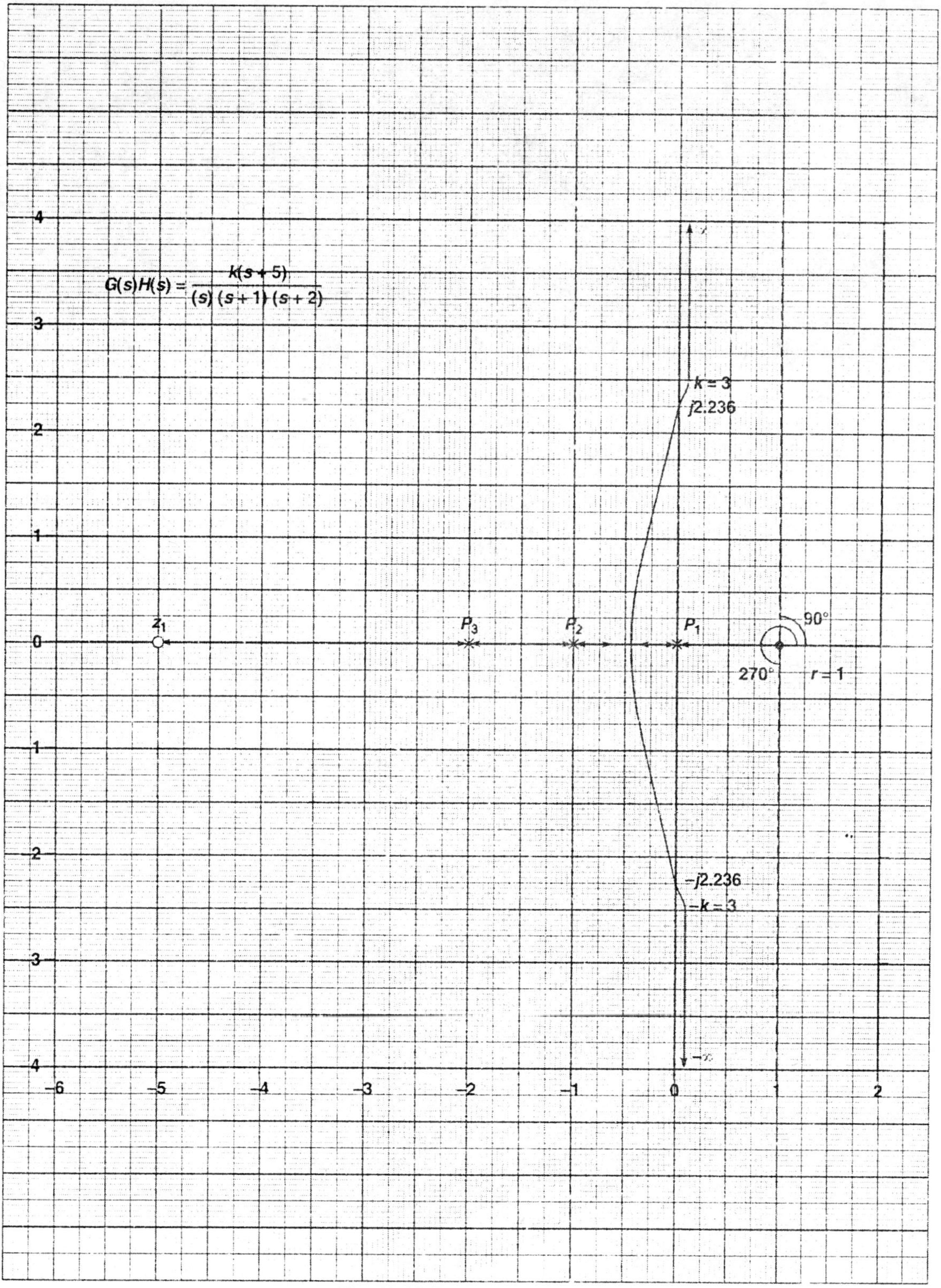

Fig. 13.14 Root locus diagram.

(v) *Breakaway point* : From the characteristic equation, we have

$$1+G(s)H(s)=0$$

$$1+\frac{k(s+5)}{(s)(s+1)(s+2)}=0$$

$$\Rightarrow \quad s^3+3s^2+2s+k(s+5)=0 \qquad \text{...(13.49)}$$

$$\Rightarrow \quad k(s+5)=-(s^3+3s^2+2s)$$

$$\Rightarrow \quad k=-\frac{(s^3+3s^2+2s)}{(s+5)} \qquad \text{...(13.50)}$$

Differentiating equation (13.50) w.r.t. '*s*' and equalizing to zero, we get

$$\frac{dk}{ds}=0$$

$$-\left[\frac{(s+5)(3s^2+6s+2)-(s^3+3s^2+2s).1}{(s+5)^2}\right]=0$$

$$\Rightarrow \quad 2s^3+18s^2+30s+10=0$$

$$\Rightarrow \quad s^3+9s^2+15s+5=0$$

By trial and error method, we get

$$s=-0.447, \quad -6.343; \quad -1.609$$

Substituting the value of '*s*' in equation (13.50), we get

when $s=-0.447$; $k=0.084$

$s=-6.343$; $k=-109.59$

$s=-1.609$; $k=-0.113$

At $s=-6.343$ and $s=-1.609$, the values of k are negative. Therefore, these values are not the breakaway point. Where as at $s=-0.447$, the value of k is positive. Therefore, it is the breakaway point.

(vi) *Intersection point 'ω' on imaginary axis* : From the polynomial characteristic equation, we have

$$s^3+3s^2+2s+k(s+5)=0 \qquad \text{...(13.49)}$$

Letting, $s=j\omega$ in equation (13.49), we get

$$(j\omega)^3-3(j\omega)+2(j\omega)+k(j\omega+5)=0$$

$$\Rightarrow \quad -j\omega^3-3\omega^2+2j\omega+kj\omega+5k=0 \qquad \text{...(13.51)}$$

On equalizing the imaginary part of equation (13.51) to zero, we get

$$-j\omega^3+2j\omega+kj\omega=0$$

$$\Rightarrow \quad -\omega^2+2+k=0$$

$$\Rightarrow \quad k=\omega^2-2 \qquad \text{...(13.52)}$$

On equalizing the real part of equation (13.51) to zero, we get

$$-3\omega^2 + 5k = 0$$

$$k = 3\omega^2/5 \qquad \ldots(13.53)$$

Comparing the equations (13.52) and (13.53), we get

$$\omega^2 - 2 = 3\omega^2/5$$

$$\Rightarrow \qquad 5\omega^2 - 10 = 3\omega^2$$

$$\Rightarrow \qquad 2\omega^2 = 10$$

$$\Rightarrow \qquad \omega^2 = 5$$

$$\Rightarrow \qquad \omega = 2.236 \qquad \ldots(13.54)$$

Substituting equation (13.54) in equation (13.53), we get

$$k = 3\,(2.236)^2/5$$

$$\Rightarrow \qquad k = 3$$

■ **Example 13.31** The open loop transfer function of a negative feed back control system is given by

$$G(s)H(s) = \frac{k(s+1)}{(s)\,(s-1)\,(s^2+4s+16)}$$

Sketch the root locus for $0 < k < \infty$ indicating all the relevant points.

Solution : Transfer function of the control system is given by

$$G(s)H(s) = \frac{k(s+1)}{(s)\,(s-1)\,(s^2+4s+16)}$$

$$\Rightarrow \qquad G(s)H(s) = \frac{k(s+1)}{(s)\,(s-1)\,(s+2-j3.464)\,(s+2+j3.464)}$$

(i) *No. of poles and zeros :*

$n = 4$; $p_1 = 0$

$p_2 = 1$

$p_3 = -2 + j3.464$

$p_4 = -2 - j3.464$

$m = 1$; $z = -1$

(ii) *No. of asymptotes :* $n - m = 4 - 1 = 3$

(iii) *Angle of asymptotes :* $\alpha = \dfrac{\pi(2k+1)}{(n-m)}$

where, $k = 0, 1, 2$

$$\Rightarrow \qquad \alpha_1 = 60°$$

$$\Rightarrow \qquad \alpha_2 = 180°$$

$$\Rightarrow \qquad \alpha_3 = 300°$$

(iv) *Centre of gravity* :

$$r = \frac{\sum_{j=1}^{n} p_j - \sum_{i=1}^{m} z_i}{(n-m)} = \frac{(0)+(1)+(-2+j3.464)+(-2-j3.464)-(-1)}{(4-1)} = -0.667$$

(v) *Breakaway point* : From the characteristic equation, we have

$$1 + G(s)H(s) = 0$$

$$\Rightarrow \quad 1 + \frac{k(s+1)}{(s)(s+1)(s^2+4s+16)} = 0$$

$$\Rightarrow \quad (s)(s-1)(s^2+4s+16) + k(s+1) = 0$$

$$\Rightarrow \quad s^4 + 3s^3 + 12s^2 - 16s + k(s+1) = 0 \qquad \text{...(13.55)}$$

$$k(s+1) = -(s^4 + 3s^3 + 12s^2 - 16s)$$

$$k = -\frac{(s^4 + 3s^3 + 12s^2 + 16s)}{(s+1)} \qquad \text{...(13.56)}$$

Differentiating equation (13.56) w.r.t. 's' and equalizing to zero, we get

$$\frac{dk}{ds} = 0$$

$$-\left[\frac{(s+1)(4s^3+9s^2+24s-16) - (s^4+3s^3+12s^2-16s).1}{(s+1)^2}\right] = 0$$

$$\Rightarrow \quad 3s^4 - 10s^3 + 21s^2 + 24s - 16 = 0$$

By trial and error method, we get $s = 0.45$

Therefore, $(s - 0.45)$ is a factor of equation

$$3s^4 + 10s^3 + 21s^2 + 24s - 16 = 0$$

$$\Rightarrow \quad (s-0.45)(3s^3 + 11.35s^2 + 26.11s + 35.75) = 0$$

Consider, $\quad (3s^3 + 11.35s^2 + 26.11s + 35.75) = 0$

By trial and error method, we get $s = -2.26$.

Therefore, $(s + 2.26)$ is a factor of equation

$$3s^3 + 11.35s^2 + 26.11s + 35.75 = 0$$

$$\Rightarrow \quad (s+2.26)(3s^2 + 4.57s + 15.79) = 0$$

$$\Rightarrow \quad 3s^2 + 4.57s + 15.79 = 0$$

The equation $3s^2 + 4.57s + 15.79 = 0$ is of the form of quadratic equation. It can be solve using the formula

$$s = \frac{-b \pm \sqrt{b^2 - 4ac}}{2a}$$

$$\Rightarrow \qquad s = \frac{-4.57 \pm \sqrt{(4.57)^2 - 4 \times 3 \times 15.79}}{2 \times 3}$$

$$\Rightarrow \qquad s = \frac{-4.57 \pm \sqrt{j^2 (168.6)}}{2 \times 3}$$

$$\Rightarrow \qquad s = -0.76 + j2.16, \quad -0.76 - j2.16$$

when, $s = 0.45$; $k = 3.072$

$s = -2.26$; $k = 70.569$

$s = -0.76 + j2.16$; k is complex

$s = -0.76 - j2.16$; k is complex

Therefore, $s = 0.45$ and -2.26 are valid breakaway points.

(vi) *Intersection point 'ω' on imaginary axis* : From the polynomial characteristic equation, we have

$$s^4 + 3s^3 + 12s^2 - 16s + k(s+1) = 0 \qquad \text{...(13.55)}$$

Letting $s = j\omega$ in equation (13.55), we get

$$\Rightarrow \qquad (j\omega)^4 + 3(j\omega)^3 + 12(j\omega)^2 - (16j\omega) + k(j\omega + 1) = 0$$

$$\Rightarrow \qquad \omega^4 - 3j\omega^3 - 12\omega^2 - 16j\omega + kj\omega + k = 0 \qquad \text{...(13.57)}$$

On equalizing the imaginary part of equation (13.57) to zero, we get

$$\Rightarrow \qquad -3j\omega^3 - 16j\omega + kj\omega = 0$$

$$\Rightarrow \qquad -3\omega^2 - 16 + k = 0$$

$$\Rightarrow \qquad k = 3\omega^2 + 16 \qquad \text{...(13.58)}$$

On equalizing the real part of equation (13.57) to zero, we get

$$\Rightarrow \qquad \omega^4 - 12\omega^2 + k = 0$$

$$\Rightarrow \qquad k = -\omega^4 + 12\omega^2 \qquad \text{...(13.59)}$$

Comparing equations (13.58) and (13.59), we get

$$\Rightarrow \qquad 3\omega^2 + 16 = -\omega^4 + 12\omega^2$$

$$\Rightarrow \qquad \omega^4 - 9\omega^2 + 16 = 0 \qquad \text{...(13.60)}$$

Letting, $\omega^2 - x$ in equation (13.60), we get

$$x^2 - 9x + 16 = 0 \qquad \text{...(13.61)}$$

The equation $x^2 - 9x + 16 = 0$ is of the form of quadratic equation. It can be solve using the formula

$$x = \frac{-b \pm \sqrt{b^2 - 4ac}}{2a}$$

$$x = \frac{-(-3) \pm \sqrt{(-3)^2 - 4 \times 1 \times 16}}{2 \times 1}$$

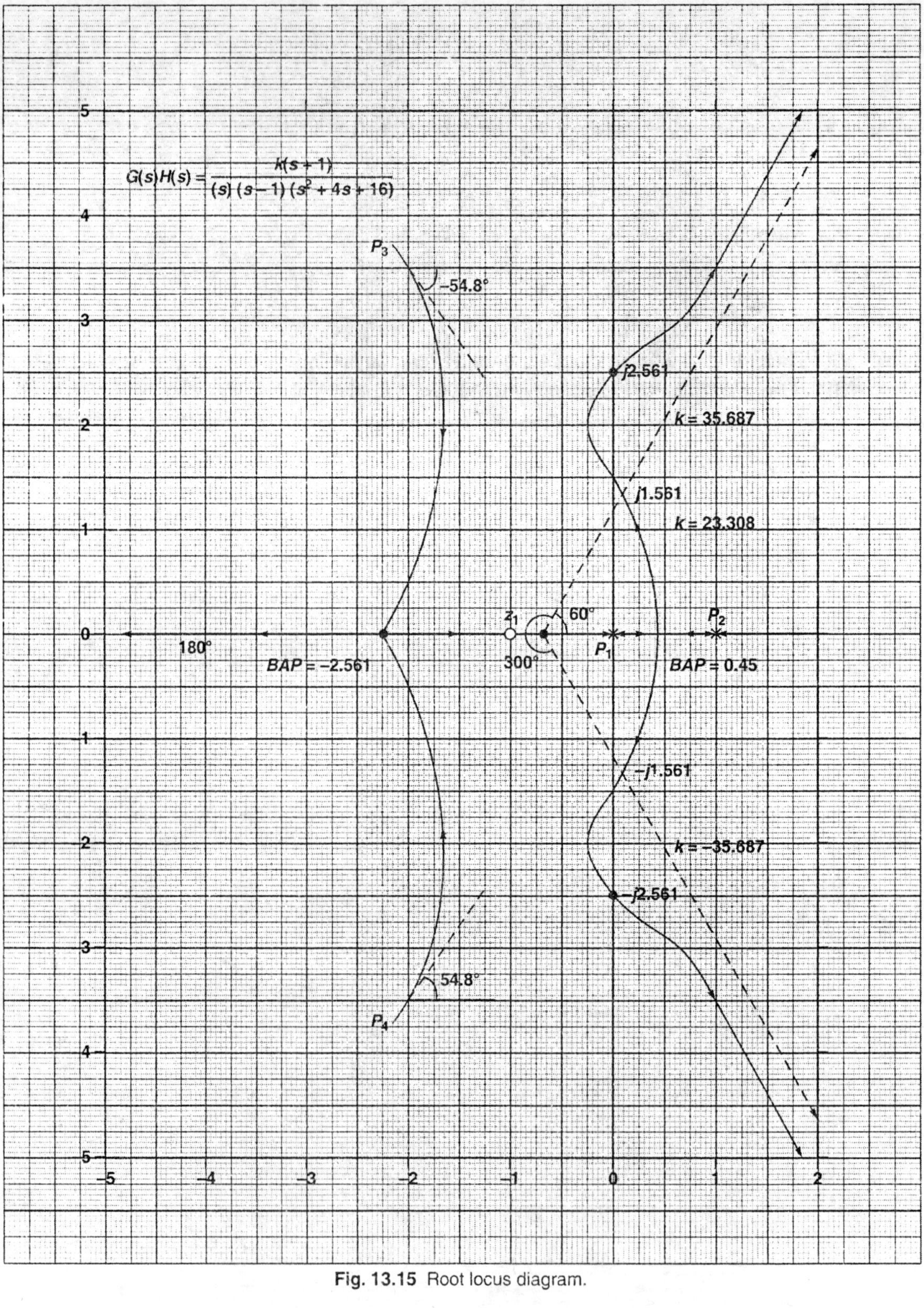

Fig. 13.15 Root locus diagram.

$$x = \frac{9 \pm \sqrt{17}}{2}$$

$$x = 6.561; \quad 2.438$$

$\Rightarrow \qquad \omega^2 = 6.561 \quad ; \quad 2.438$

$\Rightarrow \qquad \omega = 2.561 \quad ; \quad 1.561$

For $\qquad \omega = 2.561 \quad ; \quad k = 35.687$

$\qquad \omega = 1.561 \quad ; \quad k = 23.303$

Since, k is positive and real for both the values of ω, there are two pairs of imaginary points. They are, $\omega = \pm\, j2.561$ and $\omega = \pm\, j1.561$.

(vii) *Angle of departure* : From the angle of departure, we have

$$\theta_1 = 180 - \tan^{-1}(3.464/3) = 130.89°$$

$$\theta_2 = 180 - \tan^{-1}(3.464/2) = 120°$$

$$\theta_3 = 180 - \tan^{-1}(3.464/1) = 106.10°$$

Angle of departure from $s = -2 + j3.464$

$$\theta_d = 180 - (\theta_1 + \theta_2 + \theta_4) + \theta_3$$

$$= 180 - (130.89 + 120 + 90) + 106.10$$

$$= -54.8°$$

$$\theta_d^* = 54.8°$$

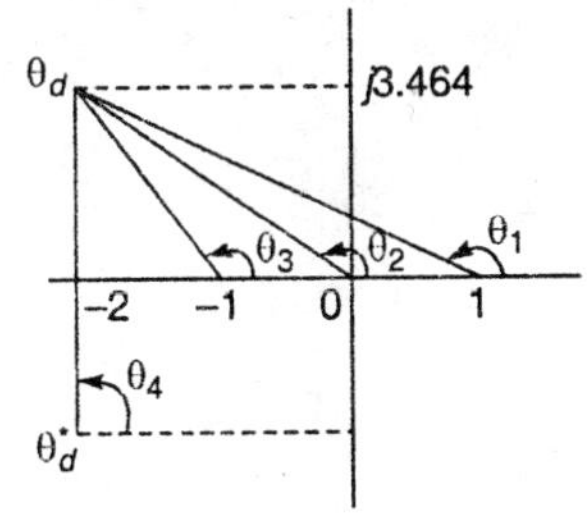

Fig. 13.16 Angle of departure.

■ **Example 13.32** Consider a negative feed back control system characterized by

$$G(s)H(s) = \frac{k(s+1)}{(s)\,(s+2)\,(s+3)\,(s+5)}$$

Sketch the root locus diagram for all values of k ranging from 0 to ∞. Also, find the gain k for which the closed-loop transfer function will have a pole on the real axis at –0.5

Solution : Transfer function of the control system is given by

$$G(s)H(s) = \frac{k(s+1)}{(s)\,(s+2)\,(s+3)\,(s+5)}$$

(i) *No. of poles and zeros* :

$n = 4 \quad ; \quad p_1 = 0$

$\qquad\qquad\quad p_2 = -2$

$\qquad\qquad\quad p_3 = -3$

$\qquad\qquad\quad p_4 = -5$

$m = 1 \quad ; \quad z_1 = -1$

(ii) *No. of asymptotes* :

$$n - m = 4 - 1 = 3$$

(iii) *Angle of asymptotes* :

$$\alpha = \frac{\pi(2k+1)}{(n-m)}$$

where, $k = 0, 1, 2$

$$\Rightarrow \qquad \alpha_1 = 60^\circ$$
$$\Rightarrow \qquad \alpha_2 = 180^\circ$$
$$\Rightarrow \qquad \alpha_3 = 300^\circ$$

(iv) *Centre of gravity* :

$$r = \frac{\sum_{j=1}^{n} p_j - \sum_{i=1}^{m} z_i}{(n-m)} = \frac{(0) + (-2) + (-3) + (-5) - (-1)}{(4-1)} = -3$$

(v) *Breakaway point* : From the characteristic equation, we have

$$1 + G(s)H(s) = 0$$
$$1 + \frac{k(s+1)}{(s)(s+2)(s+3)(s+5)} = 0$$
$$(s)(s+2)(s+3)(s+5) + k(s+1) = 0$$
$$s^4 + 10s^3 + 31s^2 + 30s + k(s+1) = 0 \qquad \ldots(13.62)$$
$$k(s+1) = -(s^4 + 10s^3 + 31s^2 + 30s)$$
$$k = -\frac{(s^4 + 10s^3 + 31s^2 + 30s)}{(s+1)} \qquad \ldots(13.63)$$

Differentiating equation (13.63) w.r.t. '*s*' and equalizing to zero, we get

$$\frac{dk}{ds} = 0$$

$$-\left[\frac{(s+1)(4s^3 + 30s^2 + 62s + 30) - (s^4 + 10s^3 + 31s^2 + 30s).1}{(s+1)^2}\right] = 0$$

$$3s^4 + 24s^3 + 61s^2 + 62s + 30 = 0$$

By trial and error method, we get $s = -2.42$.
Therefore, $(s + 2.42)$ is a factor of equation

$$3s^4 + 24s^3 + 61s^2 + 62s + 30 = 0$$
$$\Rightarrow \qquad (s + 2.42)(3s^3 + 16.74s^2 + 20.49s + 12.41) = 0$$

By trial and error method, we get $s = -4.185$.
Therefore, $(s + 4.185)$ is a factor of equation

$$3s^3 + 16.74s^2 + 20.49s + 12.41 = 0$$
$$\Rightarrow \qquad (s + 4.185)(3s^2 + 4.18s + 3) = 0$$

The equation $3s^2 + 4.18s + 3 = 0$ is of the form of quadratic equation. It can be solve using the formula

$$s = \frac{-b \pm \sqrt{b^2 - 4ac}}{2a}$$

$$s = \frac{-4.18 \pm \sqrt{(4.18)^2 - 4 \times 3 \times 3}}{2 \times 3}$$

$$s = \frac{-4.18 \pm \sqrt{j^2 (18.53)}}{2 \times 3}$$

$$s = \frac{-4.18 \pm j^2\, 4.30}{2 \times 3}$$

$$s = -0.70 \pm j(0.72)$$

when, $s = -2.42$; $k = 1.07$

$s = -4.185$; $k = -2.77$

$s = -0.70 \pm j(0.72)$; $k = \text{complex}$

The value of k is positive and real for $s = -2.42$. Therefore, it is the breakaway point and other three roots are not breakaway point.

(vi) *Intersection point 'ω' on imaginary axis*: From the polynomial characteristic equation, we have

$$s^4 + 10s^3 + 31s^2 + 30s + k(s+1) = 0 \quad ...(13.62)$$

Letting $s = j\omega$ in equation (13.62), we get

$$(j\omega)^4 + 10(j\omega)^3 + 31(j\omega)^2 + 30(j\omega) + k(j\omega + 1) = 0$$

$$\omega^4 - 10j\omega^3 - 31\omega^2 + 30j\omega + kj\omega + k = 0 \quad ...(13.64)$$

On equalizing the imaginary part of equation (13.64) to zero, we get

$$-10j\omega^3 + 30j\omega + kj\omega = 0$$

$$\Rightarrow \quad -10\omega^2 + 30 + k = 0$$

$$\Rightarrow \quad k = 10\omega^2 - 30 \quad ...(13.65)$$

On equalizing the real part of equation (13.64) to zero, we get

$$\omega^4 - 31\omega^2 + k = 0$$

$$\Rightarrow \quad k = -\omega^4 + 31\omega^2 \quad ...(13.66)$$

Comparing equations (13.65) and (13.66), we get

$$10\omega^2 - 30 = -\omega^4 + 31\omega^2$$

$$\Rightarrow \quad \omega^4 - 21\omega^2 - 30 = 0 \quad ...(13.67)$$

Letting, $\omega^2 = x$ in equation (13.67), we get

$$x^2 - 21x - 30 = 0 \quad ...(13.68)$$

The equation $x^2 - 21x - 30 = 0$ is of the form of quadratic equation. It can be solve using the formula

$$x = \frac{-b \pm \sqrt{b^2 - 4ac}}{2a}$$

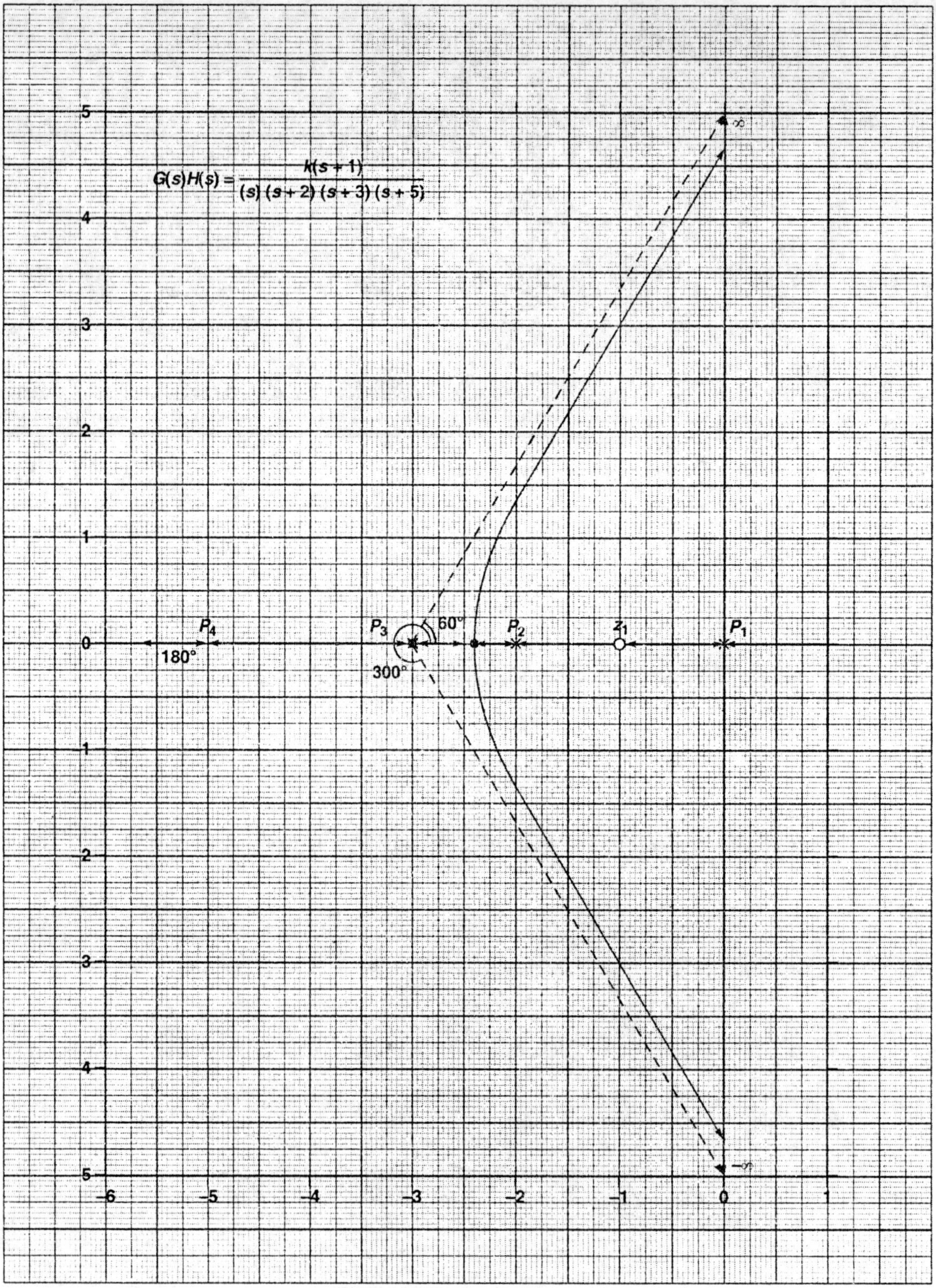

Fig. 13.17 Root locus diagram.

$$x = \frac{-(-21) \pm \sqrt{(-21)^2 - 4 \times 1 \times (-30)}}{2 \times 1}$$

$$x = \frac{21 \pm \sqrt{561}}{2}$$

$$x = 22.34; \quad -1.34$$

$\Rightarrow \quad \omega^2 = 22.34 \quad ; \quad -1.34$

$\Rightarrow \quad \omega = 4.73; \quad j(1.16)$

For $\quad \omega = 4.73; \quad k = 193$

The other value of 'ω' is complex. If ω is complex, the implication is that the root locus does not cut the $j\omega$ axis at those values. For closed loop stability, $0 < k < 193$.

Merits and Demerits

Following are the merit and demerits of root locus technique.

Merits :

(i) Root locus is a simple method to determine the stability of a control system.

(ii) It allow the rapid determination of the location of the roots of characteristic equation as a function of parameter k. In addition, it also allow relatively easy determination of the transient response for higher order system by graphical means.

Demerits :

(i) Root locus technique required both graphical as well as algebraic calculation for determination of stability.

(ii) It is difficult to apply for a system having transportation lag.

13.3.3 Frequency Response Technique

The frequency response technique is the quantitative measure of the output spectrum of a device in response to a stimulus. It is use to characterize the dynamics of the system. The following methods may be use to analyse the frequency response of a control system :

(i) Bode stability criterion

(ii) Nyquist stability criterion

13.3.3.1 Bode Stability Criterion

The bode stability criterion is a frequency response method that gives the information on the absolute and relative stability of the system. In this method required to plot bode diagram to determine the stability. Bode diagram consists of two graphs :

(i) Log (AR) vs log (ω)

(ii) φ vs log (ω)

where, AR = Amplitude Ratio

φ = Phase angle

ω = Radian frequency

Bode diagram is useful as it exhibits both low frequency (LF) and high frequency (HF) characteristics of the transfer function in one diagram. Magnifying the LF range by the use of a logarithmic scale for frequency is very advantageous since the characteristics at low frequencies are most important in practical systems.

A control system is unstable if the open loop frequency response exhibits an AR larger than 1 is at the frequency for which the phase lag is 180°. The frequency for which the phase lag is 180° known as cross over frequency. It is designated by ω_{co}. This rule is called the bode stability criterion.

Actually, this rule is not quite general. It is applicable readily to systems or which the gain and phase curve decreases continuously with frequency. Most of the process control system can be analysed with the simple bode stability criterion and therefore it finds wide application.

Application of the stability criterion requires nothing more than plotting the open loop frequency response. This may be based on the theoretical transfer function, if it is available. For example, tank level system. If the theoretical system dynamics are not known, the frequency response may be obtained experimentally. To do this the open loop system is disturbed with a sine-wave input at several frequencies. At each frequency, records of the input and output wave are to be compared to establish AR and phase lag. The results can be plotted as bode diagram.

Rules for plotting bode diagram : Consider '*n*' no. of systems in series whose overall transfer function is given by the product of individual TF

$$G(s) = G_1(s) \times G_2(s) \times \ldots \times G_n(s) \quad \ldots(13.69)$$

where, $G(s)$ = Overall transfer function

$G_1(s), G_2(s), G_n(s)$ = Individual transfer function.

Letting, $s = j\omega$ in equation (13.69), we get

$$G(j\omega) = G_1(j\omega) \times G_2(j\omega) \times \ldots \times G_n(j\omega)$$

$$\Rightarrow \quad AR = AR_1 \times AR_2 \times \ldots \times AR_n$$

$$\Rightarrow \quad \log AR = \log AR_1 + \log AR_2 + \ldots + \log AR_n$$

$$\Rightarrow \quad \phi = \phi_1 + \phi_2 + \ldots + \phi_n$$

Therefore, the following rules are applicable for plotting the bode diagram.

(i) The overall *AR* is obtained by adding the individual *AR*. For this graphical method, curve above unity *AR* is taken as positive and curve below is taken as negative.

(ii) The overall phase angle is obtained by the addition of individual phase angle.

(iii) The presence of a constant in the overall TF shift the entire *AR* curve vertically by a constant amount and has no effect on the phase angle. It is usually more convenient to include constant factor in the ordinate.

■ **Example 13.34** Draw bode diagram for the system having transfer function

$$G(s) = \frac{1}{(s+1)}$$

Solution : From the transfer function, we have

$$G(s) = \frac{1}{(s+1)}$$

Letting, $s = j\omega$, we get

$$G(j\omega) = \frac{1}{(1+j\omega)} \quad \ldots(13.70)$$

$$\Rightarrow \quad G(j\omega) = \frac{1 \times (1-j\omega)}{(1+j\omega)(1-j\omega)}$$

$$\Rightarrow \qquad G(j\omega) = \frac{(1 - j\omega)}{(1 - j^2\omega^2)}$$

$$\Rightarrow \qquad G(j\omega) = \frac{(1 - j\omega)}{(1 + \omega^2)}$$

$$\Rightarrow \qquad G(j\omega) = \frac{1}{(1 + \omega^2)} + j\frac{(-\omega)}{(1 + \omega^2)} \qquad \ldots(13.71)$$

Equation (13.71) is of the form $z = x + iy$. Therefore, from the formula, we have

$$AR = \sqrt{(x)^2 + (y)^2}$$

$$\Rightarrow \qquad AR = \sqrt{\left(\frac{1}{1+\omega^2}\right)^2 + \left(\frac{-\omega}{1+\omega^2}\right)^2}$$

$$\Rightarrow \qquad AR = \sqrt{\frac{(1+\omega^2)}{(1+\omega^2)^2}}$$

$$\Rightarrow \qquad AR = \frac{1}{\sqrt{1+\omega^2}} \qquad \ldots(13.72)$$

From the formula of phase angle, we have

$$\phi = \tan^{-1}(y/x)$$

$$\Rightarrow \qquad \phi = \tan^{-1}\left(\frac{-\omega}{1+\omega^2} \Big/ \frac{1}{1+\omega^2}\right)$$

$$\Rightarrow \qquad \phi = \tan^{-1}(-\omega) \qquad \ldots(13.73)$$

From the equations (13.72) and (13.73), we can observe that amplitude ratio and phase angle is the function of 'ω' only.

Table 13.1 : Data for plotting bode diagram.

S. No.	ω (rad/s)	AR	ϕ (deg)
1.	0.1	1	– 5.71
2.	0.2	0.98	– 11.31
3.	0.5	0.89	– 26.56
4.	1.0	0.71	– 45.00
5.	2.0	0.45	– 63.44
6.	5.0	0.20	– 78.69
7.	10.0	0.10	– 84.29
8.	20.0	0.05	– 87.14
9.	50.0	0.02	– 88.85
10.	100.0	0.01	– 89.43

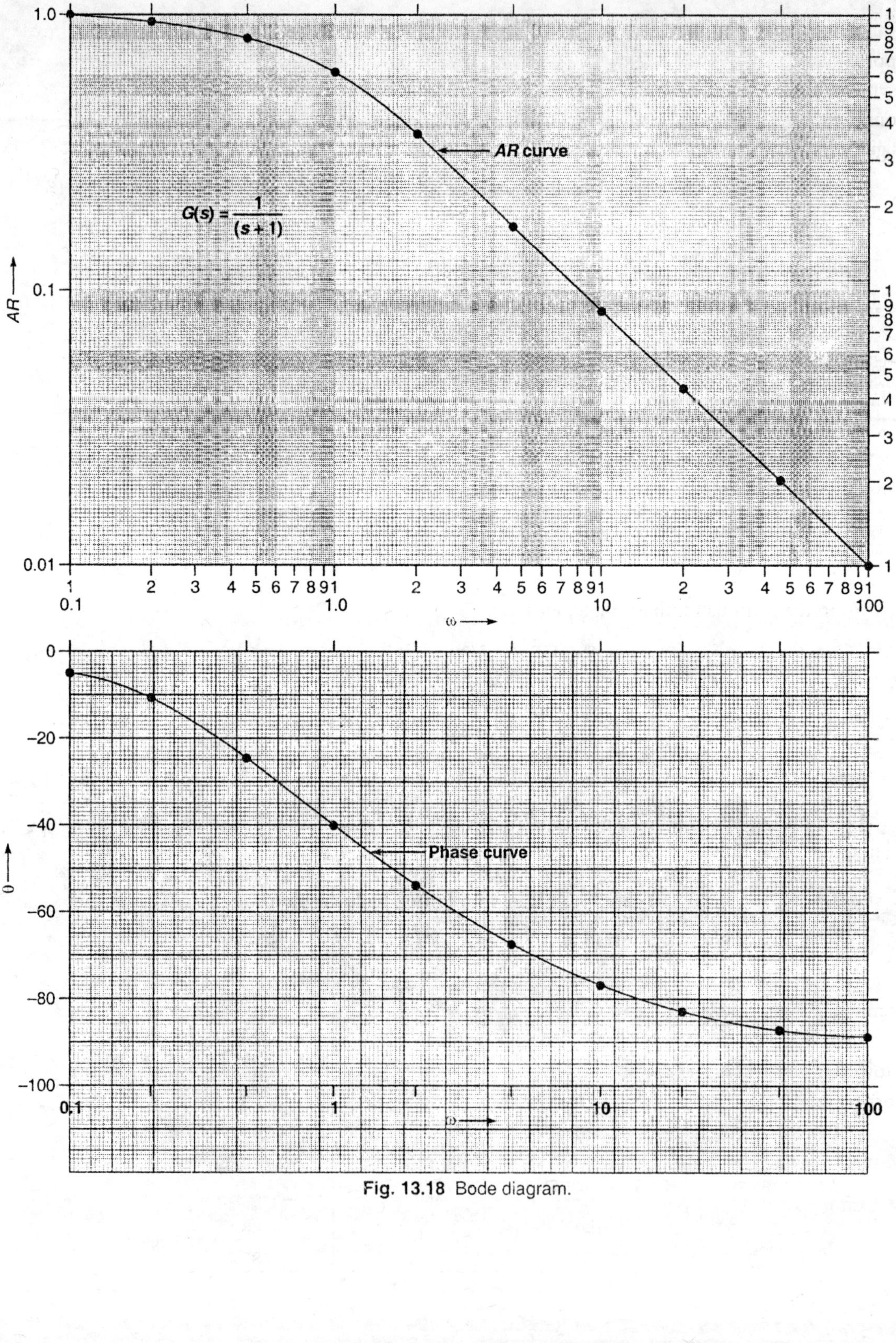

Fig. 13.18 Bode diagram.

Asymptotic consideration for bode diagram : It is observed from the Fig. 13.18 that *AR* curve approximately follows two asymptotes except at $\omega=1$, where it makes a smooth change from one asymptote to other. Practically, *AR* curve of the bode diagram is simply drawn by asymptotic consideration. For example, consider the first order system as discussed above for which amplitude ratio is given by the equation

$$AR = \frac{1}{\sqrt{1+\omega^2}} \qquad \text{...(13.72)}$$

Low Frequency Asymptotes (LFA) :

When, $\omega <<< 1$, *i.e.*, $\omega \to 0$. From the equation (13.72), we get

$$\Rightarrow \qquad AR = 1$$

$$\Rightarrow \qquad \log AR = 0$$

That means low frequency asymptote is a straight line with a slope zero passing through the points $AR = 1$ and $\omega = 0.1$ on a log-log plot.

High Frequency Asymptotes (HFA) :

When, $\omega >>> 1$, *i.e.*, $\omega \to \infty$. From the equation (13.72), we get

$$\Rightarrow \qquad AR = 1/\omega$$

$$\Rightarrow \qquad \log AR = -\log \omega$$

That means, high frequency asymptote is a straight line with a slope -1 passing through the point $AR = 1$ and $\omega = 1$ on log-log plot.

Corner Frequency : From the characteristic equation, *i.e.*, denominator of amplitude ratio, we have

$$\sqrt{1+\omega^2} = 0$$

$$\Rightarrow \qquad 1+\omega^2 = 0$$

$$\Rightarrow \qquad \omega^2 = |-1|$$

$$\Rightarrow \qquad \omega_c = 1$$

The frequency at which two asymptotes intersect each other is called corner frequency.

From the frequency response curve Fig. 13.18, we can observe that the deviation of the true *AR* curve from the asymptote is maximum at the corner frequency where the value of true $AR = 1/\sqrt{1+1} = 0.707$. At this point the deviation is around 30% and at all other frequency the deviation is much less.

For most engineering purposes, it is offer sufficient to represent the entire *AR* curve by asymptote. However, more accurate curve can be drawn easily by using the asymptote and the true value of *AR* at the corner frequency.

There are no asymptotic approximation available for phase angle curve which gives accurate results.

■ **Example 13.35** Draw bode diagram for the system having overall transfer function

$$G(s) = \frac{10}{(s+1)(s+5)}$$

Solution : From the overall transfer function, we have

$$G(s) = \frac{10}{(s+1)(s+5)}$$

$$\Rightarrow \quad G(s) = \frac{2}{(s+1)(0.2s+1)}$$

$$\Rightarrow \quad G(s) = 2 \times \frac{1}{(s+1)} \times \frac{1}{(0.2s+1)}$$

$$\Rightarrow \quad G(s) = 2 \times G_1(s) \times G_2(s)$$

where, $G_1(s) = \frac{1}{(s+1)}$; $\tau_1 = 1$

and $G_2(s) = \frac{1}{(0.2s+1)}$; $\tau_2 = 0.2$

Therefore, $AR_1 = \frac{1}{\sqrt{1+\omega^2}}$; $\phi_1 = \tan^{-1}(-\omega)$

and $AR_2 = \frac{1}{\sqrt{1+(0.2\omega)^2}}$; $\phi_2 = \tan^{-1}(-0.2\omega)$

$$\Rightarrow \quad AR = 2 \times AR_1 \times AR_2$$

$$\Rightarrow \quad \frac{AR}{2} = AR_1 \times AR_2$$

$$\Rightarrow \quad \log(AR/2) = \log AR_1 + \log AR_2$$

$$\phi = \phi_1 + \phi_2$$

$$\Rightarrow \quad \phi = \tan^{-1}(-\omega) + \tan^{-1}(-0.2\omega)$$

Low Frequency Asymptotes : From the amplitude ratio, we have

$$AR_1 = \frac{1}{\sqrt{1+\omega^2}}$$

when, $\omega <<< 1$

$$AR_1 = 1$$

$$\log AR_1 = 0$$

Table 13.2 : Data for plotting bode diagram.

S. No.	ω (rad/s)	AR_1	AR_2	$AR/2$	ϕ_1 (deg)	ϕ_2 (deg)	ϕ (deg)
1.	0.1	0.995	0.999	0.994	–5.71	–1.14	–6.85
2.	0.5	0.895	0.995	0.890	–26.56	–5.71	–32.27
3.	1.0	0.707	0.980	0.692	–45.00	–11.30	–56.30
4.	5.0	0.196	0.707	0.138	–78.69	–45.00	–123.69
5.	10.0	0.099	0.447	0.044	–84.28	–63.43	–147.71
6.	50.0	0.019	0.099	0.002	–88.85	–84.28	–173.13
7.	100.0	0.009	0.049	0.0004	–89.42	–87.13	–180.55

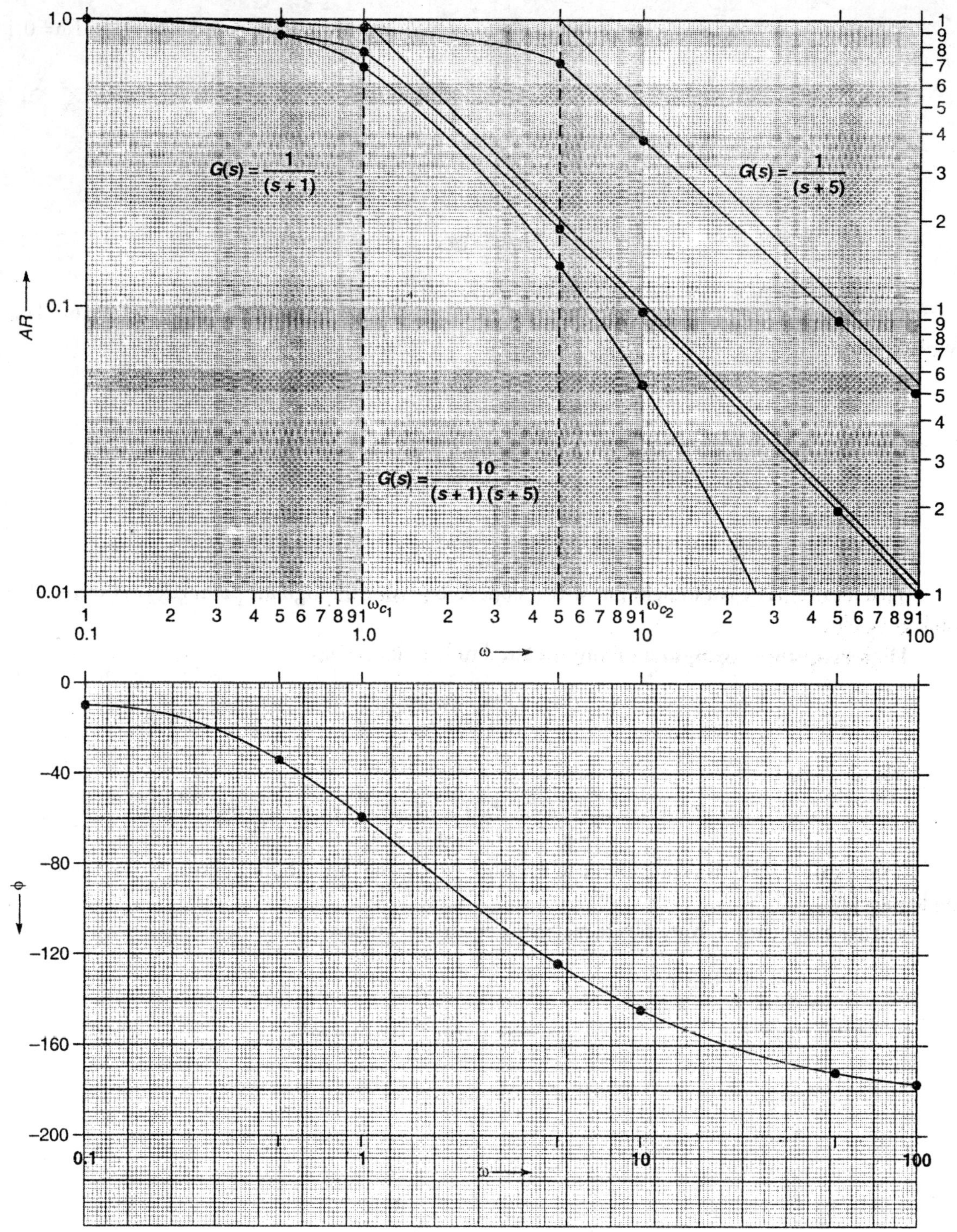

Fig. 13.19 Bode diagram.

That means LFA_1 is a straight line with a slope zero passing through the points $AR_1 = 1$ and $\omega = 0.1$ on a log-log plot.

High Frequency Asymptotes : From the amplitude ratio, we have

$$AR_1 = \frac{1}{\sqrt{(1+\omega^2)}}$$

when, $\omega >>> 1$

$$AR_1 = 1/\omega$$
$$\log AR_1 = -\log \omega$$

That means, HFA_1 is a straight line with a slope -1 passing through the point $AR_1 = 1$ and $\omega = 1$ on log-log plot.

Low Frequency Asymptote : From the amplitude ratio, we have

$$AR_2 = \frac{1}{\sqrt{1+(0.2\omega)^2}}$$

when, $\omega <<< 1$

$$AR_2 = 1$$
$$\log AR_2 = 0$$

That means LFA_2 is a straight line with a slope zero passing through the point $AR_2 = 1$ and $\omega = 0.1$ on log-log plot.

High Frequency Asymptote : From the amplitude ratio, we have

$$AR_2 = \frac{1}{\sqrt{1+(0.2\omega)^2}}$$

when, $\omega >>> 1$

$$AR_2 = 1/(0.2\omega)$$
$$\log AR_2 = -\log (0.2\omega)$$

That means, HFA_2 is a straight line with a slope -1 passing through the point $AR_2 = -1$ and $\omega = 5$ on log-log plot.

Corner Frequency : From the characteristic equation, we have

$$1+\omega^2 = 0$$
$$\omega^2 = |-1|$$
$$\omega^2 = 1$$
$$\omega_{c_1} = 1$$

From the characteristic equation, we have

$$1+(0.2\omega)^2 = 0$$
$$(0.2\omega)^2 = |-1|$$
$$(0.2\omega)^2 = 1$$
$$0.2\omega = 1$$
$$\omega_{c_2} = 5$$

■ **Example 13.36** Draw bode diagram for the system having overall transfer function

$$G(s) = \frac{2}{(s+1)(s+2)}$$

Solution : From the overall transfer function, we have

$$G(s) = \frac{2}{(s+1)(s+2)}$$

$$\Rightarrow \qquad G(s) = \frac{1}{(s+1)(0.5s+1)}$$

$$\Rightarrow \qquad G(s) = \frac{1}{(s+1)} \times \frac{1}{(0.5s+1)}$$

$$\Rightarrow \qquad G(s) = G_1(s) \times G_2(s)$$

where, $G_1(s) = \dfrac{1}{(s+1)}$; $\tau = 1$

and $G_2(s) = \dfrac{1}{(0.5s+1)}$; $\tau = 0.5$

Therefore, $AR_1 = \dfrac{1}{\sqrt{1+\omega^2}}$; $\phi_1 = \tan^{-1}(-\omega)$

and $AR_2 = \dfrac{1}{\sqrt{1+(0.5\omega^2)}}$; $\phi_2 = \tan^{-1}(-0.5\omega)$

$$\Rightarrow \qquad AR = AR_1 \times AR_2$$

$$\Rightarrow \qquad \log(AR) = \log(AR_1) + \log(AR_2)$$

$$\phi = \phi_1 + \phi_2$$

$$\Rightarrow \qquad \phi = \tan^{-1}(-\omega) + \tan^{-1}(-0.5\omega)$$

Table 13.3 : Data for plotting bode diagram.

S. No.	ω (rad/s)	AR_1	AR_2	AR	ϕ_1 (deg)	ϕ_2 (deg)	ϕ (deg)
1.	0.1	0.995	0.999	0.994	–5.71	–2.86	–8.57
2.	0.5	0.895	0.970	0.868	–26.56	–14.04	–40.60
3.	1.0	0.707	0.894	0.632	–45.00	–26.56	–71.56
4.	2.0	0.447	0.707	0.316	–63.43	–45.00	–108.43
5.	5.0	0.196	0.371	0.037	–78.69	–68.19	–146.88
6.	10.0	0.099	0.196	0.017	–84.28	–78.69	–162.97
7.	50.0	0.019	0.039	0.0007	–88.85	–87.71	–176.56
8.	100.0	0.009	0.019	0.0002	–89.42	–88.85	–178.27

Low Frequency Asymptote : From the amplitude ratio, we have

$$AR_1 = \frac{1}{\sqrt{1+\omega^2}}$$

when, $\omega <<< 1$

$$AR_1 = 1$$
$$\log AR_1 = 0$$

That means LFA_1 is a straight line with a slope zero passing through the points $AR_1 = 1$ and $\omega = 0.1$ on log-log plot.

High Frequency Asymptote : From the amplitude ratio, we have

$$AR_1 = \frac{1}{\sqrt{1+\omega^2}}$$

when, $\omega >>> 1$

$$AR_1 = 1/\omega$$
$$\log AR_1 = -\log \omega$$

That means, HFA_1 is a straight line with a slope -1 passing through the point $AR_1 = 1$ and $\omega = 1$ on log-log plot.

Low Frequency Asymptote : From the amplitude ratio, we have

$$AR_2 = \frac{1}{\sqrt{1+(0.5\omega)^2}}$$

when, $\omega <<< 1$

$$AR_2 = 1$$
$$\log AR_2 = 0$$

That means LFA_2 is a straight line with a slope zero passing through the point $AR_2 = 1$ and $\omega = 0.1$ on log-log plot.

High Frequency Asymptote : From the amplitude ratio, we have

$$AR_2 = \frac{1}{\sqrt{1+(0.5\omega)^2}}$$

when, $\omega >>> 1$

$$AR_2 = 1/(0.5\omega)$$
$$\log AR_2 = -\log(0.5\omega)$$

That means, HFA_2 is a straight line with a slope -1 passing through the point $AR_2 = -1$ and $\omega = 2$ on log-log plot.

Corner Frequency : From the characteristic equation, we have

$$1+\omega^2 = 0$$
$$\omega^2 = |-1|$$
$$\omega^2 = 1$$
$$\omega_{c_1} = 1$$

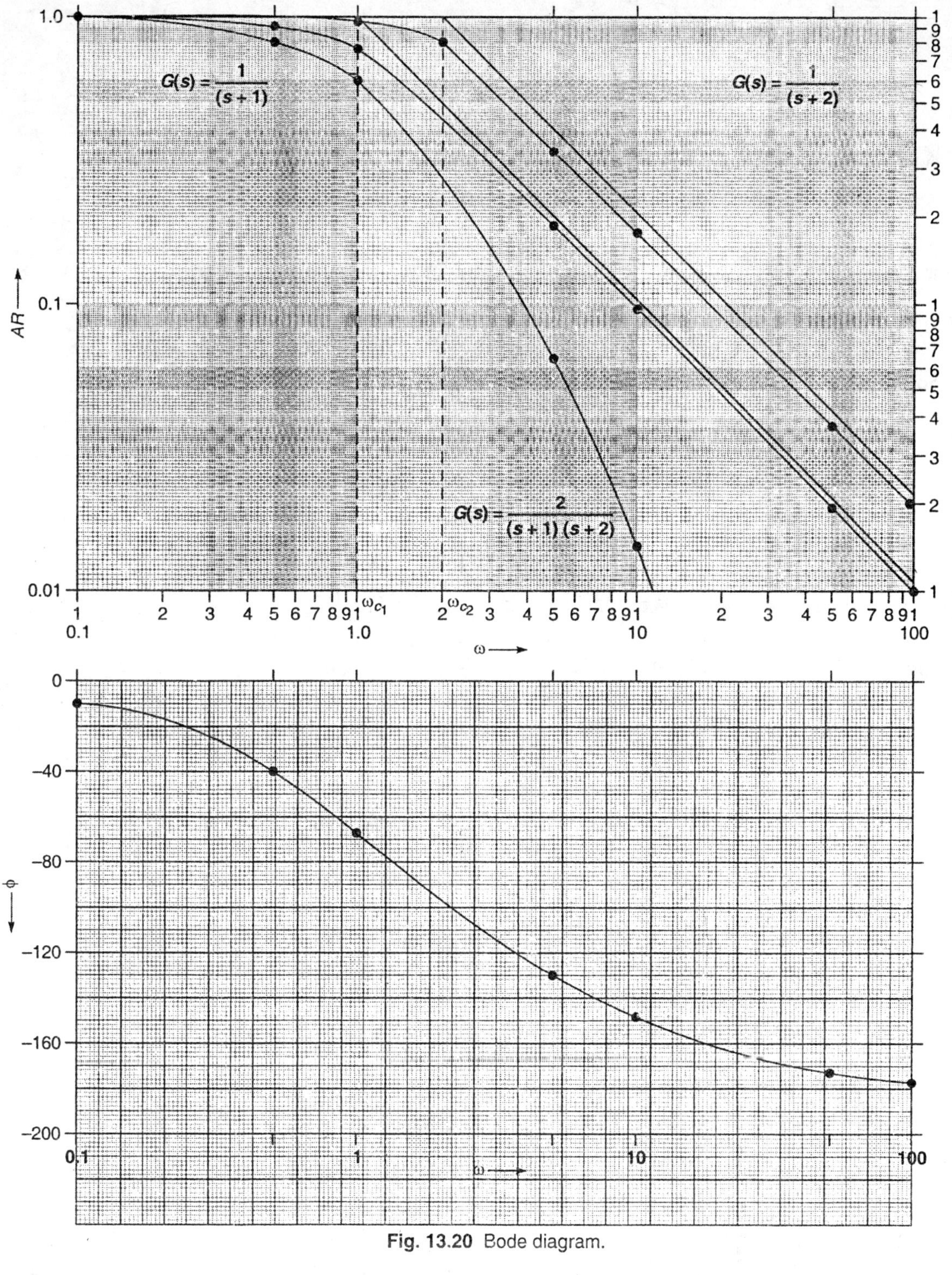

Fig. 13.20 Bode diagram.

From the characteristic equation, we have

$$1+(0.5\omega)^2 = 0$$

$$(0.5\omega)^2 = |-1|$$

$$(0.5\omega)^2 = 1$$

$$0.5\omega = 1$$

$$\omega_{c_2} = 2$$

■ **Example 13.37** Draw bode diagram for the system having overall transfer function

$$G(s) = \frac{10}{(s+2)(s+5)}$$

Solution : From the overall transfer function, we have

$$G(s) = \frac{10}{(s+2)(s+5)}$$

$$\Rightarrow \quad G(s) = \frac{1}{(0.5s+1)(0.2s+1)}$$

$$\Rightarrow \quad G(s) = G_1(s) \times G_2(s)$$

where, $G_1(s) = \dfrac{1}{(0.5s+1)}$; $\tau_1 = 0.5$

and $G_2(s) = \dfrac{1}{(0.2s+1)}$; $\tau_2 = 0.2$

Therefore, $AR_1 = \dfrac{1}{\sqrt{1+(0.5\omega)^2}}$; $\phi_1 = \tan^{-1}(-0.5\omega)$

and $AR_2 = \dfrac{1}{\sqrt{1+(0.2\omega^2)}}$; $\phi_2 = \tan^{-1}(-0.2\omega)$

$$\Rightarrow \quad AR = AR_1 \times AR_2$$

$$\Rightarrow \quad \log AR = \log(AR_1) + \log(AR_2)$$

$$\phi = \phi_1 + \phi_2$$

$$\Rightarrow \quad \phi = \tan^{-1}(-0.5\omega) + \tan^{-1}(-0.2\omega)$$

Table 13.4 : Data for plotting bode diagram.

S. No.	ω (rad/s)	AR_1	AR_2	AR	ϕ_1 (deg)	ϕ_2 (deg)	ϕ (deg)
1.	0.1	0.999	0.999	0.998	−2.86	−1.14	−4.00
2.	0.5	0.970	0.995	0.965	−14.04	−5.71	−19.75
3.	1.0	0.894	0.980	0.876	−26.56	−11.30	−37.86
4.	5.0	0.371	0.707	0.262	−68.19	−45.00	−113.19
5.	10.0	0.196	0.447	0.087	−78.69	−63.43	−142.12
6.	50.0	0.039	0.099	0.004	−87.71	−84.28	−171.99
7.	100.0	0.019	0.049	0.001	−88.85	−87.13	−176.98

Low Frequency Asymptote : From the amplitude ratio, we have

$$AR_1 = \frac{1}{\sqrt{1+(0.5\omega)^2}}$$

when, $\omega <<< 1$

$$AR_1 = 1$$

$$\log AR_1 = 0$$

That means LFA_1 is a straight line with a slope zero passing through the points $AR_1 = 1$ and $\omega = 0.1$ on log-log plot.

High Frequency Asymptote : From the amplitude ratio, we have

$$AR_1 = \frac{1}{\sqrt{1+(0.5\omega)^2}}$$

when, $\omega >>> 1$

$$AR_1 = 1/(0.5\omega)$$

$$\log AR_1 = -\log(0.5\omega)$$

That means, HFA_1 is a straight line with a slope -1 passing through the point $AR_1 = 1$ and $\omega = 2$ on log-log plot.

Low Frequency Asymptote : From the amplitude ratio, we have

$$AR_2 = \frac{1}{\sqrt{1+(0.2\omega)^2}}$$

when, $\omega <<< 1$

$$AR_2 = 1$$

$$\log AR_2 = 0$$

That means LFA_2 is a straight line with a slope zero passing through the point $AR_2 = 1$ and $\omega = 0.1$ on log-log plot.

High Frequency Asymptote : From the amplitude ratio, we have

$$AR_2 = \frac{1}{\sqrt{1+(0.2\omega)^2}}$$

when, $\omega >>> 1$

$$AR_2 = 1/(0.2\omega)$$

$$\log AR_2 = -\log(0.2\omega)$$

That means, HFA_2 is a straight line with a slope -1 passing through the point $AR_2 = 1$ and $\omega = 5$ on log-log plot.

Corner Frequency : From the characteristic equation, we have

$$1+(0.5\omega)^2 = 0$$

$$(0.5\omega)^2 = |-1|$$

$$0.5\omega = 1$$

$$\omega_{c_1} = 2$$

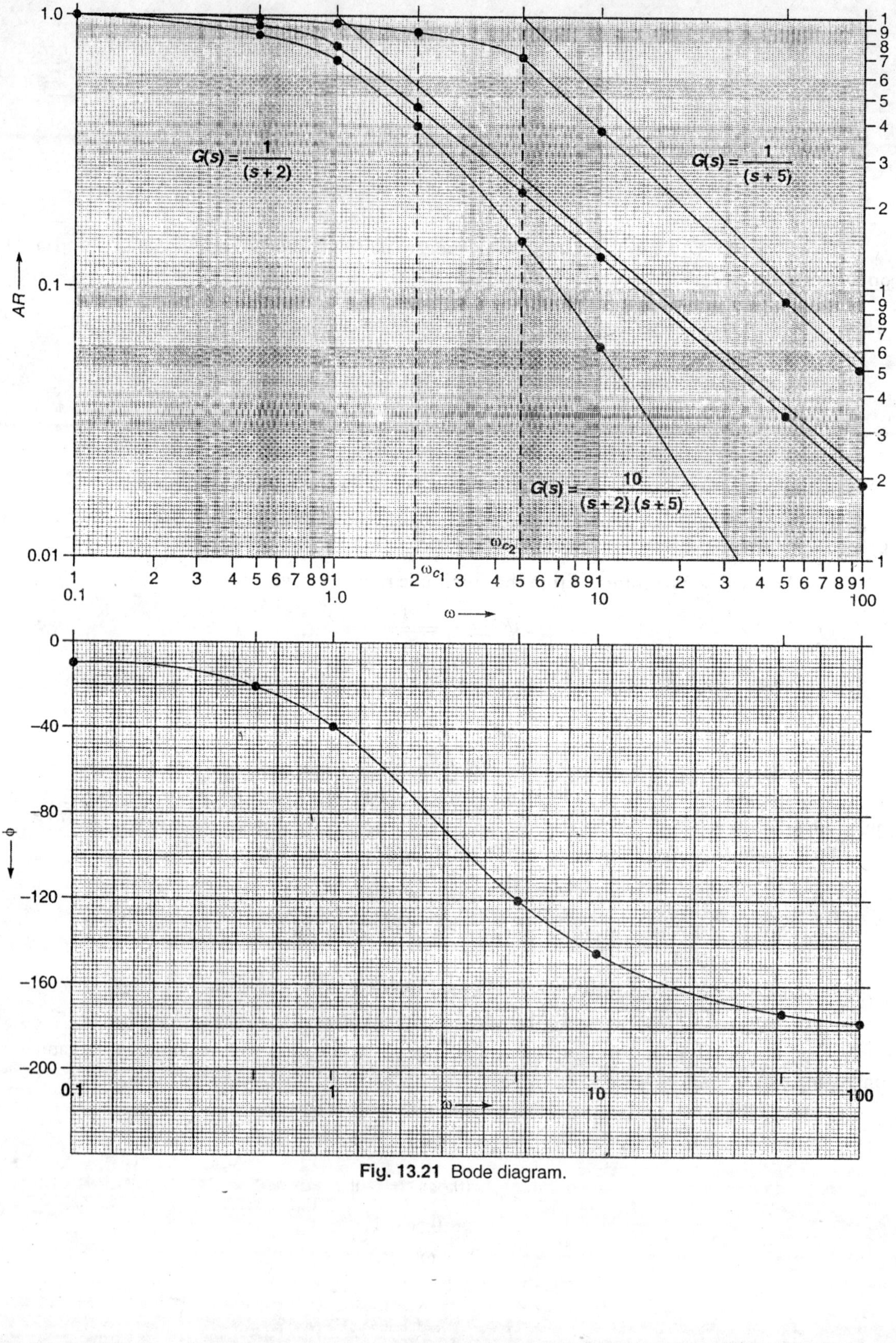

Fig. 13.21 Bode diagram.

From the characteristic equation, we have

$$1+(0.2\omega)^2 = 0$$
$$(0.2\omega)^2 = |-1|$$
$$(0.2\omega)^2 = 1$$
$$0.2\omega = 1$$
$$\omega_{c_2} = 5$$

■ **Example 13.38** Draw bode diagram for the system having overall transfer function

$$G(s) = \frac{20}{(s+2)(s+10)}$$

Solution : From the overall transfer function, we have

$$G(s) = \frac{20}{(s+2)(s+10)}$$

$$\Rightarrow \quad G(s) = \frac{1}{(0.5s+1)(0.1s+1)}$$

$$\Rightarrow \quad G(s) = G_1(s) \times G_2(s)$$

where, $G_1(s) = \dfrac{1}{(0.5s+1)}$; $\tau_1 = 0.5$

and $G_2(s) = \dfrac{1}{(0.1s+1)}$; $\tau_2 = 0.1$

Therefore, $AR_1 = \dfrac{1}{\sqrt{1+(0.5\omega)^2}}$; $\phi_1 = \tan^{-1}(-0.5\omega)$

and $AR_2 = \dfrac{1}{\sqrt{1+(0.1\omega^2)}}$; $\phi_2 = \tan^{-1}(-0.1\omega)$

$$\Rightarrow \quad AR = AR_1 \times AR_2$$

$$\Rightarrow \quad \log AR = \log(AR_1) + \log(AR_2)$$

$$\phi = \phi_1 + \phi_2$$

$$\Rightarrow \quad \phi = \tan^{-1}(-0.5\omega) + \tan^{-1}(-0.1\omega)$$

Table 13.5 : Data for plotting bode diagram.

S. No.	ω (rad/s)	AR_1	AR_2	AR	ϕ_1 (deg)	ϕ_2 (deg)	ϕ (deg)
1.	0.1	0.999	0.999	0.998	–2.86	–0.57	–3.43
2.	0.5	0.970	0.998	0.968	–14.04	–2.86	–16.90
3.	1.0	0.894	0.995	0.889	–26.56	–5.71	–32.27
4.	2.0	0.707	0.981	0.694	–45.00	–11.31	–56.31
5.	5.0	0.371	0.894	0.332	–68.19	–26.56	–84.75
6.	10.0	0.196	0.707	0.139	–78.69	–45.00	–123.69
7.	50.0	0.039	0.196	0.008	–87.71	–78.69	–166.40
8.	100.0	0.019	0.099	0.002	–88.85	–84.29	–173.14

Low Frequency Asymptote : From the amplitude ratio, we have

$$AR_1 = \frac{1}{\sqrt{1+(0.5\omega)^2}}$$

when, $\omega <<< 1$

$$AR_1 = 1$$
$$\log AR_1 = 0$$

That means LFA_1 is a straight line with a slope zero passing through the points $AR_1 = 1$ and $\omega = 0.1$ on log-log plot.

High Frequency Asymptote : From the amplitude ratio, we have

$$AR_1 = \frac{1}{\sqrt{1+(0.5\omega)^2}}$$

when, $\omega >>> 1$

$$AR_1 = 1/(0.5\omega)$$
$$\log AR_1 = -\log(0.5\omega)$$

That means, HFA_1 is a straight line with a slope -1 passing through the point $AR_1 = 1$ and $\omega = 2$ on log-log plot.

Low Frequency Asymptote : From the amplitude ratio, we have

$$AR_2 = \frac{1}{\sqrt{1+(0.1\omega)^2}}$$

when, $\omega <<< 1$

$$AR_2 = 1$$
$$\log AR_2 = 0$$

That means LFA_2 is a straight line with a slope zero passing through the point $AR_2 = 1$ and $\omega = 0.1$ on log-log plot.

High Frequency Asymptote : From the amplitude ratio, we have

$$AR_2 = \frac{1}{\sqrt{1+(0.1\omega)^2}}$$

when, $\omega >>> 1$

$$AR_2 = 1/(0.1\omega)$$

That means, HFA_2 is a straight line with a slope -1 passing through the point $AR_2 = 1$ and $\omega = 10$ on log-log plot.

Corner Frequency : From the characteristic equation, we have

$$1+(0.5\omega)^2 = 0$$
$$(0.5\omega)^2 = |-1|$$
$$(0.5\omega)^2 = 1$$
$$0.5\omega = 1$$
$$\omega_{c_1} = 2$$

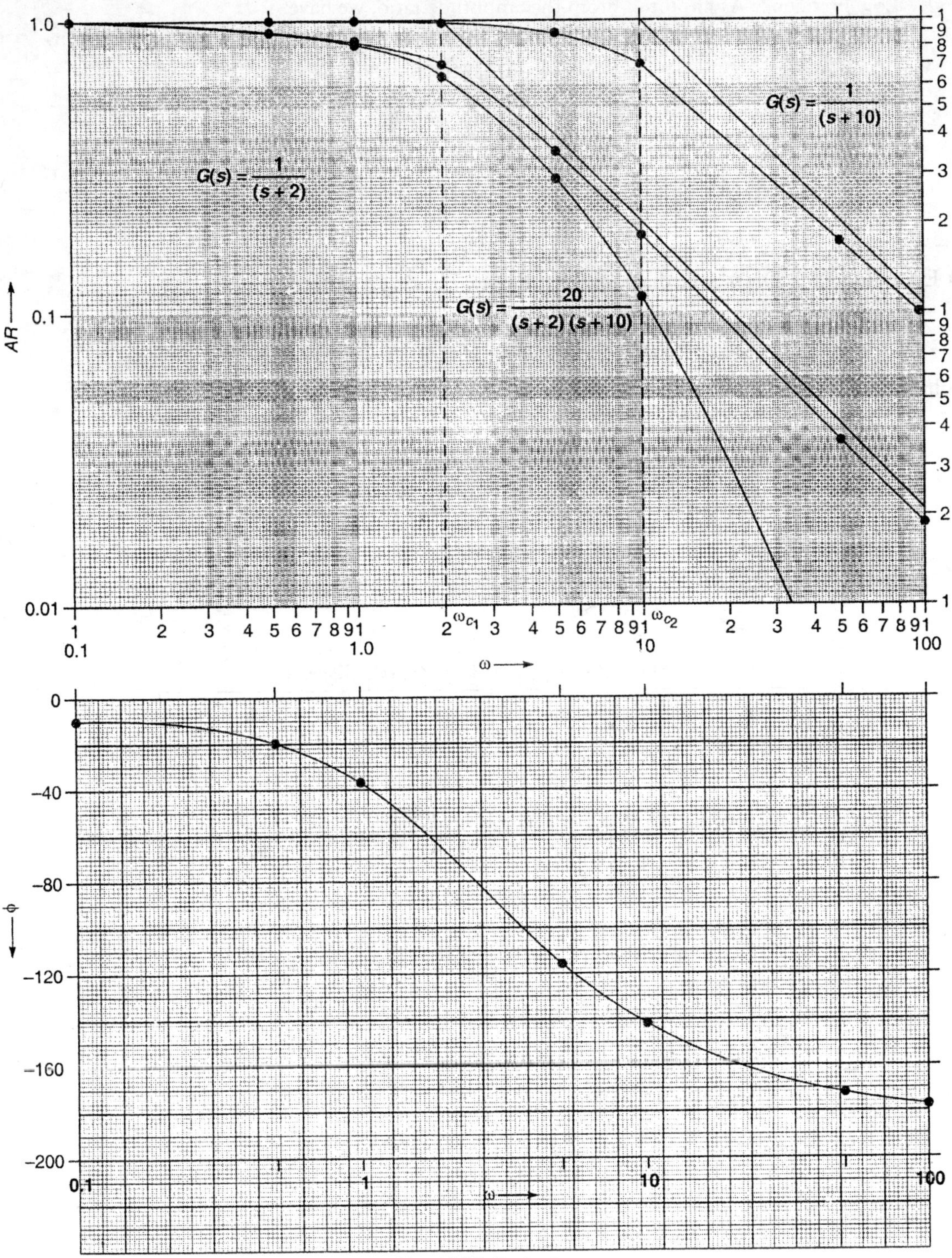

Fig. 13.22 Bode diagram.

Corner Frequency : From the characteristic equation, we have

$$1+(0.1\omega)^2=0$$

$$(0.1\omega)^2=|-1|$$

$$(0.1\omega)^2=1$$

$$0.1\omega=1$$

$$\omega_{c_2}=10$$

■ **Example 13.39** Draw bode diagram for the system having transfer function

$$G(s)=\frac{2}{(s+2)}$$

Solution : From the transfer function, we have

$$G(s)=\frac{2}{(s+2)}$$

$$\Rightarrow \quad G(s)=\frac{1}{(0.5s+1)} \quad ; \quad \tau=0.5$$

Therefore,

$$AR=\frac{1}{\sqrt{1+(0.5\omega)^2}} \quad ; \quad \phi=\tan^{-1}(-0.5\omega)$$

Table 13.6 : Data for plotting bode diagram.

S. No.	ω (rad/s)	*AR*	φ (deg)
1.	0.1	0.998	–2.86
2.	0.2	0.995	–5.71
3.	0.5	0.970	–14.03
4.	1.0	0.894	–26.56
5.	2.0	0.707	–45.00
6.	5.0	0.371	–68.19
7.	10.0	0.196	–78.69
8.	20.0	0.099	–84.29
9.	50.0	0.039	–87.71
10.	100.0	0.019	–88.85

Low Frequency Asymptote : From the amplitude ratio, we have

$$AR=\frac{1}{\sqrt{1+(0.5\omega)^2}}$$

when, $\omega <<< 1$

$$AR=1$$

$$\log AR=0$$

That means LFA is a straight line with a slope zero passing through the point $AR=1$ and $\omega=0.1$ on log-log plot.

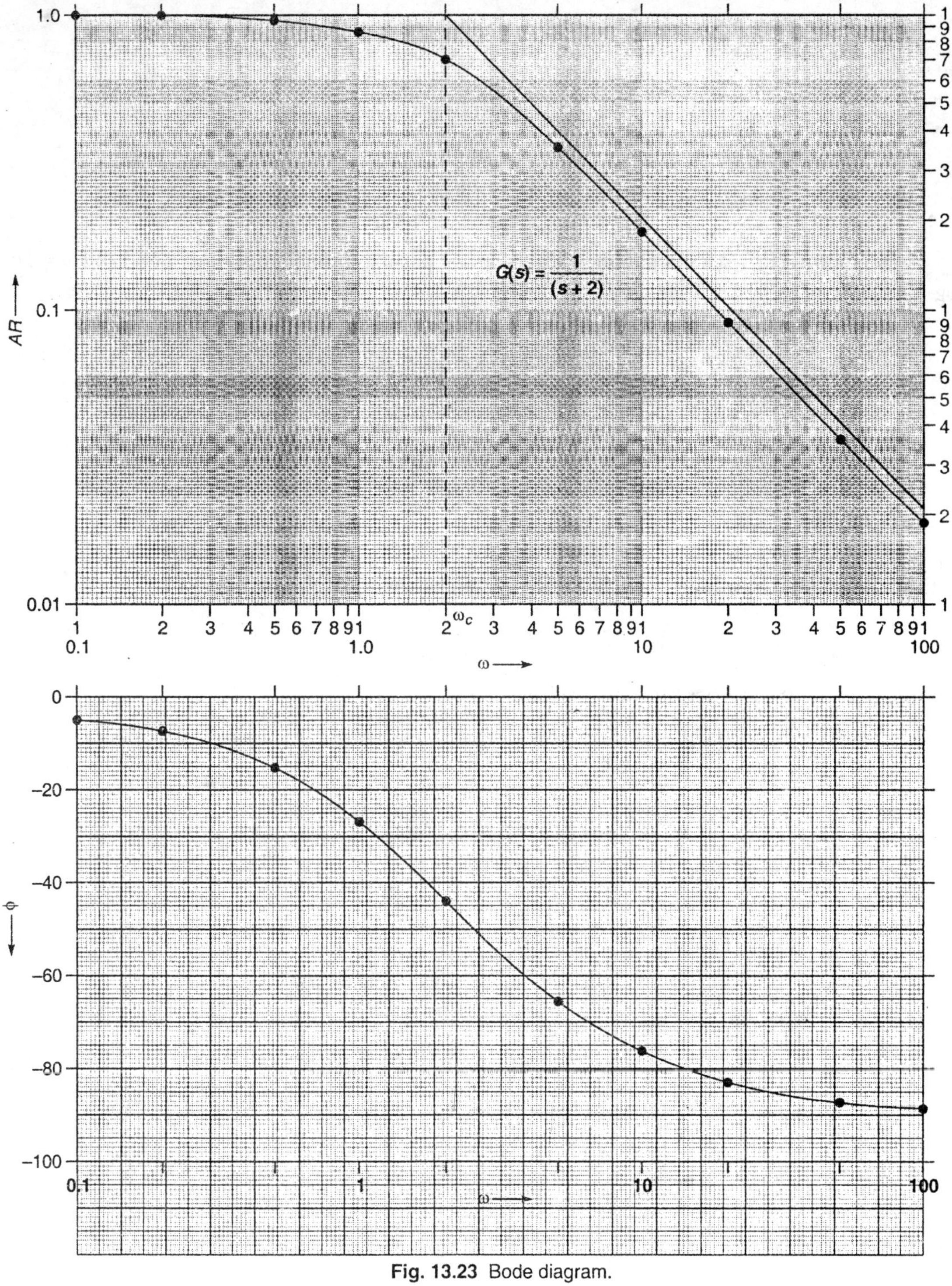

Fig. 13.23 Bode diagram.

High Frequency Asymptote : From the amplitude ratio, we have

$$AR = \frac{1}{\sqrt{1+(0.5\omega)^2}}$$

when, $\omega >>> 1$

$$AR = 1/(0.5\omega)$$
$$\log AR = -\log(0.5\omega)$$

That means, HFA is a straight line with a slope -1 passing through the point $AR = 1$ and $\omega = 2$ on log-log plot.

Corner Frequency : From the characteristic equation, we have

$$1+(0.5\omega)^2 = 0$$
$$(0.5\omega)^2 = |-1|$$
$$(0.5\omega)^2 = 1$$
$$0.5\omega = 1$$
$$\omega_c = 2$$

■ **Example 13.40** Draw bode diagram for the system having transfer function

$$G(s) = \frac{5}{(s+5)}$$

Solution : From the transfer function, we have

$$G(s) = \frac{5}{(s+5)}$$

$$\Rightarrow \quad G(s) = \frac{1}{(0.2s+1)} \quad ; \quad \tau = 0.2$$

Therefore, $$AR = \frac{1}{\sqrt{1+(0.2\omega)^2}} \quad ; \quad \phi = \tan^{-1}(-0.2\omega)$$

Table 13.7 : Data for plotting bode diagram.

S. No.	ω (rad/s)	*AR*	ϕ (deg)
1.	0.1	0.999	–1.15
2.	0.2	0.999	–2.29
3.	0.5	0.995	–5.71
4.	1.0	0.981	–11.31
5.	2.0	0.928	–21.80
6.	5.0	0.707	–45.00
7.	10.0	0.447	–63.43
8.	20.0	0.243	–75.96
9.	50.0	0.099	–84.29
10.	100.0	0.049	–87.14

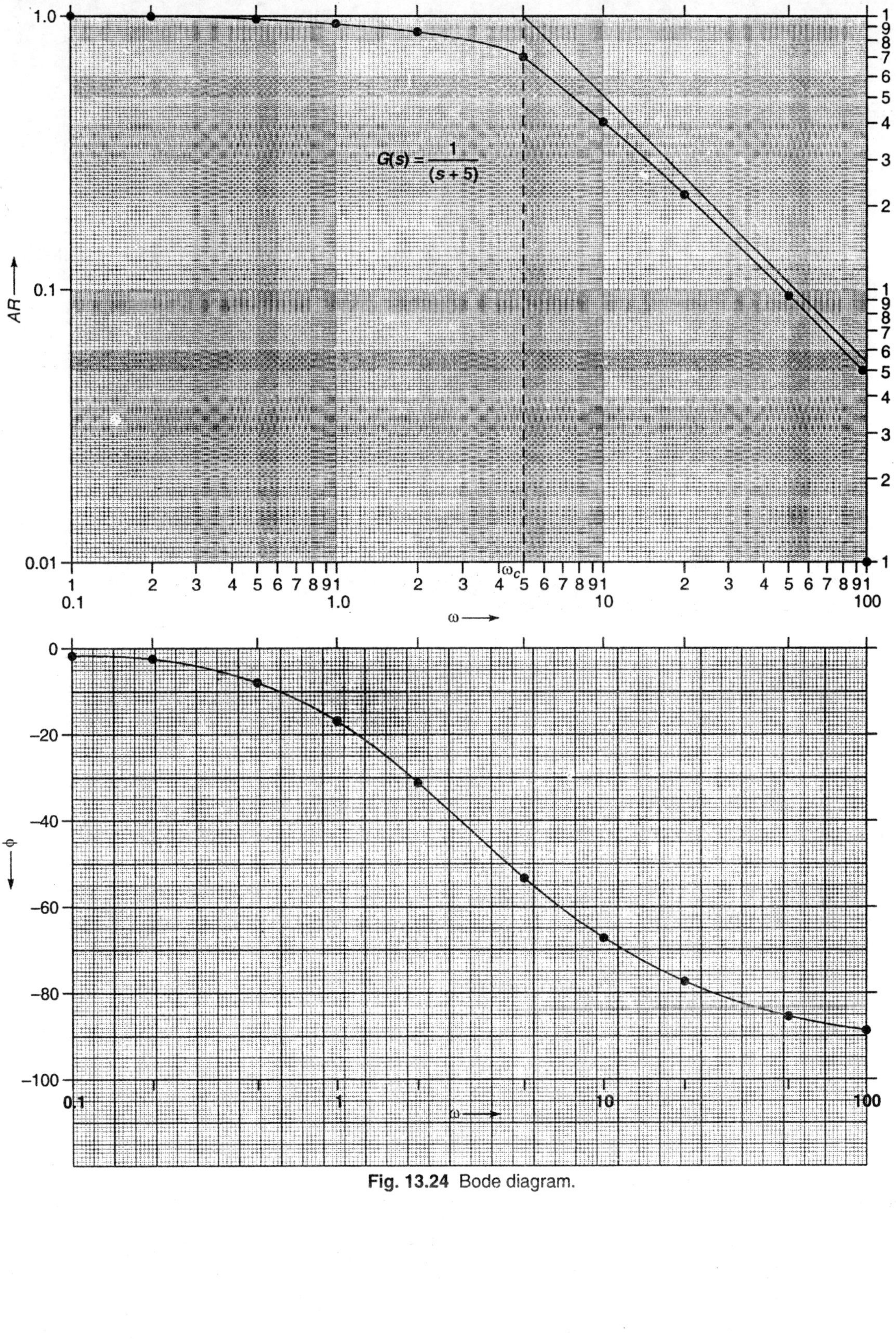

Fig. 13.24 Bode diagram.

Low Frequency Asymptote : From the amplitude ratio, we have

$$AR = \frac{1}{\sqrt{1+(0.2\omega)^2}}$$

when, $\omega <<< 1$

$$AR = 1$$

$$\log AR = 0$$

That means LFA is a straight line with a slope zero passing through the point $AR = 1$ and $\omega = 0.1$ on log-log plot.

High Frequency Asymptote : From the amplitude ratio, we have

$$AR = \frac{1}{\sqrt{1+(0.2\omega)^2}}$$

when, $\omega >>> 1$

$$AR = 1/(0.2\omega)$$

$$\log AR = -\log(0.2\omega)$$

That means, HFA is a straight line with a slope -1 passing through the point $AR = 1$ and $\omega = 5$ on log-log plot.

Corner Frequency : From the characteristic equation, we have

$$1+(0.2\omega)^2 = 0$$

$$(0.2\omega)^2 = |-1|$$

$$(0.2\omega)^2 = 1$$

$$0.2\omega = 1$$

$$\omega_c = 5$$

■ **Example 13.41** Draw bode diagram for the system having transfer function

$$G(s) = k(1+\tau_D s)$$

Solution : From the transfer function, we have

$$G(s) = k(1+\tau_D s)$$

$$\Rightarrow \quad \frac{AR}{k} = \sqrt{1+(\omega\tau_D)^2} \quad ; \quad \phi = \tan^{-1}(\omega\tau_D)$$

Low Frequency Asymptote : From the amplitude ratio, we have

$$\frac{AR}{k} = \sqrt{1+\omega^2} \quad ; \quad \tau_D = 1$$

when, $\omega <<< 1$

$$\frac{AR}{k} = 1$$

$$\log(AR/k) = 0$$

That means LFA is a straight line with a slope zero passing through the point $(AR/k) = 1$ and $\omega = 1$ on log-log plot.

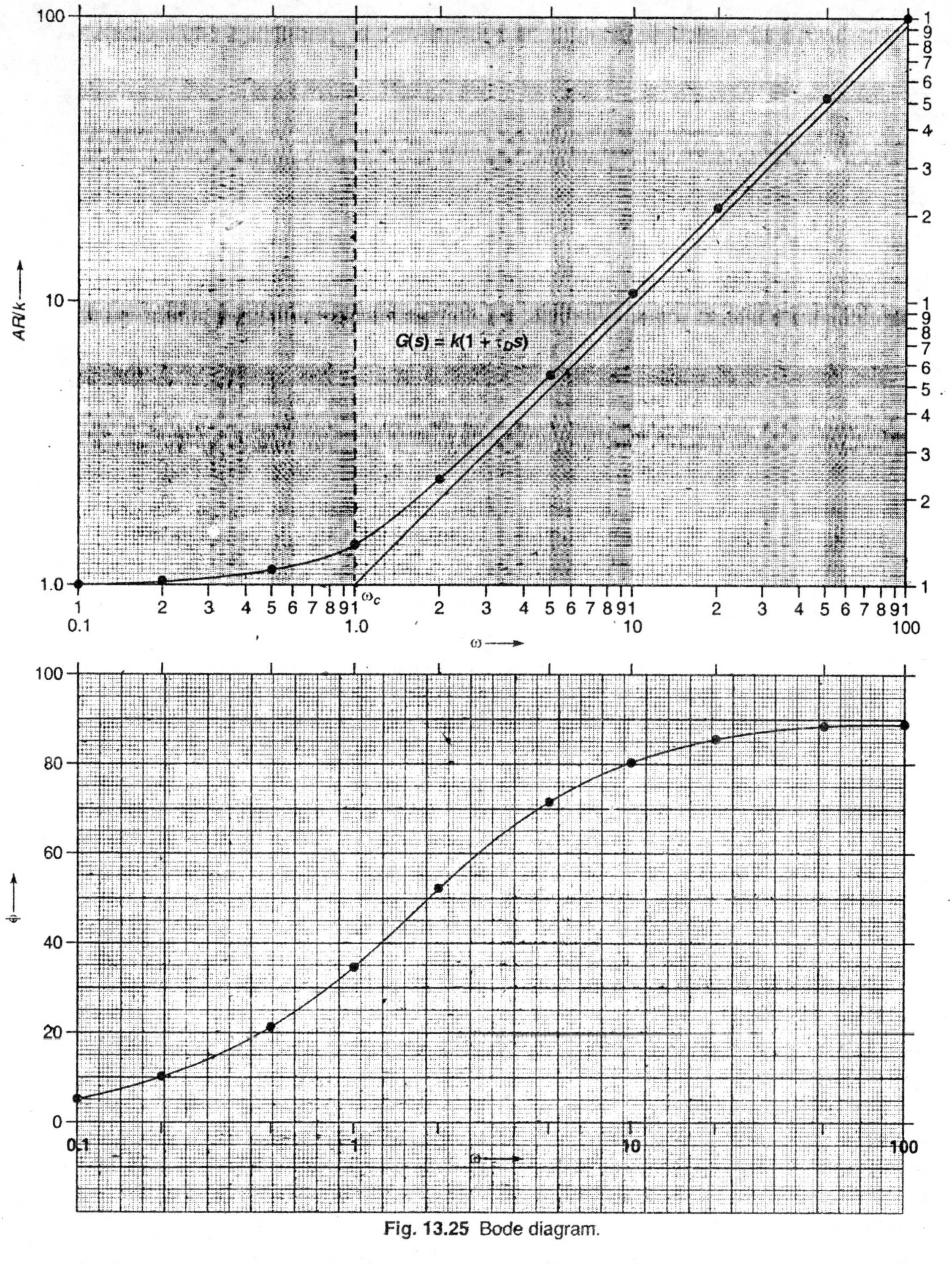

Fig. 13.25 Bode diagram.

Table 13.8 : Data for plotting bode diagram.

S. No.	ω (rad/s)	*AR*	φ (deg)
1.	0.1	1.004	5.71
2.	0.2	1.019	11.31
3.	0.5	1.118	26.56
4.	1.0	1.414	45.00
5.	2.0	2.236	63.43
6.	5.0	5.099	78.69
7.	10.0	10.049	84.29
8.	20.0	20.025	87.13
9.	50.0	50.009	88.85
10.	100.0	100.005	89.43

High Frequency Asymptote : From the amplitude ratio, we have

$$\frac{AR}{k} = \sqrt{1+\omega^2}$$

when, $\omega >>> 1$

$$\frac{AR}{k} = \omega$$

$$\log (AR/k) = \log \omega$$

Corner Frequency : From the characteristic equation, we have

$$1+(\omega\tau_D)^2 = 0$$

$$(\omega\tau_D)^2 = |-1|$$

$$(\omega\tau_D)^2 = 1$$

$$\omega\tau_D = 1$$

$$\omega_c = 1/\tau_D$$

$$\omega_c = 1$$

Gain Margin and Phase Margin : If the open loop *AR* is unity when φ is –180° then the close loop system will oscillate with constant amplitude, *i.e.,* the system will be on the verge of instability. Greater the difference between the open loop *AR* at cross over frequency and $AR = 1$, most stable will be the system. This concept is known as gain margin.

The cross over frequency, at which the phase lag is 180° indicated by ω_{co} on the bode diagram. At this frequency, the *AR* is represented by the symbol *A*. If *A* exceed unity, then the system is unstable, *i.e.,* selection of G_c is not proper. It is always assumed that '*A*' is less than unity and therefore the system is stable. It is also necessary to ascertain to what degree the system is stable. Intuitively, if *A* is only slightly less than unity then the system is almost unstable and may be expected to behave in a highly oscillatory manner even though it is theoretically stable. The constant *A* can be determined using physical parameters of the system, such as time constant. This may be only estimated and may actually change slowly with time because of wear or corrosion. Hence, a design for which *A* is close to unity does not have an adequate safety factor. To assign some indicative measure to the considerations, the concept of gain margin is introduced. Generally, the gain margin is calculated using the equation

$$\text{Gain Margin} = 1/A \qquad \text{...(13.73)}$$

Typical specification required for design is that the gain margin should be greater than 1.7. That means, *AR* at cross over frequency ω_{c0} could increase by a factor 1.7 over the design value before the system become unstable. The design value of the gain margin is really a safety factor. As such, its value varies considerably with the applications. A negative gain margin indicates an unstable system. The relation between gain margin and phase margin is shown in Fig. 13.26.

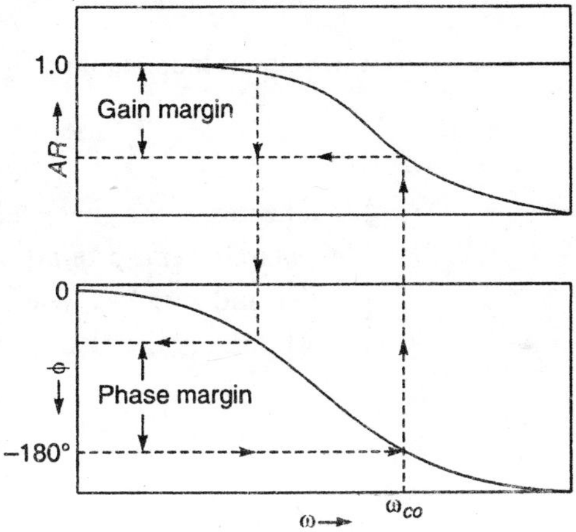

Fig. 13.26 Gain margin and phase margin.

Difference between 180° and the phase lag at the frequency for which the *AR* is unity is called phase margin. The phase margin therefore represents the additional amount of phase lag required to destablize the system, just as the gain margin represents the additional gain for the destablization. A typical design specification is that the phase margin must be greater than 30°. A negative phase margin indicates an unstable system.

13.3.3.2 Nyquist Stability Criterion

Nyquist stability criteria is a frequency response method that gives the information about absolute and relative stability of a system. It also gives the information on the degree of instability of a system. It allows to build up the information on the stability of a system with time delay that can not be determined using R–H criteria and also that are difficult to analyze using root locus technique. The stability of a system can also be improved by this method.

When the open loop transfer function has no poles on the RHS of the imaginary axis, the Nyquist plot should not encircle the point (–1, 0) for the system to be stable. When the open loop transfer function has '*P*' poles on the RHS of *s*-plane the Nyquist plot should encircle the point (–1, 0) '*p*' times in the direction opposite to the Nyquist path for the system to be stable. Generally, the open loop transfer function is used for plotting Nyquist path. The path followed is a clockwise root which will lead to anticlockwise rotation in polar plot and no root should lie to the RHS of imaginary axis.

Thus, it can be concluded that in the Nyquist plot, there should not be any anticlockwise rotation such that it encircle the point (–1, 0) for the system to be stable when no pole lie on the RHS of *s*-plane. If there is no zero in the loop, the point (–1, 0) should not be enclosed in the plot.

■ **Example 13.42** The open loop transfer function of unity feed back system is given by

$$G(s) = \frac{k}{(s)(s+1)(s+2)}$$

Sketch the Nyquist plot and find the range of *k*.

Solution :

(i) *No. of poles on RHS of s-plane* : From the transfer function we have,

$$G(s) = \frac{k}{(s)(s+1)(s+2)}$$

No. of poles,

$$n = 3 \quad ; \quad p_1 = 0$$
$$p_2 = -1$$
$$p_3 = -2$$

From the closed loop stable system, we have

$$z = N + P \qquad \text{...(13.74)}$$

where,

z = No. of zeros of the characteristic equation on the right half of s-plane.

N = Net encirclements around the point $(-1 + j0)$. Clockwise encirclements are taken as positive and anticlockwise encirclements as negative.

P = No. of poles of the characteristic equation on the right half of the s-plane.

$$\Rightarrow \quad z = 0$$

(ii) *Net encirclement* : From the equation (13.73), we have

$$z = N + P$$
$$\Rightarrow \quad o = N + P$$
$$\Rightarrow \quad N = -P = 0$$

(iii) *Nyquist path* : From the transfer function, we have

$$G(s)H(s) = \frac{k}{(s)(s+1)(s+2)} \qquad \text{...(13.75)}$$

Letting $s = j\omega$ in equation (13.75), we get

$$G(j\omega)H(j\omega) = \frac{k}{(j\omega)(j\omega+1)(j\omega+2)}$$

$$G(j\omega)H(j\omega) = \frac{k(-j\omega)\times(1-j\omega)\times(2-j\omega)}{(j\omega)(-j\omega)\times(1+j\omega)(1-j\omega)\times(2+j\omega)(2-j\omega)}$$

$$G(j\omega)H(j\omega) = \frac{k(-j\omega)(2-3j\omega-\omega^2)}{\omega^2(1+\omega^2)(4+\omega^2)}$$

$$G(j\omega)H(j\omega) = \frac{k(-2j\omega+3j^2\omega^2+j\omega^3)}{\omega^2(1+\omega^2)(4+\omega^2)}$$

$$G(j\omega)H(j\omega) = \frac{-3k\omega^2}{\omega^2(1+\omega^2)(4+\omega^2)} + \frac{-j\omega k(2-\omega^2)}{\omega^2(1+\omega^2)(4+\omega^2)} \qquad \text{...(13.76)}$$

Equalizing the imaginary part of equation (13.76) to zero, we get

$$\omega k(2-\omega^2) = 0$$
$$\Rightarrow \quad 2 - \omega^2 = 0$$
$$\Rightarrow \quad \omega^2 = 2$$
$$\Rightarrow \quad \omega = 1.414 \text{ rad/s}$$

From the real part, we get

$$\frac{-3k\omega^2}{\omega^2\,(1+\omega^2)\,(4+\omega^2)} = \frac{-6k}{(2)\,(3)\,(6)} = \frac{-k}{6}$$

For stability, $N = 0$, *i.e.*, point $(-1, 0)$ should be left side of X.

Therefore, $$|OX| < 1$$

$\Rightarrow$ $$\left|\frac{-k}{6}\right| < 1$$

$\Rightarrow$ $$k < 6$$

Therefore, range of k is $0 < k < 6$.

From the angle criterion, we have

$$\alpha = \angle G(j\omega)H(j\omega)$$

$$\alpha = -90° - \tan^{-1}(\omega) - \tan^{-1}(\omega/2)$$

$$\lim_{\omega \to 0} G(j\omega)H(j\omega) = \infty \angle -90°$$

$$\lim_{\omega \to \infty} G(j\omega)H(j\omega) = 0 \angle -270°$$

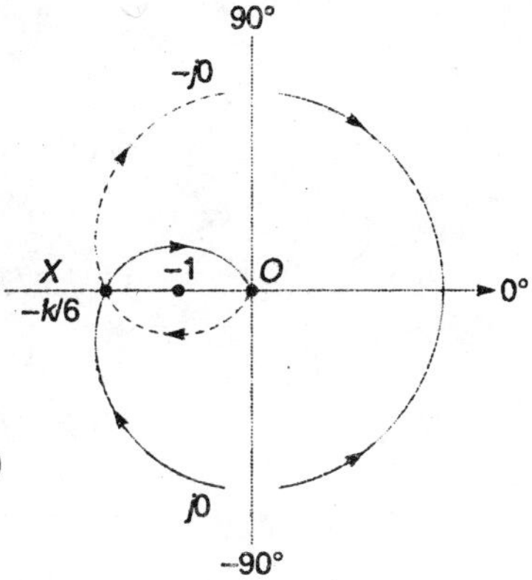

Fig. 13.27 Nyquist plot.

■ **Example 13.43** A negative feed back control system has transfer function

$$G(s) = \frac{k}{(s+1)\,(s+4)}$$

and $$H(s) = \frac{1}{s}$$

Draw the Nyquist plot and discuss the stability of the system.

Solution : From the overall transfer function, we have

$$G(s)H(s) = \frac{k}{(s)\,(s+1)\,(s+4)}$$

(i) *No. of poles on the RHS of s-plane* :

No. of poles,

$$n = 3 \quad ; \quad p_1 = 0$$
$$p_2 = -1$$
$$p_3 = -4$$

Therefore, $P = 0$

(ii) *No. of zeros on the RHS of s-plane* :

$\Rightarrow$ $$z = 0$$

(iii) *Net encirclement* : From the closed loop stable system, we have

$$z = N + P$$

$\Rightarrow$ $$O = N + P$$

$\Rightarrow$ $$N = -P = 0$$

(iv) *Nyquist path* : From the overall transfer function, we have

$$G(s)H(s) = \frac{k}{(s)(s+1)(s+4)} \qquad \text{...(13.77)}$$

Letting, $s = j\omega$ in equation (13.77), we get

$$G(j\omega)H(j\omega) = \frac{k}{(j\omega)(j\omega+1)(j\omega+4)}$$

$$G(j\omega)H(j\omega) = \frac{k(-j\omega)\times(1-j\omega)\times(4-j\omega)}{(j\omega)(-j\omega)\times(1+j\omega)(1-j\omega)\times(4+j\omega)(4-j\omega)}$$

$$G(j\omega)H(j\omega) = \frac{k(-j\omega)\times(4-5j\omega-\omega^2)}{\omega^2(1+\omega^2)(4+\omega^2)}$$

$$G(j\omega)H(j\omega) = \frac{k(-4j\omega-5\omega^2+j\omega^3)}{\omega^2(1+\omega^2)(4+\omega^2)}$$

$$G(j\omega)H(j\omega) = \frac{-5k\omega^2}{\omega^2(1+\omega^2)(16+\omega^2)} + \frac{-j\omega k(4-\omega^2)}{\omega(1+\omega^2)(16+\omega^2)} \qquad \text{...(13.78)}$$

Equalizing the imaginary part of equation (13.78) to zero, we get

$$\omega k(4-\omega^2) = 0$$

$$4-\omega^2 = 0$$

$$\omega^2 = 4$$

$$\omega = 2 \text{ rad/s}$$

Equalizing the real part of equation (13.78) to zero, we get

$$X = \frac{-5k\omega^2}{\omega^2(1+\omega^2)(16+\omega^2)} = \frac{-20k}{(4)(5)(20)} = \frac{-k}{20}$$

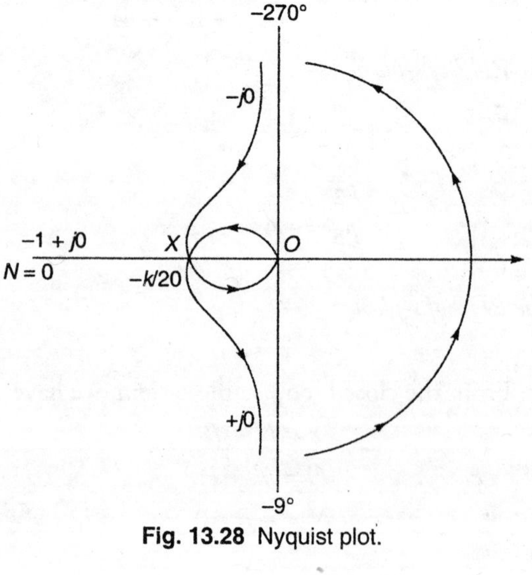

Fig. 13.28 Nyquist plot.

For stability, $N = 0$,

$$|OX| < 1$$

$$\Rightarrow \qquad \frac{k}{20} < 1$$

$$\Rightarrow \qquad k < 20$$

For $k < 20$ system will be stable

For $k = 20$ system will be be marginally stable

For $k > 20$ system will be be unstable.

Therefore, range of k is $0 < k < 20$.

From the angle criterion, we have

$$\alpha = \angle G(j\omega)H(j\omega)$$

$$\alpha = -90° - \tan^{-1}(\omega) - \tan^{-1}(\omega/4)$$

$$\lim_{\omega \to 0} G(j\omega)H(j\omega) = \infty \angle -90°$$

$$\lim_{\omega \to \infty} G(j\omega)H(j\omega) = 0 \angle -270°$$

REVIEW QUESTIONS

13.1 Define stability.

13.2 Define relative stability.

13.3 Define conditional stability.

13.4 Explain the stability criterion.

13.5 Name the various method of analysis of stability of control system.

13.6 Explain the step by step procedure for determination of stability using Routh Hurwitz criteria.

13.7 What do you mean by frequency of sustained oscillation? Explain.

13.8 Explain the merits and demerits of Routh Hurwitz criteria.

13.9 Explain the step by step procedure for determination of stability using root locus technique.

13.10 Explain the rules for plotting root locus diagram.

13.11 Explain the merits and demerits of root locus technique.

13.12 Explain the bode stability criteria.

13.13 Explain the rules for plotting bode diagram.

13.14 Explain the concept of low and high frequency asymptotes.

13.15 Explain the concept of corner frequency.

13.16 Explain the concept of gain margin and phase margin.

13.17 What do you mean by cross over frequency?

13.18 Explain the step by step procedure for determination of stability using Nyquist criteria.

EXERCISES

13.1 For the control system described by the following characteristic equation

$$s^4 + 4s^3 + 6s^2 + 5s + 3 = 0$$

Check the stability. Find the no. of roots having +ve real parts.

13.2 For the control system described by the following characteristic equation

$$s^4 + 2s^3 + 2s^2 + 6s + 3 = 0$$

Check the stability. Find the no. of roots having +ve real parts.

13.3 Determine the stability of a control system whose characteristic equation is

$$s^5 + s^4 + 4s^3 + 4s^2 + 5s + 6 = 0$$

Find the no. of roots with +ve real parts.

13.4 Discuss the stability of control system represented by the characteristic equation

$$2s^6 + 4s^5 + 16s^4 + 24s^3 + 40s^2 + 32s + 32 = 0$$

13.5 The characteristic equation of a control system is given by

$$1 + \frac{k}{(s)(s+1)(s+2)(s+4)} = 0$$

Determine the value of k for which the system is stable.

13.6 The characteristic equation of a control system is given by

$$1 + \frac{k}{(s)(s+1)(s+3)(s+5)} = 0$$

Determine the value of k for which the system is stable.

13.7 For the control system described by the characteristic equation

$$s^3 + 5s^2 + 14s + 16 = 0$$

Find the no. of roots having real parts, more negative than -1.

13.8 The open loop transfer function of a single loop unity feed back control system is given by

$$G(s) = \frac{k(s+11)}{(s)(s+3)(s+5)}$$

Find the value of k for which the closed loop system is stable.

13.9 The open loop transfer function of a chemical process is given by

$$G(s) = \frac{k}{(s+1)(50s^2 + 12s + 1)}$$

Find the range of k for which this system is stable. What will be the frequency of sustained oscillation?

13.10 The characteristic equation of a control system is given by

$$1 + \frac{k}{(s)(s+2)(s+4)(s+6)} = 0$$

Determine the value of k for which this control system is stable.

13.11 The open loop transfer function of a unity feed back control system is given by

$$G(s) = \frac{k}{(s+1)(s+2)(s^2+5s+6)}$$

By applying Routh Hurwitz criteria discuss the stability of the closed loop system as a function of k. Determine the value of k which will cause sustained oscillation in the closed loop system. What are the corresponding oscillation frequencies?

13.12 A feed back control system has an open loop transfer function

$$G(s) = \frac{ke^{-s}}{(s)(s^2+5s+6)}$$

Determine the maximum value of k for which the closed loop system is stable.

13.13 Plot the root locus diagram for the open lop transfer function

$$G(s) = \frac{k}{(s+2)(s+3)}$$

13.14 Plot the root locus diagram for the open loop transfer function

$$G(s) = \frac{k}{(s)(s+2)(s+3)}$$

13.15 Plot the root locus diagram for the open loop transfer function

$$G(s) = \frac{k}{(s)(s+2)(s+3)(s+4)}$$

13.16 A feed back control system has operating transfer function

$$G(s) = \frac{k}{(s)(s^2+5s+11)}$$

Sketch the root locus by determining the following :

(i) No. of asymptotes and asymptotes angle
(ii) Centre of gravity
(iii) Break away point, if any
(iv) Angle of departure of root loci from the poles
(v) The value of k and the frequency at which the root loci crosses imaginary axis.

13.17 Draw bode diagram for the system having transfer function

$$G(s) = \frac{1}{(0.2s+1)}$$

13.18 Draw bode diagram for the system having transfer function

$$G(s) = \frac{20}{(s+4)(s+5)}$$

13.19 Draw bode diagram for the system having transfer function

$$G(s) = \frac{10}{(s+1)(s+2)(s+5)}$$

13.20 Draw bode diagram for the system having transfer function

$$G(s) = \frac{10}{(s+1)\,(0.1s+1)^2}$$

13.21 Draw bode diagram for the system having transfer function

$$G(s) = k\left[1 + \frac{1}{\tau_I s}\right]$$

13.22 Draw bode diagram for the system having transfer function

$$G(s) = \frac{1}{(1+s)\,(1+0.2s)\,(1+0.5s)}$$

13.23 Draw the complete Nyquist plot of the system having transfer function

$$G(s) = \frac{1}{(s)\,(s+5)\,(s+10)}$$

and hence determine whether the system is stable or not.

□□□

Chapter 14

Advanced Control Techniques

14.1 INTRODUCTION

Advanced control techniques are the basic tools normally used in chemical industries in addition to basic process control, which include food processing, oil and mineral refining, petrochemicals and pharmaceuticals. These control techniques bring a higher level of computational capability to the host valuable rather than critical advanced control applications.

14.2 MODES OF ADVANCED CONTROL TECHNIQUES

The different modes of advanced control techniques are as follows :

(i) Cascade control
(ii) Feed forward control
(iii) Feed back control
(iv) Ratio control
(v) Selective control
(vi) Adaptive control
(vii) Simple alarm
(viii) Smith predictor

14.2.1 Cascade Control

A multiloop control system in which the setpoint of one controller is adjusted by the output of another controller is called cascade control. A typical cascade control is shown in Fig. 14.1.

For example, consider a water tank with inflow and outflow. The water present in the tank is required to heat at a given temperature (setpoint) by supplying steam through the tube. The input setpoint temperature is given at controller-1. The tank water temperature is sensed by a temperature sensor and pass on the information to the controller-1.

Now, the output of the controller-1 act as steam flow setpoint for the controller-2. The steam flowrate inside the tube is sensed by a flow sensor and the information is pass on to the controller-2. Based on this information the steam valve operates to supply the steam through the heating tube so that the tank water get heated up at a given setpoint temperature. Here, the setpoint of controller-2 is

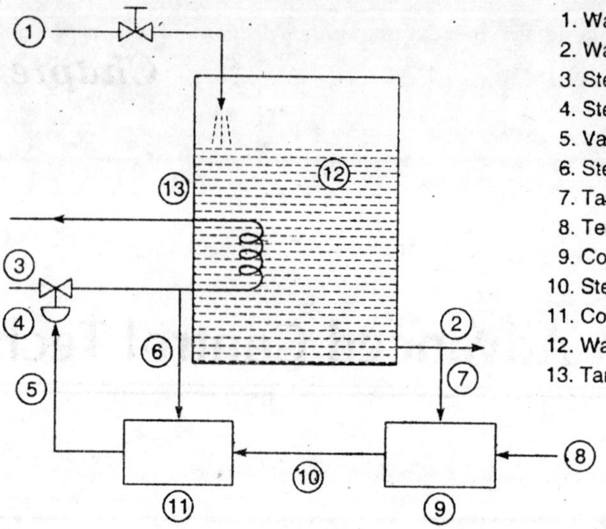

Fig. 14.1 Cascade control.

adjusted by the output of controller-1. Hence, the system is called cascade control. It is possible that the system will become slower if one controller is used. This problem is overcome in the case of cascade control. A black diagram of cascade control is shown in Fig. 14.2.

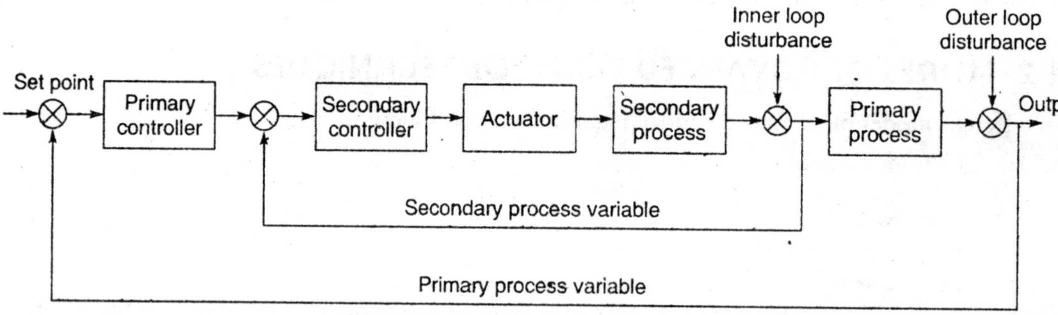

Fig. 14.2 Block diagram of cascade control.

14.2.2 Feed Forward Control

A control system which does not require the feed back information regarding the controlled variable is called feed forward control. It is also known as open loop control system.

In feed forward control system no comparison is made between the input and output of the system. Therefore, the error can not be estimated. As a result of this, there is no control over the output of the system. A block diagram of feed forward control is shown in Fig. 14.3.

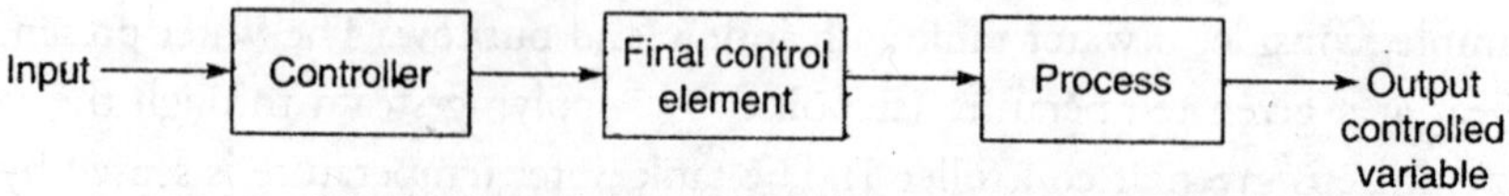

Fig. 14.3 Block diagram of feed forward control.

For example, consider a room heating system. It consists of heating element and the room. The objective is to control the actual room temperature. The input to the system is the desired temperature setting and the output of the system is the actual temperature. As there is no controller, no comparison is made between the output and the input. Hence, there is no control over the output of the system. A feed forward control system is unreliable and inaccurate. Proper judgement and experience is required to control the system.

14.2.3 Feed Back Control

A control system which require the feed back information regarding the controlled variable is called feed back control. In other words a feed back control system compares the output with the input and uses the differences between the two to drive the output into close correspondence with the input. It is this comparison that makes a feed back control system so effective. A block diagram of feed back control is shown in Fig. 14.4.

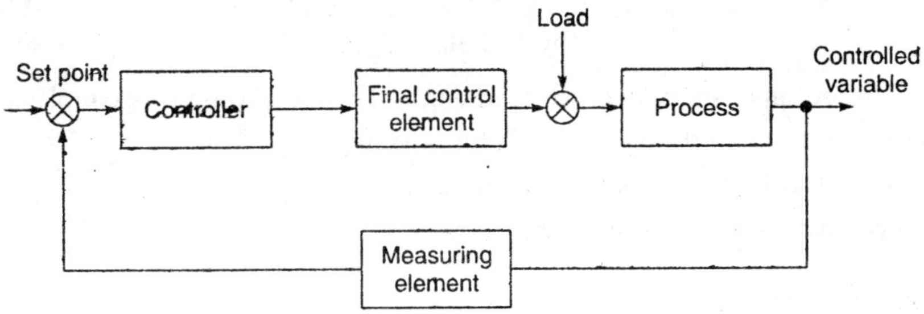

Fig. 14.4 Block diagram of feed back control.

For example, consider driving of an automobile. In this example, the driver is a controller, the steering wheel with its mechanism is the controlled agent and the front two wheels are the controlled element. The input of the system is the idea or safe position of the automobile on the road.

The output of the system is the actual position of the front two wheels on the road. The driver constantly observes the actual position of the front two wheels and compares with actual position with his idea of safe position and uses the difference between the two to drive the output into close correspondence with the input by activating the corrective action proportional to the error. In this process of driving, the driver acts as a controller initiating a corrective action and thus maintaining the error.

A feed back control system is more reliable and accurate than feed forward control system. The major advantages of a feed back control system is that it can adopt itself to the variation in the environmental conditions. Less experience and unskilled person can easily operate a feed back control system and it requires no judgement to operate the system.

14.2.4 Ratio Control

The process in which two constituents are mixed together in a constant ratio to each other is called ratio control. It is a special type of feed forward control. In this control system flow rates of both the stream are measured but only one is controlled. The flow rates of the stream which is not under control is known as wild stream. It is generally used to control the ratio of flow rates of two stream. A typical ratio control is shown in Fig. 14.5.

The flow from wild stream and controllable stream are measured using sensors and allowed to pass into the divider. The ratio of these two stream is measured. Then, the measured ratio and desired ratio entered as input to the summing junction. The difference between these two is measured and data is sent to the ratio controller. Based on this data the control valve operates to maintain the desired ratio.

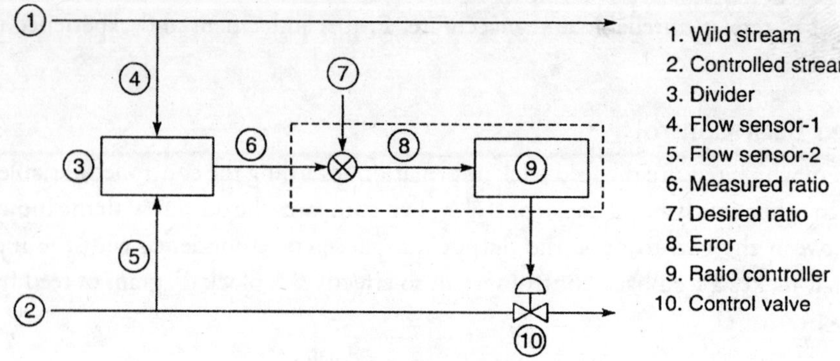

Fig. 14.5 Ratio control.

It has widespread applications in process industries. The most common examples are as follows :

(i) Burner to maintain the air-fuel ratio for burning.

(ii) Distillation column to maintain the reflux ratio.

(iii) Reactor to control the ratio of two reactants.

14.2.5 Selective Control

A control system in which one manipulated variable can select one control action from one control output to another controlled output according to the need is called selective control. The selective control are two types :

(i) Override control

(ii) Auctioneering control

14.2.5.1 Override Control

The control method that is used to select the most relevant information from the set of indicators to control the output is called override control. This situation may arise when more than one variable in a system have to be controlled so that they do not surpass the maximum or minimum limit. For example, consider the control of steam pressure and water level in a boiler. A typical override control is shown in Fig. 14.6.

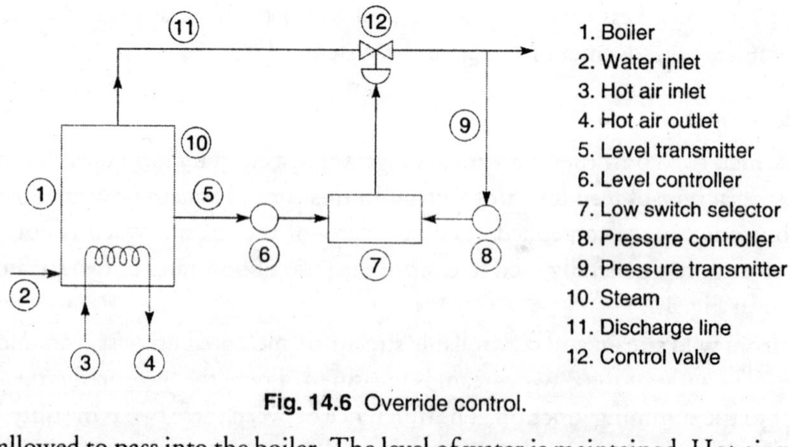

Fig. 14.6 Override control.

Water is allowed to pass into the boiler. The level of water is maintained. Hot air pass through the tube merged in water. This in tern heat the water and generate the high pressure steam. High pressure steam distributed from the discharge line. Water level is sensed using a level sensor and information is pass on to the low switch selector (LSS). Also, steam distributed from the control valve is sensed using a pressure sensor and the information is pass on to the LSS. Based on these informations the control valve operates to discharge the steam between the maximum and minimum limit.

The applications of override control are as follows :

(i) Protecting process equipment by maintaining operating variable within their limit.

(ii) Automatic startup or shut down.

14.2.5.2 Auctioneering Control

The control method that is used to select the most critical measurement condition from the multiple process measurement for a particular variable that need to be regulated through a single control action is called auctioneering control. Example of auctioneering control is tubular reactor.

14.2.6 Adaptive Control

A self optimizing control system which can adjust its parameter automatically in such a way that it can compensate disturbances within itself or in its surrounding by modifying its own parameter is called adaptive control. For example, consider an aircraft. When the aircraft flies fuel burns and it mass slowly decreases. As a result of fuel consumption a control law is needed that can adopt itself such changing scenario.

Therefore, aircraft in flying condition is an example of adaptive control. There are many different types of adaptive control. They differs only in the way the parameters of the controller are used. The adaptive control may be broadly classified as :

(i) Programmed adaptive control

(ii) Self adaptive control

14.2.6.1 Programmed Adaptive Control

It is also known as scheduled adaptive control.

Consider a well known process for which an adequate mathematical model is available. If the auxiliary process variable correlate with the change in process dynamic which in tern manipulate the best value of the controller parameters. Consequently, by measuring the auxiliary variable the adaptation of the controller parameter can be programmed suitably. A typical programmed adaptive control system is shown in Fig. 14.7.

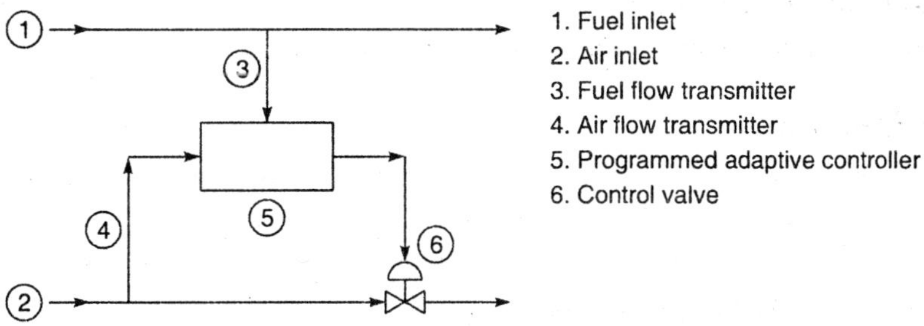

Fig. 14.7 Programmed adaptive control.

The flow of fuel and air are measured using flow transmitter and allowed to pass into the programmed adaptive controller. Where, the programmed adaptive controller optimize the air fuel ratio as per the desired value and sent the information to the control valve. Based on this information the control valve operates to maintain the optimum air fuel ratio.

Generally, this technique is used in burner of furnaces to maintain the optimum air fuel ratio.

14.2.6.2 Self Adaptive Control

The self adaptive control system composed of two loops, inner and outer. The inner loop consists of a process and a feed back controller where as the outer loop consists of a recursive parameter estimator and an adjustment mechanism for the controller parameters. The adjustment mechanism can find best value for the controller parameters using simple linear model for the process.

14.2.7 Simple Alarm

It is a device which produces a loud sound at desired circumstances. A few examples of simple alarm are as follows :

(i) Clock alarm

(ii) Door alarm

(iii) Burgler alarm

(iv) Security alarm

The working operation of a clock alarm is discussed here. A typical circuit diagram of a clock alarm is shown in the Fig. 14.8.

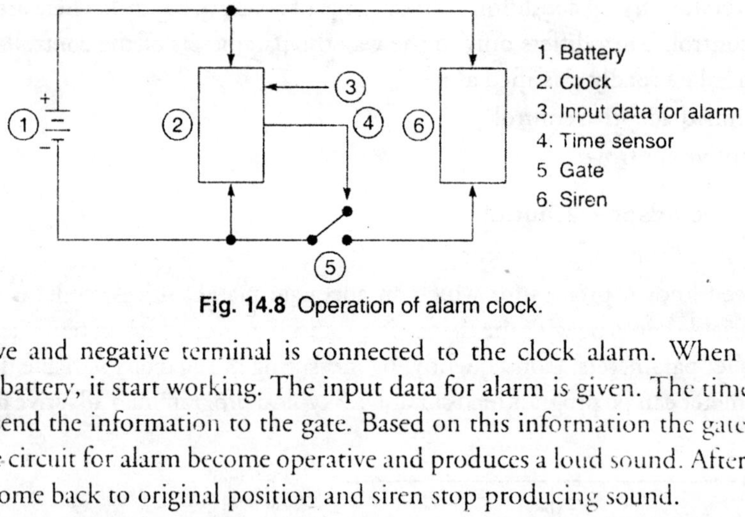

Fig. 14.8 Operation of alarm clock.

The positive and negative terminal is connected to the clock alarm. When direct current is supplied using a battery, it start working. The input data for alarm is given. The time sensor sense the alarm time and send the information to the gate. Based on this information the gate operates. By the action of gate the circuit for alarm become operative and produces a loud sound. After a particular time period the gate come back to original position and siren stop producing sound.

14.2.8 Smith Predictor

A system in which a model based predictive controller with pure time delay is used to control is called Smith predictor. This is one of the advanced control strategies developed in 1957 by OJM Smith but not used because of a lack of practical hardware to implement it. That is due to the reason that it is difficult to implement on analogue hardware.

The Smith predictors popularly used in industrial applications are as follows :

(i) Model based control

(ii) Multiloop control

14.2.8.1 Model Based Control

The control system in which computer use a process model to make control decision is called model based control. It use the intelligent control strategy which is comparable with PID controller approach A typical model based control is shown in Fig. 14.9.

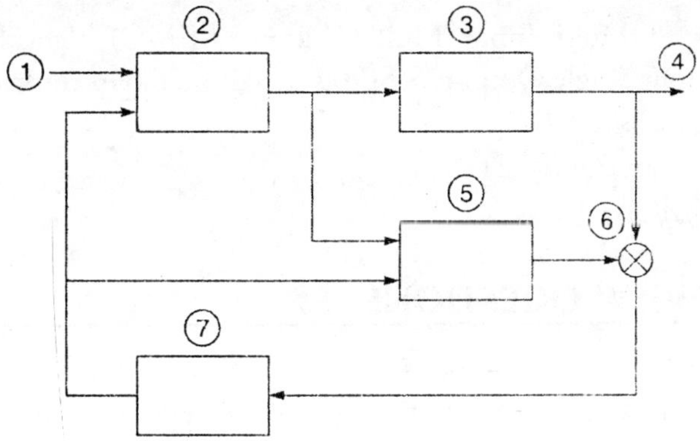

1. Set point
2. Inverse model controller
3. Process
4. Output
5. Model controller
6. Summing junction
7. Adjustment model controller

Fig. 14.9 Model based control.

Here, in model based control inverse model controller and adjustment model controller work simultaneously for the process. The difference between the process output and model controller is monitored and the information is pass on to the adjustment model controller. Then, the adjustable parameter is either bias to the setpoint or model controller. This makes the intelligent control to take decision about the structure and control law which lead the variation in model based control. Accordingly model based control are classified based on following model approaches :

(i) Transfer function model
(ii) Time series model
(iii) Non-linear phenomenological model

Based on these model approaches, the following model controller are familiar :

(i) Internal Model Control (IMC)
(ii) Predictive Model Control (PMC)
(iii) Process Model Based Control (PMBC)

Internal Model Control : A control system in which representation of the process encapsulate either implicitly or explicitly is called internal model control. A typical internal model control is shown in Fig. 14.10.

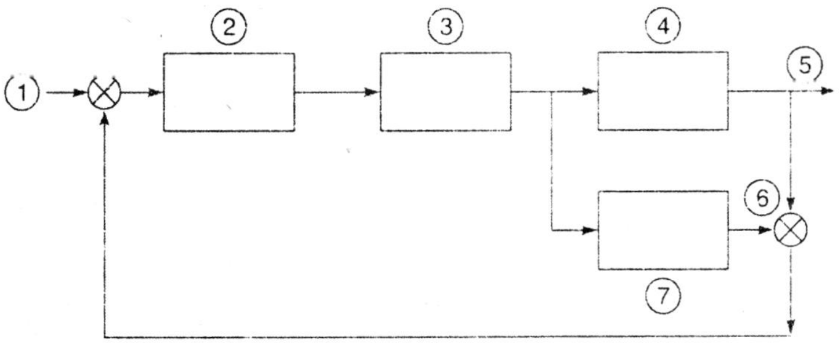

1. Set point
2. Filter controller
3. Inverse controller
4. Process
5. Output
6. Summing junction
7. Model controller

Fig. 14.10 Internal model control.

In internal model control filter controller, inverse controller and model controller work simultaneously for the process. The difference between the process output and model controller is monitored and the information is pass on to the setpoint. Based on these information the controllers operate to maintain the output variation. The main drawback of internal model control is that when the time constant changes the model based calculations either become aggressive or sluggish.

14.2.8.2 Multiloop Control

A system which provide the services to Multi Input Multi Output (MIMO) algorithms and simultaneously opposed to the Single Input Single Output (SISO) is called multiloop control system. The popular multiloop control variable are as follows :

(i) Identification and Command Control (ICC)

(ii) Dynamic Matrix Control (DMC)

REVIEW QUESTIONS

14.1 Explain the application of advanced control techniques.

14.2 Define cascade control. With a neat diagram explain the construction and working of cascade control.

14.3 Define feed forward control. With a neat diagram explain the construction and working of feed forward control.

14.4 Define feed back control. With a neat diagram explain the construction and working of feed forward control.

14.5 What is ratio control? With a neat diagram explain the construction and working of ratio control.

14.6 Define selective control. With a neat diagram explain the construction and working of selective control.

14.7 Define adaptive control. With a neat diagram explain the construction and working of adaptive control.

14.8 What is simple alarm? With a neat diagram explain the construction and working of simple alarm.

14.9 What is Smith predictor? With a neat diagram explain the construction and working of Smith predictor.

14.10 What is internal model control? With a neat diagram explain the construction and working of internal model control.

□□□

Chapter 15

Control Structure of Unit Operations

15.1 INTRODUCTION

A control structure is a block of programming that analyzes the variables of the system and choose a direction in which it has to move based on given parameters. It has the ability of an effective control of any unit operations to reduce the utility cost and to increase the capacity or key component recovery.

15.2 UNIT OPERATIONS

The physical changes occurs in the system under consideration while operations carried out in chemical process industry regarded as unit operations. The popular unit operations carried out in process industry are as follows :

(i) Distillation column

(ii) Heat exchanger

(iii) Evaporator

(iv) Dryer

15.2.1 Distillation Column

Distillation column is a major part of many processes, each of them has a unique set of operating objectives and constraints. It is for this reason that no single distillation control system suffice for all applications.

The major objective of a control system for any process is the stability. Stability, when related to distillation column operation means the ability to prevent disturbances which can result from changes in the independent variable which in tern affect the distillation column operation. A typical distillation column is shown in Fig. 15.1.

The most stable distillation operation results when product withdrawl and energy input/output are manipulated independently to produce the specified separation. This requires one product flow rate to be set in a manner which in turn maintain the material balance. Then reflux or heat is manipulated to regulate the amount of energy which resides within the process.

The overhead and bottom products are set at a rate which will maintain the level in the column. This arrangement of control promote instability because of the reason that products rate is the function of the amount of energy within the system. Energy disturbances occur due to the change in feed, reflux

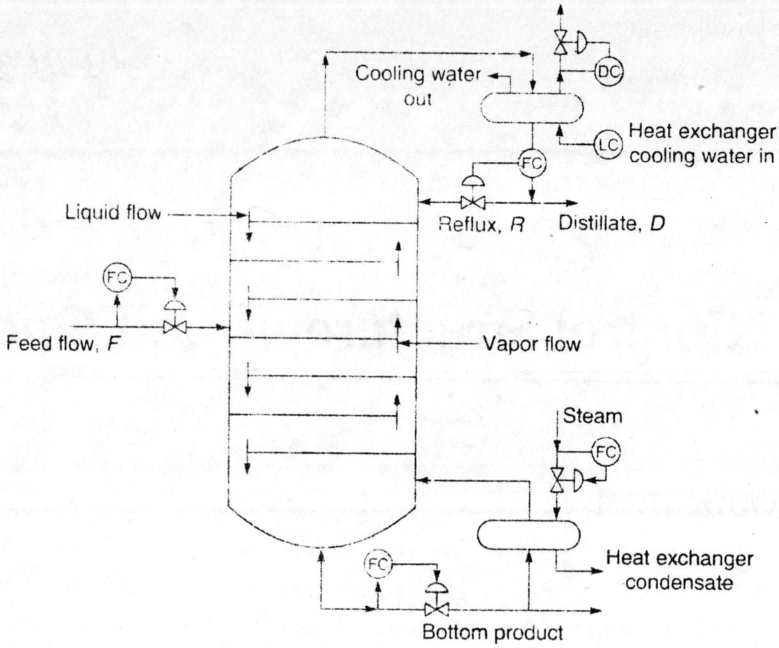

Fig. 15.1 Control of distillation column.

and heating medium enthalpy which in turn has an effect on product distillation. Thus, large variation in product quality result in such an operation because of the reason that the product quality is most sensitive to distillation. The disturbances make the recovery process slow. Therefore, the operator is forced to manipulate both reflux and heat input independently for product quality control. This further complicates the distillation operations. That is because of the reason that the amount of energy needed to produce a required separation is a function of many variables which must be taken into consideration by the operator.

In the case of acceptable control of distillation column one of the product flow rate is directly manipulated for product distribution. Reflux rate is manipulated variable for energy control when the bottom product is set from the material balance. Then, heat is manipulated for energy control when the flow of overhead product is set with respect to the material balance. Thus, the control of distillation column tend to be a self regulating system.

15.2.1.1 Classification of Distillation Variable

The distillation variable can be broadly classified as :

(i) Input (Independent) Variable :
   (a) Feed composition
   (b) Feed rate
   (c) Feed enthalpy
   (d) Reflux temperature
   (e) Heating medium enthalpy

(ii) Manipulated (Input) Variable :
   (a) Reflux rate'
   (b) Boil up rate

(c) Distillate rate

(d) Bottom product rate

(iii) Output (Dependent) Variable :

(a) Distillate composition

(b) Bottom product composition

The algorithms to design a self regulating control of distillation column are as follows :

Step 1 : Start

Step 2 : Read $F, x_F, x_D, x_W, \alpha, q, RR$

Step 3 : Calculate D and W using

$$D = F(x_F - x_W)/(x_D - x_w)$$

$$W = F - D$$

Step 4 : Calculate L and V of rectifying and stripping sections using

$$L_r = RR + D$$

$$V_r = L_r + D$$

$$L_s = L_r + qF$$

$$V_s = L_s - W$$

Step 5 : Calculate y_w corresponding to x_w using the equation

$$y_w = \alpha x_w / (1 + (\alpha - 1)x_w)$$

Step 6 : Calculate the liquid mole fraction for each tray starting from bottom most tray using

$$x_{n+1} = (Wx_w + V_s y_n)/L_s$$

Also, calculate mole fraction of vapour for each tray using

$$y_{n+1} = \alpha\, x_{n+1} / (1 + (\alpha - 1)x_{n+1})$$

Step 7 : Repeat step 6 till $x_{n+1} < x_F$ and count the number of iterations which gives the number of trays in stripping section.

Step 8 : Similarly calculate the liquid mole fraction for each tray from rectifying section using

$$x_{n+1} = (V_r y_n + wx_w - Fx_F)/L_r$$

Also, calculate mole fraction of vapour using

$$y_{n+1} = \alpha x_{n+1} / (1 + (\alpha - 1)x_{n+1})$$

Step 9 : Repeat step 8 till $y_{n+1} < x_D$ and count the number of iterations which intern gives number of trays in rectifying section.

Step 10 : Add the number of trays from stripping and rectifying segment which will be the total number of trays required for desired separation including the reboiler and display the result.

Step 11 : Stop.

15.2.2 Heat Exchanger

It is a device which used to exchange the heat between hot and cold fluid. The various heat exchanger industrially used are :

(i) Double pipe heat exchanger

(ii) Shell and tube heat exchanger

(iii) Packed bed heat exchanger

(iv) Extended surface heat exchanger

In this chapter, the structure of shell and tube heat exchanger is discussed. A typical 1-2 shell and tube heat exchanger is shown in Fig. 15.2.

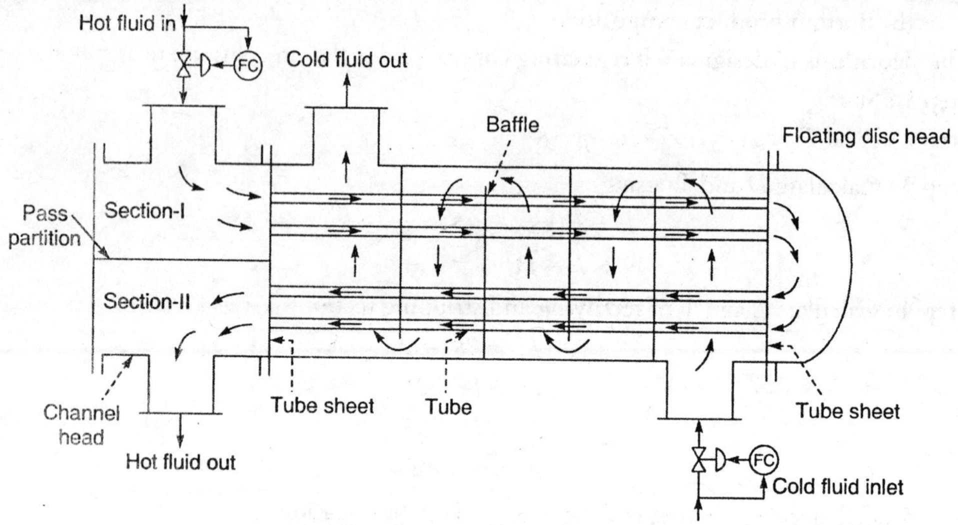

Fig. 15.2 Shell and tube heat exchanger.

It consist of shell, tubes, tube sheets, baffles, cold fluid inlet nozzle, cold fluid outlet nozzle, channel head, pass partition, hot fluid inlet, hot fluid outlet, floating disc head.

At one end of the heat exchanger a channel head is attached whereas on other end a floating disc head. The channel head is divided into two parts using a tube side pass partition. The tubes are fastened on tube sheet at both end of heat exchanger. The turbulent flow of cold fluid is achieved by providing number of baffles.

The hot fluid is allowed to pass into Section-I of the channel head from where it travel the distance of tube length and get accumulated in the floating disc head. Then, from the floating head it travel the same distance of tube length and move on to the discharge line via Section-II of the channel head. Where as the cold fluid enters from the right hand side of the shell and move on to the discharge line from left hand side of the shell. During the process heat transfer takes place from hot fluid to cold fluid.

The major user of regulated control shell and tube heat exchanger are as follows :

(i) Petroleum refinery

(ii) Fertilizer industry

In the case of acceptable control of shell and tube heat exchanger one of the flow rate either hot fluid or cold fluid is manipulated for optimum heat transfer. Algorithms to design a self regulating control of shell and tube heat exchanger are as follows :

Step 1 : Start.

Step 2 : Calculate the average temperature for hot and cold fluid.

Step 3 : Read the physical properties of hot and cold fluid at average temperature.

Step 4 : Calculate the mass flow rate using energy balance equation

$$m_h C_{ph} (T_{h1} - T_{h2}) = m_c C_{pc} (t_{c2} - t_c 1)$$

Step 5 : Declare which flow rate is flowing through inside the tube.

Step 6 : Calculate LMTD using the equation

$$\Delta T_{\text{LMTD}} = \frac{(T_{h1} - t_{c2}) - (T_{h2} - t_{c1})}{\ln\left(\dfrac{T_{h1} - t_{c2}}{T_{h2} - t_{c1}}\right)}$$

Step 7 : Calculate the N_{Re} and N_{Pr} for hot and cold fluid using the equations

$$N_{Re} = \frac{DV\rho}{\mu}$$

$$N_{Pr} = \frac{C_p\mu}{k}$$

Step 8 : Calculate the heat transfer coefficient of tube side and shell side fluid using Dittus Bolter equation

$$\frac{hD}{k} = 0.023\,(N_{Re})^{0.8}\,(N_{Pr})^{0.33}$$

Step 9 : Calculate the clear overall heat transfer coefficient using the equation

$$\frac{1}{U_c} = \frac{1}{h_o} + \frac{1}{h_{io}}$$

Step 10 : Calculate the Dirt heat transfer coefficient using the equation

$$\frac{1}{U_D} = \frac{1}{U_C} + R_d$$

Step 11 : Calculate the heat transfer surface area using the equation

$$A = \frac{Q}{U_D\,(\Delta T_{\text{LMTD}})}$$

Step 12 : Calculate the number of tubes using the equation

$$N = \frac{A}{\pi d_o L}$$

Step 13 : Calculate the number of hair pins using the equation

$$\text{Hair pins} = \frac{N - 1}{2}$$

Step 14 : Calculate the pressure drop of tube side and shell side using the equation

$$\Delta P = \frac{2 f \rho L V^2}{g d_i}$$

Step 15 : If total pressure is greater than 70,000 N/m^2 then redesign.

Step 16 : Stop.

15.2.3 Evaporator

It is a device which used to remove relatively large amount of moisture from the solution under boiling condition. This process is called evaporation. The process may be of single effect or multiple effect. It is the unit operation process generally used in process industry for concentrating weak liquor to produce thick liquor by evaporating a portion of the solvent. Some industrially used evaporators are as follows :

1. Vertical tube evaporator
2. Horizontal tube evaporator
3. Steam jacketted kettle

In this chapter, the structure of vertical tube evaporator is discussed. It is made up of stainless steel in the form of vertical cylinder. The top and bottom of the evaporator is closed with semi-spherical dish head. The bottom of the evaporator also may be conical. There are two tube sheet, one attached at the top and other at the bottom of the liquor compartment. A number of vertical tubes are fastened between these two tube sheets. A typical vertical tube evaporator is shown in Fig. 15.3.

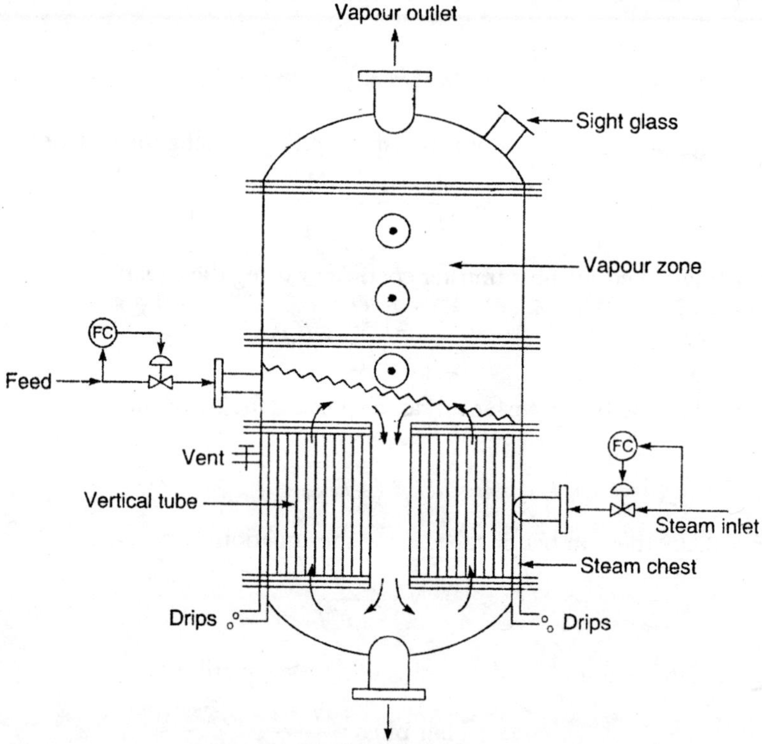

Fig. 15.3 Vertical tube evaporator.

Steam is introduced into the steam chest outside the tube. As the steam flows in, it washes the non-condensed gases and withdrawn from the vent attached at the top of the steam chest. It also washes the condensate ahead of it and withdrawn from the drips attached at the bottom of the steam chest. The feed inlet may be attached to the body of the evaporator above the tube sheets at any convenient point. The outlet for the product is made usually at the centre of the bottom dish. To view the operation of the process, most of the evaporators are provided with sight glasses. As the steam supply heat, the liquor boils, it loses density, become lighter and move upward through the tubes. On the other hand as the liquor comes at the upper surface of the tube, it contacts with vapour, loses temperature becomes denser and returns through the central down take. The vapour is withdrawn from the top of the evaporator. This process is continued till the desired thickness of the product is obtained.

The vertical tube evaporator is used for both viscous and non-viscous solution. In the case of acceptable control of vertical tube evaporator generally the steam pressure is manipulated for removal of moisture from the solution. Algorithm to design a self regulating vertical tube evaporator can be written as in the case of distillation column.

15.2.4 Dryer

It is a device which used to remove relatively small amount of moisture from the solid surface by the application of heat. This process is known as drying. In drying process, dry solid is the product. Drying of wet solid substances is possible when the surrounding environment is unsaturated with the water vapour. Hence, the humidity present in the environment is an important factor for drying solids. The various dryer industrially used are as follows :

1. Tray dryer
2. Drum dryer
3. Spray dryer
4. Fluidised bed dryer
5. Vacuum dryer
6. Freeze dryer

In this chapter the structure of tray dryer is discussed. It consists of air inlet, heater, trays, exhaust fan and air outlet. The trays are placed in the drying chamber and the wet solids are loaded into it. Fresh air is introduced through the inlet, which passes over the heater. It heats the air introduced at a required temperature. The hot air is circulated surrounding the wet substances by means of a exhaust fan. The water molecules present in the solid substances move towards the upper surface of the solids and is picked up by the circulated air. This event occurs in a single pass of air and latter it moves out from the air outlet. The time of contact between wet surface of the solid and hot circulated air is short. Hence, the amount of water picked up from the wet solid surface in a single pass is very small.

Therefore, 80% of the discharged air which is having some amount of heat circulated back using the exhaust fan and 20% fresh air after heating introduce to it. In this way, uniform airflow is maintained and constant temperature is achieved for drying. The process is continued until the desired moisture content in solid is achieved. A typical tray dryer is shown in Fig 15.4.

Tray dryer is used for drying both granules and sticky materials. In the case of acceptable control of dryer air inlet flow rate is manipulated for removal of moisture from the solid substance. Algorithms to design a self regulating dryer can be written as in the case of heat exchanger.

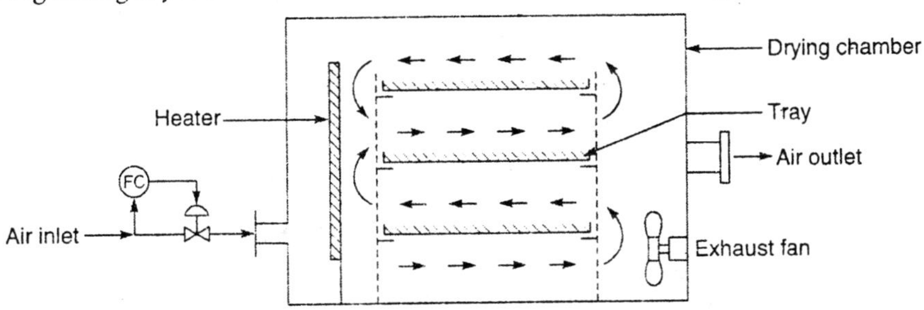

Fig. 15.4 Tray Dryer.

REVIEW QUESTIONS

15.1 What do you mean by control structure of unit operations?
15.2 Describe the control structure of distillation column with a neat diagram.
15.3 Explain the control structure of heat exchanger with a neat diagram.
15.4 Describe the control structure of evaporator with a neat diagram.
15.5 Explain the control structure of a dryer with a neat diagram.

□□□

Answers to Exercise Problems

CHAPTER – 6

6.1 (i) $\dfrac{s}{s^2+k^2}$

(ii) $\dfrac{(s+k)}{(s+k)^2+k^2}$

6.3 $x(t)=\dfrac{1}{3}-e^{-t}+e^{-2t}-\dfrac{1}{3}e^{-3t}$

6.5 $x(t)=-\dfrac{1}{6}+\dfrac{1}{2}e^{t}-\dfrac{1}{2}e^{2t}+\dfrac{1}{6}e^{3t}$

6.7 $\underset{t\to 0}{Lt}\ x(t)=0;\quad \underset{t\to\infty}{Lt}\ x(t)=1/4$

6.9 $z=(8/13)+i(1/13)$

CHAPTER – 7

7.1 $\dfrac{Y(s)}{X(s)}=\dfrac{1}{\tau s+1}$

7.3 $Y(t)=A[-\tau^2+t+\tau e^{-t/\tau}]$

7.5 $Y(t)=\left(\dfrac{A\omega\tau}{1+\omega^2\tau^2}\right)e^{-t/\tau}+\dfrac{A}{\sqrt{1+\omega^2\tau^2}}\sin\,[\omega t-\tan^{-1}(\omega\tau)]$

7.7 $t=2.77$ s

7.9 $\tau=2.39$ s

7.11 $\tau=3.87$ s

7.13 $\tau=7.5$ s

7.16 $\dfrac{H(s)}{Q(s)}=\dfrac{1}{As};\quad H(t)=t/A$

7.17 $\tau=10$ s

CHAPTER – 8

8.1 $\dfrac{H_2(s)}{Q(s)}=\dfrac{R_2}{(\tau_1 s+1)(\tau_2 s+1)}$

8.2 $$\frac{H_2(s)}{Q(s)} = \frac{R_2}{\tau_1\tau_2 s^2 + (\tau_1 + \tau_2 + A_1 R_2)s + 1}$$

8.3 $$\frac{H(s)}{Q(s)} = \prod_{i=1}^{n}\left[\frac{1}{\tau_i s + 1}\right]$$

8.5 $$H_2(t) = \frac{1}{(\tau_1 - \tau_2)}[e^{-t/\tau_1} - e^{-t/\tau_2}]$$

8.7 $$H_2(t) = 0.45[e^{-0.382t/\tau} - e^{-2.718t/\tau}]$$

8.9 $H_1(t) = 0.625$ m; $t = 0$ s

$H_2(t) = 0.338$ m; $t = 1.22$ s

8.11 $\tau_1 = 39$ s; $\tau_2 = 31$ s

CHAPTER – 9

9.1 $$\frac{Y(s)}{X(s)} = \frac{k'}{\tau^2 s^2 + 2\tau\xi s + 1}$$

9.3 $$H(t) = 1 - e^{-\xi t/\tau}\left[\cosh\left(\sqrt{(\xi^2 - 1)}\right)t/\tau + \frac{\xi}{\sqrt{(\xi^2 - 1)}}\sinh\left(\sqrt{(\xi^2 - 1)}\right)t/\tau\right]$$

9.4 $$H(t) = 1 - e^{-\xi t/\tau}\left[\cos\left(\sqrt{(1-\xi^2)}\right)t/\tau + \frac{\xi}{\sqrt{(1-\xi^2)}}\sin\left(\sqrt{(1-\xi^2)}\right)t/\tau\right]$$

9.5 $$H(t) = 1 - e^{-t/\tau}\left[1 + \frac{t}{\tau}\right]$$

9.7 $$H(t)_{\text{impulse}} = \frac{e^{-\xi t/\tau}}{\tau\sqrt{(1-\xi^2)}}\sin\left(\sqrt{(1-\xi^2)}\right)t/\tau$$

9.9 $\text{Overshoot} = e^{-\pi\xi/\sqrt{(1-\xi^2)}}$

9.10 $R_1/R_2 = 0.5$

$\tau_1 = 36$ s; $\tau_2 = 36$ s; $t = 83$ s

9.11 % O.S. = 16.3; $Y(t)_{\text{ultimate}} = 10$

CHAPTER – 11

11.1 $$\frac{P(s)}{\varepsilon(s)} = k_c$$

11.3 $$\frac{P(s)}{\varepsilon(s)} = \frac{k_c}{\tau_I s}$$

11.4 $$\frac{P(s)}{\varepsilon(s)} = k_c\left[1 + \frac{1}{\tau_I s}\right]$$

11.8 $\dfrac{P(s)}{\varepsilon(s)} = k_c[1+\tau_D s]$

11.9 $\dfrac{P(s)}{\varepsilon(s)} = k_c\left[1+\dfrac{1}{\tau_I s}+\tau_D s\right]$

11.11 $k_c = 4$ Psi/inch; $\tau_I = 40$ s

CHAPTER – 12

12.5 $\dfrac{C(s)}{R(s)} = \dfrac{6}{(s+1)(s+6)}$

12.6 $\dfrac{C(s)}{R(s)} = \dfrac{20}{(s+2)(s+10)}$

12.7 $G(s) = \dfrac{20}{(s)(s+12)}$

12.8 $C(t) = 3.5[1-2e^{-0.5t}+e^{-t}]$

12.9 $C(t) = 0.25[e^{-t}-e^{-2t}]$

12.10 $k_c = 3.9$

12.11 $\dfrac{C}{L} = \dfrac{k_c k_v k_p}{(\tau_1 s+1)(\tau_2 s+1)(\tau_m s+1)+k_c k_v k_p}$

12.12 $C(t) = 0.125[1-2e^{-t}+e^{-2t}]$

12.13 $C(t) = 0.23$

CHAPTER – 13

13.1 Stable; No root is having +ve real parts.

13.2 Unstable; 2 roots having +ve real parts.

13.3 Unstable; 3 roots having +ve real parts.

13.5 $0 < k < 12.4$

13.6 $0 < k < 26.25$

13.7 All the 3 roots have more negative than –1. The values are –2, –4 and –6.

13.8 $0 < k < 40$

13.9 $-1 < k < 15.12$; $f = 0.081$ cycle/s

13.10 $0 < k < 84$

13.11 $-12 < k < 68.5$

13.12 $0 < k < 5$

□□□

Appendix

Aptitude Test in Instrumentation and Process Control

1. A system is said to be first order if it can be represented by
   (a) linear first order differential equation
   (b) non-linear first order differential equation
   (c) both 'a' and 'b'
   (d) none of these
2. What is the Laplace transform of sin $2t$?
   (a) $\frac{s}{s^2+4}$ (b) $\frac{2}{s^2+4}$
   (c) $\frac{s}{s^2-4}$ (d) $\frac{2}{s^2-4}$
3. A control system is said to be stable
   (a) if it exhibits an unbounded response to a bounded input
   (b) if its output response is unbounded for all unbounded input
   (c) if its output response is bounded for all bounded input
   (d) none of these
4. Servo problem in control system is one in which
   (a) there is no change in load variable
   (b) there is no change in set point
   (c) there is no change in controlled variable
   (d) none of these
5. Phase margin is equal to
   (a) 180° + phase lag (b) 180° – phase lag
   (c) 120° + phase lag (d) 120° – phase lag
6. When the damping coefficient of the second order system is equal to 1, it is called
   (a) over damped system (b) under damped system
   (c) oscillatory system (d) critically damped system
7. What is dead zone?
   (a) it is the maximum change in the variable that does not change in reading of the instrument
   (b) it is same as transportation lag
   (c) it is same as time constant
   (d) none of these

8. Choose the correct value of transfer function of transportation lag

(a) $e^{\tau s}$ (b) $e^{-\tau s}$

(c) $\frac{1}{\tau s+1}$ (d) none of these

9. When two first order systems are connected in series, the overall transfer function is

(a) the product of the individual transfer function

(b) the sum of the individual transfer function

(c) the ratio of the individual transfer function

(d) none of these

10. A mercury in glass thermometer of first order characteristics takes 5 seconds to indicate 70% of the final temperature. What is the time constant of thermometer?

(a) 4.15 seconds (b) 3.15 seconds

(c) 2.15 seconds (d) 1.15 seconds

11. Pickout the second order system from among the following

(a) damped vibrator (b) U-tube manometer

(c) jacketted kettle (d) all 'a', 'b' and 'c'

12. The transfer function of a first order system is

(a) $\frac{s}{\tau s+1}$ (b) $\frac{1}{\tau s+1}$

(c) $\frac{1}{s+\tau}$ (d) none of these

13. A system is said to be second order if it can be represented by

(a) linear second order differential equation

(b) non-linear second order differential equation

(c) both 'a' and 'b'

(d) none of these

14. Laplace transform of t^n is

(a) $\frac{n}{s^{n+1}}$ (b) $\frac{n}{s^{n-1}}$

(c) $\frac{n!}{s^{n+1}}$ (d) $\frac{n!}{s^{n-1}}$

15. A single tank system is a

(a) first order system (b) second order system

(c) third order system (d) none of these

16. $\underset{t\to\infty}{Lt}\ f(t) = \underset{s\to 0}{Lt}\ s\,f(s)$ is a

(a) initial value theorem (b) final value theorem

(c) Gauss Siedel theorem (d) none of these

17. The Laplace transform of $\delta(t)$ is
(a) 0 (b) 1/2
(c) 1 (d) 2

18. The transfer function of a second order system is
(a) $\frac{1}{\tau^2 s^2 + 2\xi s + 1}$ (b) $\frac{1}{\tau^2 s^2 + 2\tau s + 1}$
(c) $\frac{2}{\tau^2 s^2 + 2\tau\xi s + 1}$ (d) $\frac{1}{\tau^2 s^2 + 2\tau\xi s + 1}$

19. What is the overall transfer function (C/R) of the following block diagram if $G = G_c . G_1 . G_2$ and $H = H_1 . H_2$

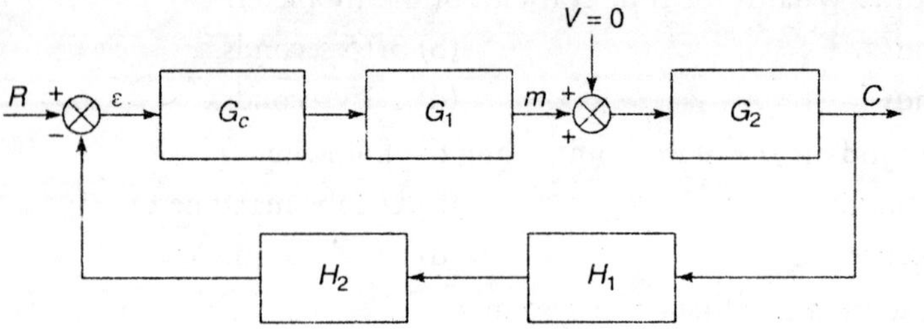

(a) $\frac{1}{1+GH}$ (b) $\frac{G}{1+GH}$
(c) $\frac{H}{1+GH}$ (d) $\frac{G}{1-GH}$

20. Phase lag of the frequency response of a first order system is
(a) $\leq 30°$ (b) $\leq 90°$
(c) 120° (d) 180°

21. Regulator problem in control system is one in which
(a) load is constant (b) set point is constant
(c) both set point and load are constant (c) neither load nor set point is constant

22. When the damping coefficient of the second order system is greater than 1, it is called
(a) under damped system (b) over damped system
(c) critically damped system (d) unstable system

23. Pickout the first order system from among the following
(a) two tank interacting system
(b) two tank non-interacting system
(c) mercury in glass thermometer kept in hot water
(d) U-tube manometer

24. Response of a system to a step input is called
(a) step response (b) unit step response
(c) impulse response (d) frequency response

25. What is time constant?
    (a) it is the time taken by the load variable to reach 63.2% of its full change
    (b) it is the time taken by the measured variable to reach 63.2% of its full change
    (c) it is same as dead time
    (d) it is the time taken by the controlled variable to reach 63.2% of its full change

26. Laplace transform of $t^n . e^{-at}$ is
    (a) $\frac{n}{(s+a)^{n+1}}$ (b) $\frac{n!}{(s+a)^{n+1}}$
    (c) $\frac{-a}{(s-a)^{n+1}}$ (d) $\frac{a}{(s+a)^{n+1}}$

27. Pneumatic control valve generally designed for pressure upto
    (a) 1500 psi (b) 2500 psi
    (c) 150 psi (d) 250 psi

28. When the damping coefficient of the second order system is less than 1, it is called
    (a) under damped system
    (b) over damped system
    (c) critically damped system
    (d) vibrating system

29. A gas tank is supplied air by a compressor at a rate of 0.12 m^3/min. The normal pressure in the tank is 3 kg/cm^2 and the volume of the tank is 0.3 m^3. The gas is then throttled through a valve to a process downstream. The normal pressure of the downstream process is 2.5 kg/cm^2. What is the time constant of the tank?
    (a) 12 minutes (b) 9 minutes
    (c) 6 minutes (d) 8 minutes

30. Gain margin is equal to
    (a) amplitude ratio
    (b) gain of steady state process
    (c) reciprocal of amplitude ratio
    (d) gain of P-D controller

31. Offset of a control system
    (a) varies with time (b) does not varies with time
    (c) varies exponentially with time (d) none of these

32. Laplace transform of e^{-at} is
    (a) $\frac{1}{s+a}$ (b) $\frac{1}{(s-a)}$
    (c) $\frac{-1}{s+a}$ (d) $\frac{-1}{s-a}$

33. What is the overall transfer (C/V) of the following block diagram if $G = G_c . G_1 . G_2$ and $H = H_1 . H_2$

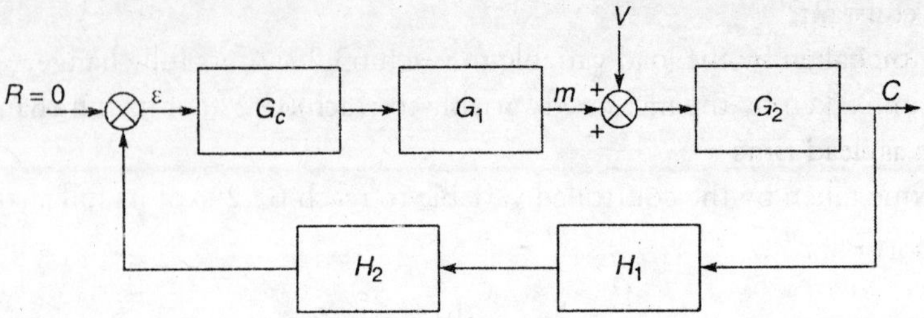

(a) $\dfrac{C}{V} = \dfrac{G_1}{1+GH}$ (b) $\dfrac{C}{V} = \dfrac{G_c}{1+GH}$

(c) $\dfrac{C}{V} = \dfrac{G_2}{1+GH}$ (d) $\dfrac{C}{V} = \dfrac{H_1}{1+GH}$

34. When the damping coefficient of the second order system is less than zero, it is called
(a) unstable system (b) critically damped system
(c) over damped system (d) under damped system

35. $\underset{t \to 0}{Lt}\ f(t) = \underset{s \to 0}{Lt}\ s\,f(s)$
(a) final value theorem (b) initial value theorem
(c) Gauss Siedel theorem (c) none of these

36. Phase lag of sinusoidal response of a second order system
(a) is 60° (b) is atleast 90°
(c) is atmost 120° (d) is approaches 180°

37. A mercury thermometer having a time constant of 0.1 minute is at a steady state temperature of 30°C. At time $t = 0$, the thermometer is placed in a temperature both maintained at 45°C. What is the time needed for the thermometer to read 40°C?
(a) 0.1098 minute (b) 0.2098 minute
(c) 0.3098 minute (d) 0.4098 minute

38. If a control system having transportation lag its stability can be best analysed by
(a) frequency response methods (b) root locus methods
(c) Routh test (d) none of these

39. When two interacting first order systems are connected in series, the overall transfer function is
(a) first order (b) second order
(c) zero order (d) none of these

40. What is the Laplace transform of sin ωt?
(a) $\dfrac{s}{s^2+\omega^2}$ (b) $\dfrac{s}{s^2-\omega^2}$

(c) $\dfrac{\omega}{s^2+\omega^2}$ (d) $\dfrac{\omega}{s^2+\omega^2}$

41. When the damping coefficient of the second order system is equal to zero, it is called
(a) oscillatory system (b) over damped system
(c) under damped system (d) critically damped system

42. Bode diagram is a plot of
(a) log AR vs ω and log ϕ vs ω
(b) AR vs log ω and ϕ vs log ω
(c) log AR vs log ω and ϕ vs log ω
(d) none of these

43. When there is a change or disturbance in the input variables, then the output variable will also
(a) change
(b) not change
(c) both 'a' and 'b'
(d) none of these

44. When there is a change in the system variable, it can be corrected by adjusting some
(a) constant
(b) variables
(c) both 'a' and 'b'
(d) none of these

45. Response of a system to a impulse input is called
(a) pulse response
(b) frequency response
(c) impulse response
(d) unit impulse response

46. A single tank system having $l = 0.45$ m, $b = 0.30$ m and $R = 150$ s/m^2. What is its time constant?
(a) 23.25 seconds
(b) 22.25 seconds
(c) 21.25 seconds
(d) 20.25 seconds

47. Laplace transform of $\cos \omega t$ is
(a) $\frac{s}{s^2 + \omega^2}$
(b) $\frac{s}{s^2 - \omega^2}$
(c) $\frac{\omega}{s^2 + \omega^2}$
(d) $\frac{\omega}{s^2 - \omega^2}$

48. An automatic control system
(a) cannot control itself
(b) required an additional controller
(c) control itself
(c) none of these

49. The improvement in uniformity of the product can be achieved by
(a) simple control
(b) automatic control
(c) both 'a' and 'b'
(d) none of these

50. A control action which does not require the feed back of information regarding the controlled variable is called
(a) open loop control system
(b) closed loop control system
(c) manipulated control system
(c) none of these

51. An instrument used to measure the controlled variable and pass the information to the comparator is called
(a) comparator
(b) measuring element
(c) both 'a' and 'b'
(d) none of these

52. A sustained (prolonged) deviation of the controlled variable from the set point is called
(a) deviation
(b) error
(c) offset
(d) none of these

53. In process control the error is the difference between
(a) set point and measured variable
(b) load and measured variable
(c) both 'a' and 'b'
(d) none of these

54. The process, energy or material which is manipulated to hold the controlled medium at its desired value is called
(a) controlled variable (b) manipulated variable
(c) controlled agent (d) none of these

55. Response of a system to a sinusoidal input is called
(a) frequency response (b) unit step response
(c) impulse response (d) none of these

56. Laplace inverse of $\frac{1}{s+2}$ is
(a) e^{-2t} (b) e^{2t}
(c) $e^{-2/t}$ (d) $e^{2/t}$

57. A mixing process in a tank having constant hold up of $5\ m^3$. If volumetric flow rate of the solution flowing is $0.5\ m^3/s$. What is the time constant of mixing process?
(a) 40 seconds (b) 30 seconds
(c) 20 seconds (d) 10 seconds

58. To obtain the response of control system free of offset and oscillation, the most suitable controller is
(a) proportional controller (b) proportional integral controller
(c) proportional derivative controller (d) proportional integral derivative controller

59. In second order under damped system
(a) overshoot $= e^{-\pi\xi/\sqrt{1-\xi^2}}$ (b) overshoot $= e^{\pi\xi/\sqrt{1-\xi^2}}$
(c) overshoot $= e^{-n\pi\xi/\sqrt{1-\xi^2}}$ (d) overshoot $= e^{n\pi\xi/\sqrt{1-\xi^2}}$

60. The mechanism of the control system which manipulate the control agent according to the signal from the controller is called
(a) measuring element (b) final control element
(c) control agent (d) manipulated variable

61. The unit of damping coefficient is
(a) m^2 (b) m^3
(c) dimensionless (d) m

62. The ratio of the transform of the response in deviation form to the transform of the input in deviation form is called
(a) forcing function (b) transfer function
(c) summing junction (d) none of these

63. A mercury thermometer bulb is 12 mm long and its diameter is 3.5 mm. It is kept in hot water. If the heat transfer coefficient of process medium is 2500 w/m^2 °C and specific heat of thermometric fluid is 502 J/kg°C. What is the time constant of mercury thermometer?
(a) 2.39 seconds (b) 3.39 seconds
(c) 4.39 seconds (d) none of these

64. Pickout the first order system from among the following
(a) liquid filled thermometer
(b) liquid level tank
(c) CSTR with first order chemical reaction
(d) all 'a', 'b' and 'c'

65. A mercury in glass thermometer works on the principle of
(a) volumetric expansion of air
(b) volumetric expansion of water
(c) volumetric expansion of mercury
(d) volumetric expansion of carbon tetrachloride

66. The time constant of mercury thermometer is given by
(a) hA/mC_p (b) mC_p/hA
(c) mC_v/hA (d) hA/mC_v

67. The Laplace transform of $e^{-at}\sin\omega t$ is
(a) $\frac{\omega}{(s+a)^2+\omega^2}$ (b) $\frac{s}{(s+a)^2+\omega^2}$
(c) $\frac{\omega}{(s-a)^2+\omega^2}$ (d) $\frac{s}{(s-a)^2-\omega^2}$

68. The unit of resistance R is
(a) m^2/s (b) s/m^2
(c) s/m^3 (d) m^3/s

69. An automatic control system in which various control units are linked in sequence, each control unit regulating the operation of the next control unit in line is called
(a) cascade control (b) ratio control
(c) adaptive control (d) none of these

70. The two tank non-interacting system has the following overall order
(a) zero (b) first
(c) second (d) none of these

71. The time constant of single tank system is
(a) AR (b) v/q
(c) mC_p/hA (d) none of these

72. The two tank interacting system has the following overall order
(a) first (b) zero
(c) second (d) none of these

73. The two tank non-interacting system has the transfer function
(a) $\frac{Q_2(s)}{Q(s)}=\frac{1}{\tau_1\tau_2 s^2+(\tau_1+\tau_2)s+1}$ (b) $\frac{Q_2(s)}{Q(s)}=\frac{R_2}{\tau_1\tau_2 s^2+(\tau_1+\tau_2)s+1}$
(c) both 'a' and 'b' (d) none of these

74. The control system in which two process variables are kept at a fixed ratio, regardless of the variation of either of the variables is called
(a) selective control (b) ratio control
(c) cascade control (d) adaptive control

75. The time constant of mixing in a tank is given by
(a) AR (b) v/q
(c) q/v (d) none of these

76. Flow rates in two separate fluid conduits are held at a fixed ratio is an example of
(a) cascade control (b) selective control
(c) ratio control (d) none of these

77. The two tank interacting system has the transfer function
(a) $\frac{Q_2(s)}{Q(s)} = \frac{R_2}{\tau_1\tau_2 s^2 + (\tau_1 + \tau_2 + A_1 R_2)s + 1}$
(b) $\frac{Q_2(s)}{Q(s)} = \frac{1}{\tau_1\tau_2 s^2 + (\tau_1 + \tau_2 + A_1 R_2)s + 1}$
(c) both 'a' and 'b
(d) none of these

78. Laplace transform of $e^{-at} \cos \omega t$ is
(a) $\frac{\omega}{(s+a)^2 + \omega^2}$ (b) $\frac{\omega}{(s-a)^2 + \omega^2}$
(c) $\frac{(s-a)}{(s-a)^2 + \omega^2}$ (d) $\frac{(s+a)}{(s+a)^2 + \omega^2}$

79. Number of poles in a system with transfer function $\frac{1}{s^3 + 6s^2 + 11s + 6}$ is
(a) 2 (b) 6
(c) 1 (d) 3

80. Number of zero's in a system with transfer function $\frac{1}{s^3 + 6s^2 + 11s + 6}$ is
(a) 3 (b) 0
(c) 1 (d) 2

81. The corner frequency of first order system having transfer function $\frac{i}{\tau s + 1}$ is
(a) $1/s$ (b) $1/\tau$
(c) τ (d) 1

82. A system that modifies its control algorithm as the systems operating conditions changes so that optimal performance can be achieved is called
(a) cascade control (b) ratio control
(c) adaptive control (d) selective control

83. The transfer function of PI controller is given by

(a) $\frac{1}{\tau s+1}$ (b) $\frac{1}{\tau^2 s^2 + 2\tau\xi s + 1}$

(c) $k_c\left[1+\frac{1}{\tau_I s}\right]$ (d) none of these

84. The function of final control element is to return the controlled variable to the

(a) measured variable (b) set point

(c) process variable (d) none of these

85. The time constant of U-tube manometer is given by

(a) $\sqrt{l/2g}$ (b) $\sqrt{2g/l}$

(c) $\sqrt{2l/g}$ (d) $\sqrt{g/2l}$

86. The transfer function of PD controller is given by

(a) $k_c[1+\tau_I s]$ (b) $(1+\tau_D s)$

(c) $(1+\tau s)$ (d) $k_c[1+\tau_D s]$

87. A system with transfer function $\frac{3s}{4s+1}$ is of

(a) first order (b) zero order

(c) second order (d) none of these

88. The transfer function of P controller is given by

(a) $P(s)$ (b) $\varepsilon(s)$

(c) τ_D (d) k_c

89. The Laplace transform of cosh ωt is

(a) $\frac{s}{s^2+\omega^2}$ (b) $\frac{s}{s^2-\omega^2}$

(c) $\frac{\omega}{s^2+\omega^2}$ (d) $\frac{\omega}{s^2-\omega^2}$

90. Decay ratio of under damped second order system is given by

(a) $e^{-2\pi\xi/\sqrt{1-\xi^2}}$ (b) $e^{-\pi\xi/\sqrt{1-\xi^2}}$

(c) $e^{-3\pi\xi/\sqrt{1-\xi^2}}$ (d) none of these

91. The damping coefficient of U-tube manometer is given by

(a) $\frac{16\mu}{\rho D^2}\sqrt{2l/g}$ (b) $\frac{16\mu}{\rho D^2}\sqrt{2g/l}$

(c) $\frac{8\mu}{\rho D^2}\sqrt{2l/g}$ (d) $\frac{8\mu}{\rho D^2}\sqrt{2g/l}$

92. Force acting on the manometric liquid at any instant due to differential pressure is

(a) P/A (b) A/P

(c) F/A (d) PA

93. Force acting on the manometric liquid at any instant due to gravity is
(a) $AP\rho g$ (b) $AH\rho g$
(c) $Al\rho g$ (d) none of these

94. Force acting on the manometric liquid at any instant due to frictional drag is
(a) $\frac{16\mu LA}{D^2}\frac{dH}{dt}$ (b) $\frac{16\mu LP}{D^2}\frac{dH}{dt}$
(c) $\frac{16\mu LD}{A^2}\frac{dH}{dt}$ (d) none of these

95. Amplitude can be converted into decibels using the formula
(a) Decibel = 20 $\log_e AR$ (b) Decibel = 20 $\log_{10} AR$
(c) Decibel = 20 $\log_e (AR)^2$ (d) none of these

96. Laplace transform of sinh ωt is
(a) $\frac{\omega}{s^2+\omega^2}$ (b) $\frac{\omega}{s^2-\omega^2}$
(c) $\frac{s}{s^2+\omega^2}$ (d) $\frac{s}{s^2-\omega^2}$

97. The cross over frequency means the phase lag is
(a) 180° (b) 160°
(c) 90° (d) 30°

98. Generally, the gain margin calculated using the equation
(a) $GM = \frac{2}{A}$ (b) $GM = \frac{1}{A}$
(c) $GM = \frac{3}{A}$ (d) $GM = A$

99. A multiloop control system in which the set point of one controller is adjusted by the output of another controller is called
(a) forward control (b) backward control
(c) cascade control (d) ratio control

100. A control system in which representation of the process encapsulate either implicitly or explicitly is called
(a) selective control (b) predictive control
(c) preventive control (d) internal model control

ANSWERS

1. (a)	2. (b)	3. (c)	4. (a)	5. (b)	6. (d)	7. (a)	8. (b)	9. (a)	10. (a)
11. (d)	12. (b)	13. (a)	14. (c)	15. (a)	16. (b)	17. (c)	18. (d)	19. (b)	20. (b)
21. (b)	22. (b)	23. (c)	24. (a)	25. (d)	26. (b)	27. (c)	28. (a)	29. (d)	30. (c)
31. (b)	32. (a)	33. (c)	34. (a)	35. (b)	36. (d)	37. (a)	38. (a)	39. (b)	40. (d)
41. (a)	42. (c)	43. (a)	44. (b)	45. (c)	46. (d)	47. (a)	48. (c)	49. (b)	50. (a)
51. (b)	52. (c)	53. (a)	54. (c)	55. (a)	56. (a)	57. (d)	58. (d)	59. (a)	60. (b)
61. (c)	62. (b)	63. (a)	64. (d)	65. (c)	66. (b)	67. (a)	68. (b)	69. (a)	70. (c)
71. (a)	72. (c)	73. (c)	74. (b)	75. (b)	76. (c)	77. (c)	78. (d)	79. (d)	80. (b)
81. (b)	82. (c)	83. (c)	84. (b)	85. (a)	86. (d)	87. (a)	88. (d)	89. (b)	90. (a)
91. (c)	92. (d)	93. (b)	94. (a)	95. (b)	96. (b)	97. (a)	98. (b)	99. (c)	100. (d)

References

1. Coughnour D.R., Process System Analysis and Control, 2nd edition, McGraw Hills, New York, 1991.
2. Stephanopoulos G., Chemical Process Control, Prentice Hall, New Delhi, 1994.
3. Singh S.K., Process Control, Prentice Hall of India, New Delhi, 2010.
4. Ganesh R.D., Venkatesh K.C., Control Engineering, Sanguine Technical Publishers, Bangalore, 2011.
5. Luben, Procgess Modelling Simulation and Control for Chemical Engineers, 2nd edition, McGraw Hill, 1990.
6. Coulson and Richardson, Chemical Engineering Vol. III, 3rd edition, Pergamon Press, 1998.
7. Surekha B., Process Control Principles and Applications, Oxford University Press, 2008.
8. Sarker P.K., Advanced Process Dynamics and Control, Prentiee Hall of India, New Delhi, 2014.
9. William D., Fundamentals of Industrial Instrumentation and Process Control, McGraw Hill, 2009.
10. Johnson C., Process Control Instrumentation Technology Prentice Hall of India, New Delhi, 2006.
11. Andrew S., Hand Book of Instrumentation, Kallisti Publisher, 1994.
12. Pavel R., Modern Sensor Hand Book, ISTE Publisher, 2007.
13. Khandpur R.K., Hand Book of Biomedical Instrumentation, McGraw Hill, 2012.

□□□

INDEX

□□□